普通高等教育土木工程专业新形态教材

混凝土结构基本原理
Basic Principles of Concrete Structures

（中英双语版）

史美东　主编

清华大学出版社
北京

内容简介

本书采用中英文对照体例编写,由教学目标、导读、引例、核心词汇、正文、拓展知识、特别提示、任务、例题、小结、习题、数字资源等内容组成,配套视频、图片、拓展知识、特别提示、英文翻译等数字资源,参照《混凝土结构设计标准》(GB/T 50010—2010)、《工程结构通用规范》(GB 55001—2021)、《混凝土结构通用规范》(GB 55008—2021)和《混凝土结构施工图平面整体表示方法制图规则和构造详图(现浇混凝土框架、剪力墙、梁、板)》(22G101—1)等编写。

本书共分9章,内容包括绪论,混凝土和钢筋的材料性能,混凝土结构设计基本原则,受弯构件的正截面受弯承载力,受弯构件的斜截面承载力,受压构件的截面承载力,受拉构件的截面承载力,受扭构件的扭曲截面承载力,裂缝、变形及耐久性等。

本书既可作为土木工程专业及相关专业的教材和指导书,也可作为土木工程专业人员及相关工程技术人员的学习参考书及职业资格考试参考用书,还可作为土木工程专业人员学习理解新规范、新图集的参考用书。

版权所有,侵权必究。举报:010-62782989,beiqinquan@tup.tsinghua.edu.cn。

图书在版编目(CIP)数据

混凝土结构基本原理:汉、英 / 史美东主编.
北京:清华大学出版社,2024.8.--(普通高等教育土木工程专业新形态教材).-- ISBN 978-7-302-66806-0
Ⅰ.TU37
中国国家版本馆CIP数据核字第2024J9Z130号

责任编辑:秦 娜 赵从棉
封面设计:陈国熙
责任校对:欧 洋
责任印制:丛怀宇

出版发行:清华大学出版社
网　　址:https://www.tup.com.cn,https://www.wqxuetang.com
地　　址:北京清华大学学研大厦A座　　邮　编:100084
社 总 机:010-83470000　　邮　购:010-62786544
投稿与读者服务:010-62776969,c-service@tup.tsinghua.edu.cn
质量反馈:010-62772015,zhiliang@tup.tsinghua.edu.cn
印 装 者:涿州汇美亿浓印刷有限公司
经　　销:全国新华书店
开　　本:185mm×260mm　　印　张:23.25　　字　数:566千字
版　　次:2024年8月第1版　　印　次:2024年8月第1次印刷
定　　价:75.00元

产品编号:101123-01

我国土木工程在国际舞台上发挥着越来越重要的作用,涉外工程、科研项目等越来越多,国际化趋势日益加速。我国土木工程领域中既能熟练使用专业英语,又有过硬技术本领的人才极度匮乏,采用传统中文教材已无法满足人才培养需求,一些高校迫切需要具有足够资源、形式新颖的双语教材配套教学。本书采用中英文对照体例编写,突破了已有相关教材的编写框架,适合读者阅读、理解,使其在掌握专业知识的同时能够学习到相应的混凝土结构相关专业英语词汇和表达,为今后阅读混凝土结构相关英文科技文献,撰写英文科技论文,从事涉外设计、施工、管理等工作奠定基础。

本书编写具有以下特点:

1. 中文英文对照,便于双语教学

本书根据国内学生实际英文水平和课程难度,采用中文和英文对照同时编写,标题、表名、图名、任务、拓展知识、特别提示、例题题干及正文的中文段落都有对应的英文翻译,方便阅读和理解。在每章开始部分列出该章核心中文词汇、词组及对应的英文翻译,有利于学生对该章核心中英文词汇和英文段落的理解与掌握,便于教师开展双语教学和学生学习,强化学生双语能力的培养。

2. 体例内容丰富,结合工程实际

一般在各章开始设置教学目标、导读、引例、核心词汇等内容,引导学生对每章学习内容进行初步了解,通过工程实例激发学生学习兴趣。在每章内容中穿插拓展知识、特别提示、任务、例题、数字资源等内容,结合现行规范、图集和工程实际编写,开阔学生视野,引发深度学习。在每章结束部分引入小结、习题等,对该章主要内容进行概括、总结,勾画出知识脉络和学习路径,使学生对整章内容有全局性的认识和掌握。习题形式多样,引入填空题、选择题、判断题、思考题、计算题等题型,便于课堂练习、作业布置和复习。内容编写化难为简,以"必需""够用""能用"为原则,简化公式推导过程,重点在公式应用,内容力求深入浅出,理论联系实际,增强教材的可读性。

3. 配套数字资源,线上线下融合

配套视频、图片、拓展知识、特别提示、英文翻译等数字资源,以二维码形式提供,将教材、课堂、教学资源三者融合,实现线上线下结合的教学模式(拓展知识和特别提示汇总到每章小结后面的二维码中,只在行文中以标题形式标明对应位置)。结合工程实例和混凝土结构的最新发展编写,可以促使学生不断掌握新知识,及时了解最新成果。

本书按照高等教育土木工程专业人才培养目标、规格以及本课程的教学大纲，根据《建筑结构可靠性设计统一标准》(GB 50068—2018)、《混凝土结构设计标准》(GB/T 50010—2010)、《工程结构通用规范》(GB 55001—2021)、《混凝土结构通用规范》(GB 55008—2021)、《建筑结构荷载规范》(GB 50009—2012)、《建筑抗震设计规范》(GB 50011—2010)(2016年版)、《混凝土结构施工图平面整体表示方法制图规则和构造详图(现浇混凝土框架、剪力墙、梁、板)》(22G101—1)等国家标准、设计图集，结合编者多年教学实践经验，兼顾最新规范、图集的内容，并根据实际教学需要编写。注重内容更新，强调结构设计中应贯彻的规范条文，替换某些实际工作中不常应用及过时的内容。

本书内容包括绪论，混凝土和钢筋的材料性能，混凝土结构设计基本原则，受弯构件的正截面受弯承载力，受弯构件的斜截面承载力，受压构件的截面承载力，受拉构件的截面承载力，受扭构件的扭曲截面承载力，裂缝、变形及耐久性等。本书内容可安排50~70学时进行授课，教师可根据不同专业要求灵活安排学时。

本书由浙江水利水电学院史美东编写，并统稿和定稿。本书参考和引用了国内外诸多文献资料，在此对文献资料的作者表示衷心感谢。

由于编者水平有限，编写内容较多，编写时间较仓促，书中难免存在不足之处，恳请广大同行和读者指正。

史美东

2023年12月

于杭州

第 1 章	绪论	1
1.1	混凝土结构的概念和分类	3
1.2	混凝土结构的特点	4
	1.2.1 配筋的作用与要求	4
	1.2.2 钢筋和混凝土共同工作的原因	6
	1.2.3 钢筋混凝土结构的特点	7
1.3	混凝土结构的应用与发展	11
	1.3.1 混凝土结构的历史与应用	11
	1.3.2 混凝土结构的发展	13
1.4	课程学习目标、内容及要求	14
	1.4.1 课程的学习目标	14
	1.4.2 课程的学习内容	15
	1.4.3 课程的学习要求	16
本章小结		18
习题		18
第 2 章	混凝土和钢筋的材料性能	20
2.1	混凝土的基本性能	22
	2.1.1 混凝土的强度	22
	2.1.2 混凝土的变形	30
2.2	钢筋的基本性能	39
	2.2.1 钢筋的化学成分、品种和选用	39
	2.2.2 钢筋的强度和变形	42
	2.2.3 钢筋的本构关系	47
	2.2.4 钢筋的疲劳性能	48
	2.2.5 混凝土结构对钢筋性能的要求	49
2.3	混凝土和钢筋的黏结	51
	2.3.1 混凝土与钢筋间的黏结作用	51

2.3.2　拔出试验 ··· 53
　　2.3.3　黏结破坏机理 ··· 55
　　2.3.4　影响黏结强度的因素 ·· 56
本章小结 ··· 57
习题 ··· 57

第3章　混凝土结构设计基本原则 ·· 60
3.1　结构上的作用和荷载 ·· 61
　　3.1.1　结构上的作用、效应及结构抗力 ····································· 61
　　3.1.2　结构上的荷载 ··· 63
　　3.1.3　结构的荷载代表值 ·· 64
3.2　结构的功能 ··· 68
　　3.2.1　结构的功能要求 ·· 68
　　3.2.2　设计工作年限、设计基准期和结构安全等级 ··················· 69
3.3　结构功能的极限状态 ·· 72
　　3.3.1　承载能力极限状态 ·· 72
　　3.3.2　正常使用极限状态 ·· 74
　　3.3.3　耐久性极限状态 ·· 75
　　3.3.4　结构的设计状况 ·· 76
3.4　极限状态设计法和设计表达式 ··· 77
　　3.4.1　功能函数和极限状态方程 ·· 77
　　3.4.2　结构的可靠度、可靠概率与失效概率 ······························ 77
　　3.4.3　可靠指标与目标可靠指标 ·· 79
　　3.4.4　承载能力极限状态设计表达式 ··· 80
本章小结 ··· 86
习题 ··· 87

第4章　受弯构件的正截面受弯承载力 ·· 89
4.1　受弯构件一般构造要求 ·· 91
　　4.1.1　受弯构件的材料选择、截面形式与尺寸 ·························· 91
　　4.1.2　混凝土保护层厚度 ·· 93
4.2　受弯构件正截面受弯性能的试验研究 ·· 95
　　4.2.1　简支试验梁受弯承载力试验 ·· 96
　　4.2.2　钢筋混凝土梁正截面的三个工作阶段 ······························ 97
　　4.2.3　受弯构件正截面的破坏形态 ·· 99
4.3　正截面受弯承载力计算原理 ··· 103
　　4.3.1　受弯构件正截面承载力计算的基本假定 ························ 103
　　4.3.2　受压区等效矩形应力图形 ·· 105
　　4.3.3　相对受压区高度和界限相对受压区高度 ························ 106

    4.3.4 纵向受拉钢筋的配筋率 ·· 109
  4.4 单筋矩形截面受弯承载力计算 ·· 112
    4.4.1 基本计算公式及适用条件 ·· 113
    4.4.2 截面设计 ·· 115
    4.4.3 截面复核 ·· 125
  4.5 双筋矩形截面受弯承载力计算 ·· 127
    4.5.1 双筋截面概述 ·· 127
    4.5.2 基本计算公式及适用条件 ·· 128
    4.5.3 截面设计 ·· 130
    4.5.4 截面复核 ·· 134
  4.6 T形截面受弯承载力计算 ·· 135
    4.6.1 T形截面概述 ·· 135
    4.6.2 T形截面的分类和判别 ··· 137
    4.6.3 基本计算公式及适用条件 ·· 140
    4.6.4 截面设计 ·· 141
    4.6.5 截面复核 ·· 143
 本章小结 ··· 145
 习题 ·· 146

第5章 受弯构件的斜截面承载力 ·· 150

  5.1 概述 ·· 151
  5.2 斜裂缝、剪跨比及斜截面受剪破坏形态 ·· 152
    5.2.1 腹剪斜裂缝和弯剪斜裂缝 ·· 152
    5.2.2 剪跨比 ·· 152
    5.2.3 无腹筋梁的斜截面受剪破坏形态 ··· 154
    5.2.4 有腹筋梁的斜截面受剪破坏形态 ··· 156
  5.3 斜截面受剪承载力的计算 ·· 159
    5.3.1 斜截面受剪承载力计算原则 ·· 159
    5.3.2 无腹筋梁混凝土剪压区受剪承载力的试验结果及计算公式 ····························· 161
    5.3.3 板类受弯构件的斜截面受剪承载力计算公式 ··· 162
    5.3.4 有腹筋梁斜截面受剪承载力计算公式 ·· 163
    5.3.5 计算公式的适用范围 ·· 168
    5.3.6 斜截面受剪承载力的计算截面 ·· 172
    5.3.7 截面设计 ·· 173
    5.3.8 截面复核 ·· 175
  5.4 保证斜截面受弯承载力的构造要求 ·· 179
    5.4.1 箍筋的构造要求 ··· 180
    5.4.2 弯起钢筋的构造要求 ·· 183
    5.4.3 抵抗弯矩图 ·· 185

	5.4.4 纵向钢筋的弯起	190
	5.4.5 钢筋的锚固	191
	5.4.6 纵向钢筋的截断	203

5.5 梁、板内钢筋的其他构造要求 ... 206
    5.5.1 梁中构造钢筋的构造要求 ... 206
    5.5.2 钢筋的连接 ... 208

本章小结 ... 212
习题 ... 213

第6章 受压构件的截面承载力 ... 217

6.1 轴心受压构件正截面受压承载力计算 ... 219
    6.1.1 概述 ... 219
    6.1.2 轴心受压普通箍筋柱的正截面受力分析和破坏形态 ... 221
    6.1.3 轴心受压普通箍筋柱的正截面承载力 ... 222
    6.1.4 轴心受压构件的截面设计和截面复核 ... 224

6.2 偏心受压构件正截面受压破坏形态 ... 228
    6.2.1 偏心受压短柱的破坏形态 ... 228
    6.2.2 偏心受压长柱的破坏形态 ... 232

6.3 偏心受压构件的二阶效应 ... 233
    6.3.1 结构无侧移时由受压构件自身挠曲产生的 $P\text{-}\delta$ 二阶效应 ... 234
    6.3.2 考虑 $P\text{-}\delta$ 二阶效应后控制截面的设计弯矩 ... 236
    6.3.3 结构有侧移时偏心受压构件的 $P\text{-}\Delta$ 二阶效应 ... 240

6.4 矩形截面非对称配筋偏心受压构件正截面受压承载力计算 ... 241
    6.4.1 基本计算公式及适用条件 ... 241
    6.4.2 非对称配筋偏心受压构件截面设计 ... 246
    6.4.3 非对称配筋偏心受压构件截面承载力复核 ... 254

6.5 矩形截面对称配筋偏心受压构件正截面受压承载力计算 ... 254
    6.5.1 对称配筋偏心受压构件截面设计 ... 255
    6.5.2 对称配筋偏心受压构件截面承载力复核 ... 259

6.6 I形截面对称配筋偏心受压构件正截面受压承载力计算 ... 260
    6.6.1 I形截面对称配筋大偏心受压 ... 260
    6.6.2 I形截面对称配筋小偏心受压 ... 262

6.7 受压构件的一般构造要求 ... 263
    6.7.1 截面形式和尺寸 ... 263
    6.7.2 材料强度要求 ... 264
    6.7.3 纵向钢筋 ... 265
    6.7.4 箍筋 ... 268

本章小结 ... 271
习题 ... 272

第7章 受拉构件的截面承载力 ……………………………………………………… 276
7.1 轴心受拉构件承载力计算 ………………………………………………… 277
7.1.1 轴心受拉构件的受力特点 ………………………………………… 277
7.1.2 轴心受拉构件正截面承载力计算 ………………………………… 278
7.1.3 轴心受拉构件的构造要求 ………………………………………… 278
7.2 偏心受拉构件承载力计算 ………………………………………………… 279
7.2.1 偏心受拉构件正截面的受力特点 ………………………………… 279
7.2.2 矩形截面偏心受拉构件正截面承载力计算 ……………………… 280
本章小结 ……………………………………………………………………… 284
习题 …………………………………………………………………………… 284

第8章 受扭构件的扭曲截面承载力 ……………………………………………… 287
8.1 纯扭构件的扭曲截面承载力计算 ………………………………………… 288
8.1.1 纯扭构件的试验研究 ……………………………………………… 289
8.1.2 纯扭构件的破坏形态 ……………………………………………… 291
8.1.3 矩形截面纯扭构件的开裂扭矩 …………………………………… 292
8.1.4 按变角度空间桁架模型的扭曲截面受扭承载力 ………………… 295
8.1.5 按《混凝土结构设计标准》的纯扭构件受扭承载力 …………… 300
8.2 复合受扭构件截面承载力计算 …………………………………………… 305
8.2.1 剪扭构件承载力计算 ……………………………………………… 305
8.2.2 弯扭构件承载力计算 ……………………………………………… 310
8.2.3 弯剪扭构件承载力计算 …………………………………………… 310
8.3 受扭构件的构造要求 ……………………………………………………… 321
8.3.1 截面尺寸控制条件 ………………………………………………… 321
8.3.2 受扭钢筋的构造要求 ……………………………………………… 322
本章小结 ……………………………………………………………………… 324
习题 …………………………………………………………………………… 325

第9章 裂缝、变形及耐久性 ……………………………………………………… 328
9.1 概述 ………………………………………………………………………… 329
9.2 钢筋混凝土构件的裂缝宽度计算 ………………………………………… 332
9.2.1 混凝土构件裂缝概述 ……………………………………………… 332
9.2.2 构件裂缝宽度的验算 ……………………………………………… 334
9.3 受弯构件的挠度计算 ……………………………………………………… 340
9.3.1 受弯构件的挠度和刚度 …………………………………………… 340
9.3.2 受弯构件的挠度验算 ……………………………………………… 343
9.4 混凝土结构的耐久性 ……………………………………………………… 347
9.4.1 耐久性的一般概念 ………………………………………………… 347

  9.4.2 影响混凝土结构耐久性的因素 ··· 347
  9.4.3 混凝土结构的耐久性设计 ··· 351
本章小结 ·· 353
习题 ·· 354

参考文献 ·· 356

附录1 《混凝土结构设计标准》(GB/T 50010—2010)和《混凝土结构通用规范》
   （GB/T 55008—2021）规定的混凝土和钢筋的力学性能指标 ··················· 357

附录2 钢筋的公称直径、公称截面面积及理论质量 ·· 359

附录3 《混凝土结构设计标准》(GB/T 50010—2010)和《混凝土结构通用规范》
   （GB/T 55008—2021）的有关规定 ·· 361

绪论

Introduction

教学目标：

1. 了解各种建筑结构的材料和形式，熟悉建筑结构和混凝土结构的概念和分类；
2. 理解配筋的作用和要求，理解钢筋和混凝土共同工作的原因，了解混凝土结构的主要优点和缺点，了解混凝土结构的应用和发展概况；
3. 明确课程学习目标，了解课程学习内容，熟悉课程学习要求，了解课程学习中应注意的主要问题。

导读：

建筑结构是指组成工业与民用房屋建筑包括基础在内的承重骨架体系，为房屋建筑结构的简称。建筑结构设计是在满足安全、适用、耐久、经济和施工可行的要求下，按有关设计标准的规定对建筑结构进行总体布置、技术与经济分析、计算、构造和制图工作，并寻求优化的全过程。

作用是指能够引起结构产生内力和变形的各种因素，如荷载、地震、温度变化以及基础沉降等因素。楼盖是在房屋楼层间用以承受各种楼面作用的楼板、次梁和主梁等所组成的构件总称。板是建筑结构中直接承受楼面荷载的构件，具有较大平面尺寸，但厚度却相对较小，属于受弯构件，通过板将荷载传递到梁或墙或柱上。梁一般指承受垂直于其纵轴方向荷载的线型构件，属于受弯构件，是板与柱之间的支撑构件，承受板传来的荷载并传递到柱上。梁包括主梁和次梁，主梁是将楼盖荷载传递到柱、墙上的梁，次梁是将楼面荷载传递到主梁上的梁。柱和墙都是建筑结构中承受轴向压力的承重构件，柱是主要承受平行于其纵轴方向荷载的线型构件，截面尺寸小于高度；墙是主要承受平行于墙体方向荷载的竖向构件，并将荷载传到基础上；柱和墙都属于受压构件，有时也承受弯矩和剪力。基础是地面以下部分的结构构件，将柱和墙等传来的上部结构荷载传递给地基。

按照所用材料不同，建筑结构分为混凝土结构、砌体结构、钢结构和木结构。

混凝土是由胶凝材料（水泥或其他胶结料）、粗细骨料和水等拌和而成的先可塑后硬化的结构材料，需要时可另加掺合料或外加剂。混凝土产生于古罗马时期，现代混凝土的广泛应用开始于19世纪中期。混凝土结构是以混凝土为主要建筑材料的结构，主要包括素混凝土结构、钢筋混凝土结构和预应力混凝土结构等。随着生产的发展、理论研究以及施工技术的改进，混凝土结构逐步提升及完善，得到了迅速发展。

砌体结构是由块体（如砖、石或砌块）与砂浆或其他胶结料砌筑而成的结构，大量用于居住建筑和多层民用房屋（如办公楼、教学楼、商店、旅馆等）中，其中以砖砌体的应用最为广泛。砖、石、砂等材料具有就地取材、成本低等优点，结构的耐久性和耐腐蚀性也很好；缺点是材料强度较低、结构自重大、施工砌筑速度慢、现场作业量大等。

钢结构是以钢材为主组成的结构，钢材是结构用的型钢、钢板、钢管、带钢或薄壁型钢以及钢筋、钢丝和钢绞线等的总称。钢结构主要用于大跨度的建筑屋盖（如体育馆、剧院等）、吊车吨位很大或跨度很大的工业厂房骨架和吊车梁以及超高层建筑的房屋骨架等。钢结构的材料质量均匀、强度高、构件截面小、重量轻、可焊性好、制造工艺比较简单、便于工业化施工；缺点是钢材易锈蚀、耐火性较差、价格较贵。

木结构是以木材为主制作的结构，由于受自然条件的限制，我国木材相当缺乏，仅在山区、林区和农村地区使用。

按照组成建筑主体结构的受力体系的不同，建筑结构可分为框架结构、剪力墙结构、框架-剪力墙结构、筒体结构、混合结构、大跨结构（排架结构、网架结构、悬索结构、壳体结构、膜结构）等。高层建筑的混凝土结构可采用框架、剪力墙、框架-剪力墙和筒体等结构体系。框架结构是由梁和柱以刚接或铰接形式连接成承重体系的房屋建筑结构。剪力墙结构是指纵横向的主要承重结构用钢筋混凝土墙板来代替框架结构中的柱，形成一种有效抵抗水平作用的结构体系，又能起到对空间的分割作用，其墙身平面内的抗侧移刚度很大，抵抗变形的能力强，建筑物上大部分水平作用（风荷载和水平地震力）或水平剪力被分配到剪力墙上，这也是剪力墙名称的由来。框架-剪力墙结构也称框剪结构，是在框架结构中的适当部位布置一定数量的钢筋混凝土剪力墙，能构成灵活自由的使用空间，满足不同建筑功能的要求，足够的剪力墙有相当大的侧向刚度，承担大部分水平荷载，框架则主要承担竖向荷载。筒体结构是由竖向悬臂的筒体组成能承受竖向、水平作用的高层建筑结构，筒体分剪力墙围成的薄壁筒和由密柱框架围成的框筒等。框架筒体结构是由中央薄壁筒与外围的一般框架组成的高层建筑结构。筒中筒结构是由若干并列筒体组成的高层建筑结构。

任务1-1：观察并判断学校教学楼的建筑结构类型。

Task 1-1：Observe and judge the building structure type of the school teaching building.

核心词汇：

中文	English	中文	English
混凝土结构	concrete structure	木结构	wood structure
柱	column	纤维混凝土	fiber concrete
素混凝土结构	plain concrete structure	钢管混凝土结构	concrete-filled steel tube structure
共同工作	work together		
钢筋混凝土结构	reinforced concrete structure	智能混凝土	intelligent concrete
温度线膨胀系数	thermal linear expansion coefficient	钢筋	reinforcement
		土木工程	civil engineering
预应力混凝土结构	prestressed concrete structure	梁	beam
耐火性	fire resistance	建筑工程	architectural engineering
砌体结构	masonry structure	板	slab
轻质混凝土	light-weight concrete	设计规范	design code
钢结构	steel structure	基础	foundation
高强高性能混凝土	high strength and high performance concrete	高层建筑	high-rise building
		图集	atlas

1.1 混凝土结构的概念和分类
Concept and classification of concrete structures

任务 1-2：判断图 1-1 中的建筑各属于哪种建筑结构。
Task 1-2：Determine which building structure the buildings in Fig. 1-1 belong to.

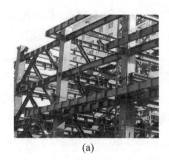

(a) (b) (c)

图 1-1 不同结构材料的建筑
(a) 北京某施工中的建筑；(b) 北京某正在浇筑混凝土的建筑；(c) 清华大学图书馆
Figure 1-1 Buildings with different structural materials
(a) A building under construction in Beijing；(b) A building undergoing concrete pouring in Beijing；
(c) Tsinghua University library

混凝土结构是指以混凝土为主要材料，并根据需要在其中合理配置钢筋、型钢、钢管及纤维，作为主要承重材料的结构，主要有素混凝土结构、钢筋混凝土结构、预应力混凝土结构、钢骨混凝土结构、钢管混凝土结构、纤维混凝土结构等，其中应用最广泛的是钢筋混凝土结构。

Concrete structure refers to the structure with concrete as the main material according to the need of the rational allocation of steel bar, section steel, steel tube and fiber as the main load-bearing material structure. There are mainly plain concrete structures, reinforced concrete structures, prestressed concrete structures, steel-reinforced concrete structures, concrete-filled steel tube structures, fiber-reinforced concrete structures, etc., among which reinforced concrete structures are the most widely used.

素混凝土结构是指无筋或不配置受力钢筋的混凝土结构，其承载力低，性质脆，使用很有限，常用于路面、受压的基础、柱墩和一些非承重结构。钢筋混凝土结构指配有受力钢筋、钢筋网或钢筋骨架的混凝土结构，是目前应用最广泛的结构形式。承重的主要构件是用钢筋混凝土建造的，钢筋承受拉力或压力，混凝土承受压力，具有坚固、耐久、防火性能好、比钢结构节省钢材和成本低等优点。

Plain concrete structure refers to the concrete structure without reinforcement or stressed reinforcement. It has low bearing capacity, brittle nature and limited use. It is often used for pavements, pressed foundations, piers and some non-load-bearing structures. Reinforced concrete structure refers to the concrete structure equipped with stressed steel bars, steel nets or steel skeletons, and is currently the most widely used structural form. The main load-bearing members are constructed of reinforced concrete.

The steel bars bear tensile force or compressive force, and the concrete bears compression. It has the advantages of firmness, durability, good fire resistance, saving steel materials and lower cost than steel structures.

预应力混凝土结构指配置受力的预应力钢筋，通过张拉预应力钢筋或其他方法建立预加应力的混凝土结构。钢骨混凝土结构是配置型钢或用钢板焊成钢骨架的混凝土结构，承载力大、延性好、刚度大、抗震性能好，但耗钢量较多，可在高层、大跨或抗震要求较高的工程中采用。钢管混凝土结构是把混凝土灌入钢管中并捣实以加大钢管混凝土的强度和刚度，主要用于以轴心受压、偏心受压构件为主的高层建筑结构。钢管混凝土的构件连接较复杂，维护费用大。纤维混凝土结构是在混凝土中掺入钢纤维、合成纤维等纤维材料构成的结构，具有抗拉强度高、抗裂性能好、抗渗性能强、抗磨损和抗冲击等优点。纤维混凝土使混凝土的性能获得飞跃发展，混凝土的拉压强度比从 1/10 提高到 1/2，并且具有早期强度高和收缩、徐变小的特性。纤维混凝土主要应用于建筑楼面、高速公路路面、机场跑道、停车场以及储液池等。

Prestressed concrete structure refers to the concrete structure configuring of load bearing prestressed steel bars, and the prestressing is established by tensioning the prestressed steel bars or other methods. The steel-reinforced concrete structure is the concrete structure with steel frame welded by steel or steel plate. It has large bearing capacity, good ductility, high rigidity and good seismic performance, but it consumes more steel. It can be used in high-rise, large-span or projects with high seismic requirements. The concrete-filled steel tube structure is to pour concrete into the steel tube and to increase the strength and rigidity of the concrete-filled steel tube. It is mainly used for high-rise building structures with axial compression and eccentric compression members. The splice of concrete-filled steel tube members is complex and the maintenance cost is high. Fiber-reinforced concrete structure is the structure composed of concrete mixed with fiber materials such as steel fiber and synthetic fiber. It has the advantages of high tensile strength, good crack resistance, strong impermeability, abrasion resistance and impact resistance. Fiber concrete has made a leap development in the properties of concrete. The tensile-compression strength ratio of concrete has been increased from 1/10 to 1/2, and it has the characteristics of high early strength, small shrinkage and small creep. Fiber concrete is mainly used in building floors, highway pavements, airport runways, car parks and liquid storage tanks etc.

 拓展知识 1-1

1.2 混凝土结构的特点
Characteristics of concrete structures

1.2.1 配筋的作用与要求
Role and requirements of reinforcement

混凝土材料最显著的特点是抗压强度高，抗拉强度低。普通混凝土的抗拉强度一般是其抗压强度的1/17~1/8,高强混凝土的抗拉强度一般是其抗压强度的1/24~1/20。大部分钢筋混凝土结构构件要承受弯矩等作用,在截面上产生拉应力,如果拉应力全部由素混凝土承担,当截面受拉边缘的拉应力达到混凝土的抗拉强度时,构件就会突然发生脆性断裂破坏,破坏前无明显预兆,属于脆性破坏,构件的承载能力将会非常低,在很小的弯矩作用下构件就会破坏。因此,不配置钢筋的素混凝土无法应用于截面具有拉应力的结构构件中。

The most notable features of concrete materials are high compressive strength and low tensile strength. The tensile strength of ordinary concrete is generally 1/17-1/8 of the compressive strength, and the tensile strength of high-strength concrete is generally 1/24-1/20 of the compressive strength. Most of the reinforced concrete structural members have to bear the action of bending moment, etc., and tensile stress is generated at the section. If the tensile stress is all borne by plain concrete, when the tensile stress on the tension side of the section reaches the tensile strength of the concrete, the member will undergo suddenly brittle failure. There is no obvious advance warning before failure occurs. It belongs to brittle failure. The bearing capacity of the member will be very small. The member will be damaged under the action of a small bending moment. Therefore, plain concrete without reinforcement cannot be applied to structural members with tensile stress on the cross section.

如果在构件受拉区边缘设置钢筋,当受拉区混凝土开裂后,混凝土上的拉应力可以转移到钢筋上,钢筋可以继续承受受拉区的拉应力。当荷载继续增加,钢筋达到屈服阶段,虽然荷载不能继续增加,但构件可以继续变形,具有足够的变形能力。受压区的混凝土受到的压应力和变形继续增加,当受压区混凝土边缘压应力达到混凝土轴心抗压强度时,混凝土被压碎而导致构件破坏。与素混凝土梁相比,混凝土的抗压强度和钢筋的抗拉强度都得到充分利用,钢筋混凝土梁的破坏荷载和承载能力大大提高,破坏前构件有明显的裂缝和变形,有显著的预兆,属于延性破坏,在实际工程中可以最大限度地减少财产损失和人员伤亡。

If steel bars are installed at the edge of the tension zone of the member, after the cracking of the concrete in the tension zone, the tensile stress on the concrete can be transferred to the steel bars, and the steel bars can continue to bear the tensile stress in the tension area. When the load continues to increase and the steel bars reach the yielding stage, although the load cannot continue to increase, the member can continue to deform and has sufficient deflection capability. The compressive stress and deformation of the concrete in the compression zone continue to increase. When the edge compressive stress of the concrete in the compression zone reaches the axial compressive strength of the concrete, the concrete fails by crushing and the member fails. Compared with plain concrete beams, both the compressive strength of concrete and the tensile strength of steel bars have been fully utilized. The failure load and bearing capacity of reinforced concrete beams have been greatly improved. There are obvious cracks and deformations in the members before failure occurs, which has significant advance warning. It belongs to ductile failure, which can minimize property losses and casualties in actual projects.

除了在受弯构件的受拉区配置钢筋,在结构构件中可能出现拉应力的位置都应该配置钢筋,如在梁中配置箍筋提高构件的抗剪承载力,配置钢筋网来防止出现温度裂缝或收缩裂缝等。还可以在构件的受压区配置受压钢筋,提高构件的承载力和延性,如采用双筋截面和在柱受压区配置钢筋等。

In addition to the arrangement of reinforcement in the tension zone of flexural members, steel bars should be arranged where tensile stress may occur in structural members. For example, stirrups are arranged in beams to increase the shear capacity of the members, and steel meshes are arranged to prevent temperature cracks or shrinkage cracks, etc. The compression reinforcement can also be configured in the compression zone of the member to improve the bearing capacity and ductility of the member, such as the use of doubly reinforced section and the configuration of steel bars in the compression zone of the column, etc.

因此,钢筋混凝土结构构件充分利用混凝土和钢材两种材料的力学特点,在荷载作用下,混凝土和钢筋共同作用,发挥各自的优点。

Therefore, reinforced concrete structural members make full use of the mechanical characteristics of the two materials of concrete and steel. Under the action of load, concrete and steel bars work together to play their respective advantages.

1.2.2 钢筋和混凝土共同工作的原因
Reasons why steel and concrete work together

在混凝土中加筋的主要目的是抵抗结构构件中产生的拉应力,为什么一般把钢筋作为加筋材料?

The main purpose of reinforcement in concrete is to resist the tensile stress generated in structural members. Why are steel bars usually used as reinforcement materials?

(1) 钢筋和混凝土之间存在很好的黏结力。混凝土结硬后,能与钢筋牢固地黏结在一起,相互传递内力。在荷载作用下,可以保证两者协调变形、共同受力、整体工作,两者的接触面不会出现滑移、开裂等现象,这是钢筋和混凝土两种性质不同的材料共同工作的基础。

(1) There is a good bond between steel and concrete. After the concrete is hardened, it can be firmly bonded with the steel and transfer internal forces to each other. Under the action of load, it can ensure that the coordinated deformation, joint force and overall work of the two. The contact surface of the two will not appear slip or crack, etc. This is the basis for the joint work of reinforcement and concrete with different properties.

(2) 钢筋与混凝土的温度线膨胀系数接近。钢筋的温度线膨胀系数为 $1.2\times10^{-5}/℃$,混凝土的温度线膨胀系数为 $(1.0\sim1.5)\times10^{-5}/℃$,温度变化时,钢筋与混凝土的黏结力不会因温度变化引起两者较大的相对变形而导致破坏。

(2) The thermal linear expansion coefficient of reinforcement and concrete is close. The thermal linear expansion coefficient of reinforcement is $1.2 \times 10^{-5}/℃$, and the thermal linear expansion coefficient of concrete is $(1.0\text{-}1.5)\times 10^{-5}/℃$. When the temperature changes, the bond force between steel and concrete will not be damaged due to the large relative deformation of reinforcement and concrete caused by temperature

changes.

（3）钢筋外的混凝土保护层包裹住钢筋，可以防止钢筋锈蚀，保证结构的耐久性；能提高钢筋的耐火性能，在遭受火灾时不致因钢筋很快软化而导致结构整体破坏。

(3) The concrete protective layer outside reinforcement wraps reinforcement to prevent steel corrosion and ensure the durability of the structure; it can improve the fire resistance of reinforcement, and will not cause overall collapse of structure due to the rapid softening of reinforcement in the event of a fire.

另外，钢筋的弹性模量是混凝土的6~10倍，在相同变形下钢筋能承担更大的应力，有利于钢筋强度的充分利用。钢筋在生产、施工、经济方面有比较显著的优势。

In addition, the elastic modulus of reinforcement is 6-10 times that of concrete, and reinforcement can bear greater stress under the same deformation, which is conducive to the full utilization of the reinforcement strength. The reinforcement has obvious advantages in production, construction and economy.

1.2.3 钢筋混凝土结构的特点
Characteristics of reinforced concrete structures

1. 钢筋混凝土结构的主要优点
Main advantages of reinforced concrete structures

钢筋混凝土结构在土木工程的各个领域得到广泛的应用，主要是因为其具有以下优点：

Reinforced concrete structures are widely used in various fields of civil engineering, mainly because of the following advantages：

（1）取材容易。混凝土所用砂、石等原材料来源广泛，易于就地取材，造价相对低廉，还可以利用建筑垃圾、工业固体废料如矿渣、粉煤灰等来制作人工骨料或胶凝材料，改善混凝土的性能，制造再生骨料混凝土。既可以废物利用，变废为宝，又有利于环境保护，实现建筑业的可持续发展。钢材的生产、加工比较简单，是用途广泛、价格相对低廉的材料。

(1) Easy to obtain materials. The raw materials such as sand and stone used in concrete have a wide range of sources, are easy to use local materials, and are relatively inexpensive. Construction waste, industrial solid wastes such as slag, fly ash, etc. can also be used to make artificial aggregates or cementitious materials to improve the performance of concrete and manufacture recycled aggregate concrete. It can not only use waste, turn waste into treasure, but also be beneficial to environmental protection and realize the sustainable development of the construction industry. The production and processing of steel are relatively simple, and it is a material with a wide range of uses and relatively low prices.

（2）合理用材。充分发挥钢筋和混凝土材料的力学性能，发挥各自的优势，两种材料的结合不需要特别的措施，解决了钢材容易失稳等方面的问题，结构具有较高的承载力，和钢结构相比可以降低造价。

(2) Reasonable use of materials. Give full play to the mechanical properties and respective advantages of reinforcement and concrete materials. The combination of two

materials requires no special measures and solves the problem of easy instability of steel. The structure has higher bearing capacity, which can reduce the cost compared with steel structure.

（3）可模性好。混凝土拌合物具有流动性、可塑性，可以利用模板将混凝土浇筑成各种形状和尺寸的构件或结构，以满足工程需要。

(3) Good mouldability. The concrete mixture has fluidity and plasticity, and concrete can be placed into various shapes and sizes of members or structures through using formwork to meet engineering needs.

（4）施工优势。对施工场地环境要求比较低，一般情况下，可以在任何需要的地方施工，对环境的破坏相对较小。施工工艺简单、成熟，施工机具和熟练工人容易获得，施工质量比较稳定。

(4) Construction advantages. The environmental requirements of the construction site are relatively low. Under normal circumstances, construction can be carried out wherever needed, and the damage to the environment is relatively small. The construction process is simple and mature, construction equipment and skilled workers are easy to obtain, and the construction quality is relatively stable.

（5）整体性好。现浇式或现浇整体式混凝土结构整体性好，具有抗震、抵抗振动和爆破冲击波、防辐射等多种用途。混凝土结构的刚度比较大，有利于变形控制。

(5) Good integrity. The cast-in-place or cast-in-place monolithic concrete structure has good integrity and has resistance to earthquake, vibration, blasting shock waves and radiation, etc. The rigidity of concrete structure is relatively large, which is conducive to deformation control.

（6）耐火性好。受到火灾等高温作用时，因混凝土传热性能比较差，钢筋外有足够厚度的混凝土保护层时不会像钢结构那样很快升温达到屈服强度而丧失承载力，从而提高结构的耐火极限。厚度为 30mm 的混凝土保护层可耐火 2h，比裸露的木结构和钢结构耐火性好。

(6) Good fire resistance. When subjected to high temperatures such as fire, the heat transfer performance of concrete is relatively poor, and there is concrete protective layer of sufficient thickness outside reinforcement, which will not heat up as quickly as steel structure and reach the yield strength, lose the bearing capacity, thereby increase the fire resistance limit of the structure. The concrete protective layer with the thickness of 30mm can resist fire for 2 hours, which is better than bare wood structure and steel structure.

（7）耐久性好。混凝土本身具有很好的化学稳定性，一般环境下，混凝土本身的性能不会退化，随着时间的增长，混凝土的强度还会有所增长，后期维护费用低。钢筋因混凝土保护层的存在而不易锈蚀。混凝土呈碱性，钢筋包裹在混凝土中，钢筋表面形成一层致密的氧化膜，能避免或延缓钢筋腐蚀，具有良好的耐久性，维修费用很低。

(7) Good durability. The concrete itself has good chemical stability. In general environment, the performance of concrete itself will not be degraded. With the increase of time, the strength of concrete will increase, and the maintenance costs in the later stage

will be low. The reinforcement is not easy to rust due to the existence of the concrete protective layer. The concrete is alkaline, the steel bars are wrapped in the concrete, and a dense oxide film is formed on the surface of reinforcement, which can avoid or delay the corrosion of reinforcement. It has good durability, maintenance costs are very low.

另外,钢筋混凝土结构比钢结构等其他结构形式造价相对便宜,性价比好。经过长时间的发展和工程应用,设计理论比较成熟可靠,结构安全性有保障。

In addition, the costs of reinforced concrete structures are relatively cheaper than steel structures and other structural forms, and the performance price ratio is good. After a long period of development and engineering application, the design theory is relatively mature and reliable, and the structural safety is guaranteed.

2. 钢筋混凝土结构的主要缺点
Main disadvantages of reinforced concrete structures

(1) 自重大。混凝土材料的重度较大,约为 $25kN/m^3$,比砌体和木材的重度都大,结构构件的截面尺寸比钢结构的大,因此结构的自重大。其自重远远超过相同跨度或高度的钢结构,不利于高层建筑结构、大跨度结构和结构抗震。目前正在大力研究与发展轻质、高强、高性能混凝土及预应力混凝土以减轻自重。

(1) Large self-weight. The weight of concrete material is relatively large, about $25kN/m^3$, which is greater than that of masonry and wood. The cross-sectional dimensions of structural members are larger than that of steel structures. Therefore, the self-weight of the structure is far more than that of the steel structure with the same span or height. It is not conducive to high-rise building structures, large-span structures and structural seismic resistance. At present, it is vigorously researching and developing lightweight, high-strength, high-performance concrete and prestressed concrete to reduce its own weight.

(2) 易开裂。混凝土材料的抗拉强度低,容易出现裂缝,普通钢筋混凝土结构在正常使用阶段往往是带裂缝工作的,如果裂缝宽度符合规范要求,就不会影响混凝土结构的正常使用。当裂缝数量较多、宽度较宽时,会给人带来不安全感。在工作条件较差的环境,影响结构的耐久性和适用性,不适用于对防渗、防漏要求较高的结构。对一些不允许出现裂缝或对裂缝宽度有严格控制的结构,采用预应力混凝土结构是解决混凝土开裂的有效途径之一,但要满足这些要求就需要提高工程造价。在混凝土中掺入适量纤维也能够提高混凝土的抗拉强度,增强混凝土的抗裂能力。

(2) Easy to crack. The tensile strength of concrete materials is low and cracks are easy to appear. Ordinary reinforced concrete structures often work with cracks in the normal service stage. If the crack width meets the specification requirements, it will not affect the normal use of concrete structures. When the number of cracks is large and the width is wide, people will feel a sense of insecurity. In an environment with poor working conditions, it affects the durability and applicability of the structure, and is not suitable for structures with high requirements for watertight and leakage. For some structures that do not allow cracks or have strict control of crack width, the use of prestressed concrete

structures is one of the effective ways to solve the cracking of the concrete, but to meet these requirements, it is necessary to increase the project cost. Adding proper amount of fiber into concrete can also improve the tensile strength of concrete and enhance crack resistance of concrete.

（3）施工复杂。混凝土结构的施工具有工序多、工期长、受季节及天气影响大等缺点。钢筋混凝土结构的建造需要经过支模板、绑钢筋、浇筑、养护、拆模等多道施工工序，工期长，湿作业多，施工质量和进度易受季节、气候等环境条件的影响。混凝土结构损伤修复比较困难，特别是某些隐蔽工程。采用早强混凝土、泵送混凝土、自密实混凝土等高性能混凝土，施工采用大模板、滑模、飞模、爬模等先进模板技术，可以大大提高混凝土工程的施工效率。

(3) The construction is complex. The construction of concrete structures has many shortcomings, such as many procedures, long construction period, and great influence by seasons and weather, etc. The construction of reinforced concrete structures requires multiple construction processes such as supporting formwork, binding steel bars, pouring, curing, and demoulding. The construction period is long and there are many wet operations. The construction quality and progress are easily affected by environmental conditions such as seasons and climate. It is difficult to repair the damage of concrete structure, especially for some concealed projects. The concrete adopts high-performance concrete such as early-strength concrete, pumped concrete, self-compacting concrete, and advanced formwork technologies such as large formwork, sliding formwork, flying formwork and climbing formwork, which can greatly improve the construction efficiency of concrete projects.

（4）承载能力有限。与钢材相比，混凝土的强度还是比较低的，用作高层建筑的底部结构时，往往需要比较大的构件尺寸，占用比较大的建筑空间。通常采用高强混凝土、钢骨混凝土或钢管混凝土等混合结构来解决这一问题。

(4) Limited carrying capacity. Compared with steel, the strength of concrete is relatively low. When used as the bottom structure of the high-rise building, it often requires relatively large member sizes and occupies relatively large building space. It is usually solved by mixed structures such as high-strength concrete, steel-reinforced concrete or concrete-filled steel tube.

（5）补强修复困难。混凝土一旦破坏，其修复、加固以及补强比较困难。采用植筋、粘贴钢板、粘贴碳纤维布、外包钢等加固技术，能够较好地对发生损坏的混凝土结构或构件进行修复。

(5) Difficulty in reinforcement and repair. Once the concrete is damaged, it is difficult to repair, strengthen and reinforce it. Reinforcement techniques such as planting steel bars, pasting steel plates, pasting carbon fiber cloth and wrapping steel outside can better repair damaged concrete structures or members.

随着科学技术的不断发展，以上缺点正逐渐被克服。

With the continuous development of science and technology, the above shortcomings are gradually being overcome.

1.3 混凝土结构的应用与发展
Application and development of concrete structures

1.3.1 混凝土结构的历史与应用
History and application of concrete structures

1. 混凝土结构的历史
History of concrete structures

钢筋混凝土是当今最主要的建筑材料之一,其历史不长,但发展迅速,使用广泛。它的发明者是一位名叫约瑟夫·莫尼埃的法国园艺师。1867年,莫尼埃移栽花时,不小心打碎了一盆花,花根四周的土却紧紧抱成一团。他从这件事中得到启发,将铁丝仿照花木根系编成网状,然后和水泥、砂、石一起搅拌做成花坛,果然十分牢固。1861年,他获得了钢筋混凝土板、管道和拱桥的专利。1849年,法国的蓝波特建造了一艘钢丝网水泥砂浆小船,并于1855年在巴黎博览会上展出。1867年,法国工程师艾纳比克在巴黎博览会上看到莫尼尔用铁丝网和混凝土制作的花盆、浴盆和水箱后,受到启发,设法把这种材料应用于房屋建筑上。1879年,他开始制造钢筋混凝土楼板,几年后,他在巴黎建造公寓大楼时采用了经过改善迄今仍普遍使用的钢筋混凝土柱、梁和楼板。

英文翻译1-1

1872年,世界上第一座钢筋混凝土结构的建筑在美国纽约落成。1875年,莫尼埃主持建造了世界上第一座钢筋混凝土大桥。1884年德国建筑公司购买了莫尼埃的专利,进行了第一批钢筋混凝土的科学实验,研究了钢筋混凝土的强度、耐火性能,钢筋与混凝土之间的黏结力等。1884年,德国威士、包辛格和康纳等提出了钢筋应配置在构件中受拉力的部位和钢筋混凝土板的计算理论,钢筋混凝土结构逐渐得到了推广应用。英国 W. D. 威尔森申请了钢筋混凝土板的专利。美国 T. 海厄特对混凝土梁进行了实验。1895—1900 年,法国用钢筋混凝土建成了第一批桥梁和人行道。1900 年,万国博览会上展示了钢筋混凝土在很多方面的应用。钢筋混凝土的发明以及 19 世纪中叶钢材在建筑业中的应用使高层建筑与大跨度桥梁的建造成为可能。1903 年,我国建成第一座钢筋混凝土建筑——上海东风饭店。1872 年,美国沃德建造了第一个无梁平板,从此钢筋混凝土构件进入工程实用阶段。1918 年,艾布拉姆发表了著名的计算混凝土强度的水灰比理论,初步奠定了混凝土强度的理论基础。此后,相继出现了轻集料混凝土、加气混凝土及其他混凝土,各种混凝土外加剂也开始使用。1922 年,广州建成国内第一座钢筋混凝土高层建筑——南方大厦(12 层)。1928 年,法国工程师弗来西奈发明了预应力混凝土,并于第二次世界大战后广泛用于建造大跨度结构、高层建筑结构以及对抗震、防裂等有较高要求的结构,大大扩展了混凝土结构的应用范围。1928 年,混凝土收缩和徐变理论在法国被提出。其后钢筋混凝土与预应力混凝土在分析、设计、施工及科研等方面迅速发展,出现了许多独特的建筑物,如美国波士顿市的 Kresge 大会堂、英国的 1951 节日穹顶、美国芝加哥市的 Marina 摩天大楼、湖滨公寓等建筑物。

20 世纪 50 年代至今,出现了轻质、高强、高性能的混凝土和高强、高延性、低松弛的钢筋与钢丝等新型结构材料。20 世纪 60 年代以来,由于广泛应用减水剂,一些高分子材料开始进入混凝土材料领域,出现了聚合物混凝土。多种纤维被用于分散配筋的纤维混凝土。现代测试技术也越来越多地应用于混凝土材料科学的研究。

2. 混凝土结构的应用

Application of concrete structures

英文翻译 1-2

混凝土结构历史不长,但发展迅速,使用广泛,我国超过 $100m$ 高的高层建筑绝大多数是混凝土结构或混凝土和钢的组合结构。混凝土结构的耐久性和耐火性都比钢结构优越,因此在建筑工程、水利工程、道路桥梁工程、给排水工程中得到了广泛应用,应用范围日益扩大。混凝土结构不仅大量应用于住宅、办公楼、工业厂房等工业与民用建筑中,还广泛应用于大跨度、高耸、重载结构中。

水利工程中的大坝、拦海闸墩、渡槽、港口等多采用混凝土结构,钢筋混凝土建造的水闸、水电站、船坞和码头已经星罗棋布,如黄河上的刘家峡、龙羊峡、小浪底水电站和长江上的葛洲坝水利枢纽工程、三峡工程等。

拓展知识 1-2

目前世界上最高的混凝土重力坝是瑞士狄克桑斯大坝,坝高 $285m$,坝长 $695m$,库容量 4 亿 m^3。世界上最大的水利工程长江三峡水利枢纽工程的大坝高 $185m$,坝体混凝土用量达 1527 万 m^3。我国龙羊峡水电站拦河大坝是混凝土重力坝,坝高 $178m$,坝顶宽 $15m$,坝底宽 $80m$,坝长 $393.34m$。

在大跨度建筑方面,预应力混凝土屋架、薄腹梁、钢筋混凝土拱以及薄壳等已得到广泛应用。法国巴黎国家工业与发展技术展览中心大厅为薄壳结构,屋顶平面为三角形,边长 $219m$,屋盖结构采用拱身为钢筋混凝土装配整体式薄壁结构的落地拱,跨度为 $206m$。意大利都灵展览馆拱顶由装配式混凝土构件组成,跨度达 $95m$。澳大利亚悉尼歌剧院的主体结构由三组巨大的壳片组成,壳片曲率半径为 $76m$,建筑涂白色,状如帆船,已成为世界著名的风光建筑。美国底特律的韦恩县体育馆屋顶为圆形平面,直径 $266m$。

钢筋混凝土和预应力混凝土桥梁也有很大的发展,如著名的武汉长江大桥引桥、福建乌龙江大桥、四川泸州大桥等,城市道路立交桥也发展迅速。跨度较大的混凝土拱桥如重庆市万州区长江大桥,为劲性骨架钢管混凝土上承式拱桥,主跨 $420m$。较大的铁路拱桥如广深港高铁广深段的骝岗涌大桥,$160m$ 的主跨采用预应力混凝土连续梁与钢管拱组合结构。上海杨浦大桥主跨 $602m$,桥塔和桥面均为混凝土结构。

钢筋混凝土结构在特种结构中也得到广泛的应用,如核电站安全壳、反应堆压力容器、热电厂的冷却塔、储水池、储气罐、上下水管道、海洋石油平台、巨型运油船、填海造地工程、地铁支护和站台工程、大吨位水压机机架、飞机场跑道、电视塔等,解决了钢结构难以解决的技术问题。

随着滑模等施工技术的发展,许多高耸建筑可以采用混凝土结构。如加拿大多伦多电视塔(高 $553.33m$)、我国上海东方明珠广播电视塔(高 $468m$),它们的主体均为混凝土结构。

任务 1-3:查找混凝土结构在建筑工程、水利工程、道路桥梁工程、给排水工程中应用的工程图片和介绍,了解其结构背景资料。

Task 1-3: Find the engineering pictures and introductions of concrete structures in architectural engineering, water conservancy project, road and bridge engineering, water supply and drainage engineering, and understand their structural background information.

1.3.2 混凝土结构的发展
Development of concrete structures

英文翻译 1-3

混凝土结构的发展可概括为：材料强度不断提高，计算理论趋于完善，施工机械化程度越来越高，建筑物向大跨、高层发展。

混凝土材料的主要发展方向是轻质、高强、高性能、耐久、复合、抗裂和环保。新型外加剂的研制与应用将不断改善混凝土的力学性能，以适应不同环境、不同要求的混凝土结构。掺入外加剂或胶凝材料除了可以提高强度，还可以显著改善和提高混凝土的施工性能、耐久性能和工作性能。随着高强度钢筋、高强高性能混凝土（强度可达 $100N/mm^2$）以及高性能外加剂和混合材料的研制使用，高强、高性能混凝土的应用范围不断扩大。具有防射线、耐磨、耐腐蚀、防渗透、保温等特殊性能的混凝土也正在研究中。

混凝土内掺纤维材料或有机材料，可以制成纤维混凝土、聚合物混凝土等，钢纤维混凝土和聚合物混凝土的研究和应用有了很大发展，用于一些特殊工程领域。利用建筑垃圾等制作再生骨料，可以制成再生混凝土。为改善混凝土自重大的缺点，各种轻质混凝土得到了开发应用，其重力密度（简称重度）一般不大于 $18kN/m^3$，如陶粒混凝土、浮石混凝土、火山渣混凝土、膨胀矿渣混凝土等。钢筋的发展方向是高强、防腐、有较好延性和高黏结锚固性能等，为提高钢筋的防腐性能，带有环氧树脂涂层的热轧钢筋已在某些有特殊防腐要求的工程中应用。

康纳于 1886 年发表了第一篇关于混凝土结构的理论与设计手稿。1887 年德国工程师科伦发表了钢筋混凝土的计算方法。1922 年英国的狄森提出了受弯构件按破损阶段的计算方法。从混凝土应用的初期直到 20 世纪 20 年代，混凝土结构设计计算采用的都是弹性理论，采用允许应力设计法，不考虑混凝土的非线性性质。这一阶段，所采用的钢筋和混凝土的强度都比较低，混凝土仅应用在简单的结构和构件中，主要用于建造中小型楼板、梁、拱和基础等构件。

1950 年苏联根据极限平衡理论制定了"塑性内力重分布计算规程"，1955 年颁布了极限状态设计法，从而结束了按破坏阶段的设计计算方法。由于第二次世界大战后许多大城市百废待兴，重建任务繁重，工程中大量应用预制构件和机械化施工以加快建造速度。继苏联提出极限状态设计法之后，1970 年英国、德国、加拿大、波兰相继采用此方法，并在欧洲混凝土委员会与国际预应力混凝土协会（CEB-FIP）第六届国际会议上提出了混凝土结构设计与施工建议，形成了设计思想方面的国际化统一准则。由于近代钢筋混凝土力学这一新的学科的科学分支逐渐形成，以统计数学为基础的结构可靠性理论已逐渐进入工程实用阶段。电算的迅速发展使复杂的数学运算成为可能，设计计算依据的是概率极限状态设计法。

预应力技术的完善与普及使建筑结构、桥梁的跨度普遍增大。预应力混凝土的应用不仅克服了钢筋混凝土易产生裂缝的缺点，又对材料强度提出更高的要求，高强混凝土及钢筋的发展反过来又促进预应力混凝土结构应用范围的不断扩大，如高层建筑、桥梁和隧道、海洋平台、压力容器、飞机跑道及公路路面等。

钢管混凝土在建筑工程中,特别是高层、超高层建筑上应用越来越多。钢管内浇筑高强混凝土可以使柱的承载力大幅提高,从而减小柱的截面尺寸。混凝土-型钢复合结构(即型钢外包裹混凝土)越来越多地在建筑结构上使用,可大幅度提高结构承载力和建筑的耐火安全性。

高强混凝土(C80～C120)主要在超高层建筑的抗压结构上使用。超高性能混凝土(UHPC,C120～C200)和超高性能钢筋混凝土在一些防爆结构、薄壳结构、大跨度结构和高耐久性结构上应用。振动台试验、拟动力试验以及风洞试验较普遍地开展,提高了抗震设计水平。计算机模拟试验大大减少了试验工作量,节约了大量人力和物力。有限元法得到广泛应用,人们还在创立和发展其他数值计算方法。

随着混凝土结构在工程中的大量应用,我国在混凝土结构的科研方面也取得较大进展。模型试验技术和计算机仿真技术日益发展,在混凝土结构基本理论和设计方法、可靠度理论、高层、大跨、特种结构、工业化建筑体系、结构抗震、现代化测试技术等方面的研究取得了很多新成果,某些方面已达到或接近国际水平。

  拓展知识 1-3

1.4 课程学习目标、内容及要求
Learning objectives, contents and requirements of the course

1.4.1 课程的学习目标
Learning objectives of the course

本课程是一门实践性很强,与现行规范、标准、图集等密切相关的课程。要求学生通过学习,熟悉混凝土和钢筋的材料性能,掌握钢筋混凝土梁、板、柱等基本构件的基本计算理论、设计方法及构造要求,熟悉变形和裂缝计算,了解混凝土结构耐久性的一般知识,具有一般钢筋混凝土结构构件的设计计算能力,为毕业后在混凝土结构领域继续学习提供坚实的基础,为将来进行钢筋混凝土楼盖结构设计、单层厂房结构设计和多层、高层结构设计等打下基础,将来在土木工程设计、施工和管理中能根据相关规范、标准、图集等对钢筋混凝土结构构件进行设计计算、校核等。

This course is a highly practical course closely related to current codes, standards, atlas, etc. Through learning, students are required to familiarize themselves with the material properties of concrete and reinforcement, master the basic calculation theory, design methods and structural requirements of basic members such as reinforced concrete beams, slabs, columns, etc., familiarized themselves with the calculation of deformation and cracks, understand the general knowledge of the durability of concrete structures, possess the design and calculation ability of general reinforced concrete structural members. It provides a solid foundation for further study in the field of concrete structure after graduation. It will lay a foundation for the future design of reinforced concrete floor structure, single-storey plant structure and multistory and high-rise structures. In the future, in civil engineering design, construction and management, reinforced concrete structural members can be designed, calculated and checked according to relevant specifications, standards, atlases, etc.

通过对《混凝土结构设计标准》(GB/T 50010—2010)等规范的学习和应用,了解土木工程专业相关的法律法规和行业标准,理解土木工程师的责任。理解并遵守土木工程师的职业道德和行为规范,培养严谨、科学的思维方式和认真、负责、细致的态度,树立责任担当、贡献国家、服务社会的职业观念。

Through the study and application of the "Standard for design of concrete structures" (GB/T 50010—2010) and other codes, understand the relevant laws, regulations and trade standards of civil engineering, and understand the responsibilities of civil engineers. Understand and abide by the professional ethics and codes of conduct of civil engineers, cultivate rigorous, scientific way of thinking and serious, responsible and meticulous attitude and establish professional concept of taking responsibility, contributing to the country, and serving the society.

任务 1-4:依据课程学习目标,制订个人学习目标。

Task 1-4: Develop personal learning goals based on the learning goals of the course.

1.4.2 课程的学习内容
Learning contents of the course

钢筋混凝土结构是由一系列受力类型不同的构件组成的,这些构件称为基本构件,按其主要受力特点的不同可以分为:受弯构件,如梁、板等;受压构件,如柱、剪力墙、屋架的压杆等;受拉构件,如屋架的拉杆、水池的池壁等;受扭构件,如带有悬挑雨篷的过梁、框架的边梁等。还有一些复合受力构件,受力比较复杂,如压弯构件、拉弯构件、弯扭构件、拉弯扭构件等。

Reinforced concrete structure is composed of a series of members with different types of forces. These members are called basic members. According to their main force characteristics, they can be divided into: flexural members, such as beams, slabs, etc; compression members, such as columns, shear wall, roof truss pressure bar, etc.; tension members, such as the tension rods of the roof truss, the pool wall of the pool, etc.; torsional members, such as the lintel with cantilevered canopy, the edge beam of the frame, etc. There are also some composite load-bearing members, the force is more complex, such as compression-flexural members, tension-flexural members, bending-torsion members, tension-bending-torsion members, and so on.

本课程主要学习钢筋和混凝土的材料性能、设计原则、受力性能、计算方法和构造要求等。研究的构件主要是受弯、受压、受拉、受扭构件。课程的学习内容主要包括:混凝土结构的概念、特点、发展及应用,混凝土和钢筋的物理力学性能,钢筋与混凝土的黏结性能,混凝土结构设计基本原则,单筋、双筋矩形截面及 T 形截面受弯构件正截面受弯承载力计算,受弯构件斜截面承载力计算,受压构件正截面承载力计算,受拉构件正截面承载力计算,受扭构件扭曲截面承载力计算,受弯构件挠度验算,构件裂缝宽度验算,混凝土结构的耐久性等内容。

This course mainly teaches the material properties of reinforcement and concrete, design principles, force performance, calculation methods and structural requirements, etc. The members studied are mainly bending, compressive, tensile and torsional members. The learning content of this course mainly includes: the concept, characteristics, development and application of concrete structures, the physical and mechanical properties of concrete and steel bars, the bonding properties between steel and concrete, the basic principles of concrete structure design, singly reinforced, doubly reinforced rectangular section, and T-shaped section calculation of flexural capacity of normal section of flexural members, calculation of bearing capacity of inclined section of flexural members, calculation of bearing capacity of normal section of compression members, calculation of bearing capacity of normal section of tension members, calculation of bearing capacity of torsion section of torsion members, deflection calculation of flexural members, crack width checking calculation of members, durability of concrete structures, etc.

1.4.3 课程的学习要求
Learning requirements of the course

1. 正确理解和使用公式

Correctly understand and use the formula

课程中各类公式、符号较多,有些公式比较复杂,参数较多。与力学中的公式有所不同,理论力学中研究的材料是刚性材料,材料力学、结构力学中研究的是理想弹性或塑性材料;而钢筋混凝土结构研究的是非匀质、非弹性材料。混凝土结构材料自身性能较复杂,有些方面的设计理论还不够完善,公式及其参数取值还需参照试验资料的统计分析,处于半理论、半经验状态。学习时要理解公式是建立在科学实验和工程经验基础上的,虽然不完全科学严谨,却能够较好地反映结构的真实受力性能。学习公式时要了解公式的来源、公式出现的试验背景及基本假定,注意公式与试验研究结果的关系,重视受力性能分析,注意计算公式的适用条件。不要死记硬背,要结合实际,在推导、理解的基础上逐步记忆,并能正确应用公式解决工程问题。

2. 重视结构构造要求

Pay attention to the structural requirements

钢筋混凝土结构设计除了满足计算要求,还应满足各种构造要求。构造要求是在建筑结构设计中,为保证结构安全或正常使用,在构造上考虑各种难以分析计算的因素,一般指不通过计算而必须采取的各种细部措施。构造要求与计算是同等重要的,构造要求是对计算的必要补充,是根据长期的科学实验与大量的工程实践积累起来的,截面设计计算必须辅以良好的构造设计才能保证结构构件完成预定的功能。学习时要避免出现重计算、轻构造的思想,要充分重视对构造规定和要求的理解,并会具体应用。

3. 考虑结构设计的综合性
Consider the comprehensiveness of structural design

概念设计、设计计算和构造设计是结构设计不可或缺的三方面内容。设计时要考虑适用、经济、安全、施工可行，还要进行方案比较、构件选型、强度和变形的计算并满足构造要求。结构设计是综合性问题，设计时要多方面比较，设计结果不是唯一的。应对各种方案进行比较，综合考虑可行性，才能确定较合适的设计结果。

4. 重视理论联系实际
Emphasis on linking theory with practice

本课程是一门实践性很强的课程，学习时要避免出现重理论、轻实践的思想。除学习设计理论外，还要完成必要的作业、讨论、实验、课程设计等。混凝土结构的基本理论是以实验为基础的，除课堂理论学习外，学生还要通过实验，掌握钢筋混凝土矩形截面梁的正截面、斜截面承载力实验方法、测试手段、仪表识读，了解挠度变化及裂缝出现和发展过程，掌握受弯构件正截面的开裂荷载和极限承载力的测定方法，掌握受弯构件斜截面极限承载力的测定方法。通过识读、绘制结构施工图，加强动手能力的培养，培养工程素质和职业能力。通过参观、实习等深入工程现场，增加感性认识，积累工程经验，加深对知识的理解和应用，做到理论联系实际。

5. 能正确理解和应用规范、标准、规程、图集等
Able to correctly understand and apply codes, standards, regulations, atlases, etc.

混凝土结构的设计、建设必须按照国家颁布的法规进行，设计人员必须遵照各种设计规范、标准、规程、图集等进行设计。结构设计规范、标准、规程、图集等是国家或各地区、行业颁布的关于结构设计计算和构造要求的技术规定和标准，规范条文尤其是强制性条文是设计中必须遵守的带有约束性和法律性的技术文件，目的是使工程结构在符合国家经济政策的条件下，保证设计的质量和工程项目的安全可靠，这些规范和标准等都具有立法性和约束性，设计、施工等工程技术人员都应遵循，认真执行。在学习过程中，要逐步熟悉和正确运用我国颁布的和课程相关的设计规范、标准、规程和图集等。

和本课程密切相关的规范、标准有《建筑结构荷载规范》(GB 50009—2012)、《建筑结构可靠性设计统一标准》(GB 50068—2018)、《工程结构通用规范》(GB 55001—2021)、《混凝土结构通用规范》(GB 55008—2021)、《混凝土结构设计标准》(GB/T 50010—2010)、《建筑抗震设计规范》(GB 50011—2010)(2016年版)等，学习时要注意熟悉规范条文，并正确应用规范。混凝土结构设计规范中的方法来源于设计基本原理，结构设计中应遵循设计规范的具体要求，但更要遵循设计基本原理，有关基本理论的应用最终都要落实到规范的具体规定。

现行国家标准设计图集中和本课程密切相关的图集是《混凝土结构施工图平面整体表示方法制图规则和构造详图(现浇混凝土框架、剪力墙、梁、板)》(22G101—1)。学习过程中应在学习规范、图集的基础上多看工程图纸，结合实际工程，理论联系实际。

任务 1-5：查找、借阅相关规范、标准、图集等课程学习资料。

Task 1-5: Find and borrow relevant codes, standards, atlases and other learning materials.

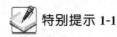 特别提示 1-1

建筑结构是指在房屋建筑中,由各种构件(屋盖、梁、板、柱等)组成的能够承受各种作用的体系。按照所用材料不同,建筑结构分为混凝土结构、砌体结构、钢结构和木结构。

按照组成建筑主体结构受力体系的不同,建筑结构可分为框架结构、剪力墙结构、框架-剪力墙结构、筒体结构、混合结构、大跨结构(排架结构、网架结构、悬索结构、壳体结构、膜结构)等。混凝土结构是以混凝土为主要建筑材料的结构,包括素混凝土结构、钢筋混凝土结构和预应力混凝土结构等。钢筋混凝土结构构件充分利用混凝土和钢材两种材料的力学特点,在荷载作用下,混凝土和钢筋共同作用,发挥各自的优点。

混凝土结构历史不长,但发展迅速,在土木工程的各个领域得到广泛的应用。因为它的耐久性和耐火性都比钢结构优越,因此在建筑工程、水利工程、道路桥梁工程、给排水工程中都有广泛应用,应用范围日益扩大。混凝土结构的发展可概括为:材料强度不断提高,计算理论趋于完善,施工机械化程度越来越高,建筑物向大跨、高层发展。近年来,混凝土结构从新材料、新技术的研究、开发和推广应用,到工程结构的建造,取得了惊人的巨大成就,创造了一个个新的纪录。

本课程是一门理论联系实际,与现行规范、标准、规程、图集等紧密相关的课程。课程主要学习钢筋和混凝土的材料性能、设计原则、受力性能、计算方法和构造要求等。学习本章内容后应明确课程学习目标,了解课程学习内容,熟悉课程学习要求,会确立个人学习目标,制订课程学习计划。

第1章拓展知识和特别提示　　　　　视频:第1章小结讲解

1-1　填空题

1. 建筑结构根据所用材料的不同分为_____结构、_____结构、_____结构和_____结构等。

2. 混凝土结构分为_____结构、_____结构、_____结构等,其中应用最广泛的是_____结构。

3. 钢筋混凝土结构是以_____为主要材料,并根据需要在其中合理配置_____,作为主要承重材料的结构。

4. 预应力混凝土结构指具有_____钢筋,通过_____或其他方法建立预加应力的混凝土结构。

5. 在混凝土中配置受力钢筋的主要作用是提高结构或构件的_____和_____。

6. 混凝土结构材料的主要发展方向是_____。

7. 钢筋混凝土结构的基本构件有_____构件、_____构件、_____构件、_____构件等。

1-2 名词解释

1. 建筑结构	2. 混凝土结构	3. 素混凝土结构	4. 钢筋混凝土结构
5. 预应力混凝土结构	6. 钢骨混凝土结构	7. 钢管混凝土结构	8. 钢-混凝土组合结构
9. 轻质混凝土	10. 绿色混凝土	11. 高强高性能混凝土	12. 纤维混凝土
13. 聚合物混凝土	14. 智能混凝土	15. FRP混凝土	

1-3 简答题

1. 钢筋混凝土结构有哪些优点和缺点？如何克服这些缺点？

2. 为什么要在混凝土内配置受力钢筋？

3. 钢筋和混凝土为什么可以在一起共同工作？

4. 混凝土结构在土木工程方面有哪些应用？举例说明。

5. 本课程的学习目标和学习内容是什么？写出你学习本课程的学习目标。

6. 学习本课程应注意哪些问题？谈谈你的想法。

7. 与本课程密切相关的规范和图集有哪些？

第 2 章

混凝土和钢筋的材料性能

Material properties of concrete and reinforcement

教学目标：

1. 掌握混凝土的立方体抗压强度、轴心抗压强度和轴心抗拉强度，理解复杂应力状态下混凝土的强度；了解荷载作用下混凝土的变形，了解混凝土的弹性模量、泊松比，理解混凝土的收缩和徐变。

2. 熟悉钢筋的形式和品种，理解短期荷载下钢筋的应力-应变关系，掌握钢筋的主要力学指标；了解钢筋的疲劳，理解钢筋的变形性能；了解钢筋混凝土结构对钢筋性能的要求。

3. 了解钢筋与混凝土之间黏结力的组成，理解钢筋与混凝土黏结的重要性和黏结机理，为后续内容的学习奠定基础。

导读：

混凝土，简写为"砼"，拼音为 tong，声调第 2 声。混凝土的出现可以追溯到古老的年代，其所用的胶凝材料为黏土、石灰、石膏、火山灰等。考古发现 5000 年前的凌家滩人已懂得"挖槽填烧土，木骨撑泥墙"的建筑工艺，这和如今的钢筋混凝土非常相似。1884 年，德国建筑公司进行了第一批钢筋混凝土的科学实验，研究了钢筋混凝土的强度、钢筋与混凝土的黏结力等。1900 年，万国博览会上展示了钢筋混凝土在很多方面的使用，在建材领域引起了一场革命。1918 年，艾布拉姆发表了著名的计算混凝土强度的水灰比理论。自 19 世纪 20 年代出现波特兰水泥，用它配制成的混凝土具有工程所需要的强度和耐久性，钢筋混凝土开始成为改变世界的重要材料。

混凝土具有原料丰富、价格低廉、生产工艺简单等特点，同时还具有抗压强度高、耐久性好等优点，混凝土抗拉强度较低，在结构中主要起抗压作用。混凝土已成为土木工程中用途最广、用量最大的一种建筑材料。

混凝土一般用在结构的承重部位。如果混凝土强度不足，可能导致结构出现裂缝、渗水、隔音效果不好等情况，严重的会导致整栋楼倒塌。出现混凝土强度不足的原因主要有两个方面，一种可能是选择了不合格产品，另一种可能是混凝土搅拌时的配比出现了问题，如水泥和砂的配比错了。如果混凝土强度不达标，建筑物的使用寿命就会减少，可能未到设计工作年限就不能使用了。另外，混凝土强度不达标还会导致建成的房屋抗震能力不够。在实际工程中,不少工程项目混凝土试件的制作、养护不符合标准的规定,给混凝土强度的评定带来一定难度。从养护上讲,作为混凝土强度的验收,必须实行标准养护。

钢筋是配置在钢筋混凝土及预应力混凝土构件中的钢条或钢丝的总称。钢筋种类很多,主要起抗弯、抗折作用。

混凝土和钢筋的力学性能对结构受力有很大影响,直接影响结构的正常和安全使用。强度和变形是本章的主要学习内容,强度是混凝土硬化后最重要的力学性能,是指混凝土抵抗压、拉、弯、剪等应力的能力。混凝土在荷载或温湿度作用下会产生变形,主要包括弹性变形、塑性变形、收缩、膨胀和温度变形等。本章是本课程的重要基础,必须很好地掌握。

引例:

江苏省某单层厂房屋面薄腹梁,11 榀屋面梁中仅 2 榀的混凝土强度超过设计值,4 榀梁与混凝土试块按现行标准不可作评定依据,1 榀梁无试块,其余 4 榀梁混凝土强度仅 20.9N/mm^2。因此,该工程因混凝土强度不足发生事故,性质较严重。

核心词汇:

复合应力	combined stress	劈裂试验	split test
标准试件	standard test specimen	温度变形	temperature deformation
混凝土立方体	concrete cube	收缩	shrinkage
换算系数	conversion coefficient	体积膨胀	volume expansion
立方体抗压强度标准值	cubic compressive strength standard value	重复荷载	repeated load
		延伸率	elongation
变形模量	deformation modulus	冷弯性能	cold bending property
弹性模量	elastic modulus	热轧钢筋	hot-rolled steel bar
混凝土强度等级	concrete strength grade	热处理钢筋	heat treated steel bar
割线模量	secant modulus	光圆钢筋	round bar
切线模量	tangent modulus	变形钢筋	deformed bar
应力-应变曲线	stress-strain curve	带肋钢筋	ribbed bar
轴心抗压强度	axial compressive strength	单轴受拉	uniaxial tension
轴心抗拉强度	axial tensile strength	软钢	mild steel
变形性能	deformation property	颈缩	necking
破坏形态	destruction form	塑性变形	plastic deformation
弹性极限	elasticity limit	比例极限	proportional limit
疲劳破坏	fatigue failure	屈服强度	yield strength
疲劳强度	fatigue strength	强化阶段	strengthen stage
长期荷载	long-term load	强度标准值	standard value of strength
徐变	creep	强度设计值	design value of strength
横向变形	lateral deformation	黏结作用	bonding effect
泊松比	Poisson's ratio	黏结应力	bond stress
棱柱体	prism	黏结机理	bonding mechanism
材料强度	material strength	拔出实验	pull-out test
强度等级	strength grade	黏结强度	bond strength

2.1 混凝土的基本性能
Basic properties of concrete

混凝土构件和结构的力学性能很大程度上取决于混凝土材料的性能。混凝土的性能包括混凝土的强度、变形、碳化、耐腐蚀、耐热、防渗等。本节主要阐述混凝土的强度和变形性能。

The mechanical properties of concrete members and structures are largely determined by the properties of concrete material. The properties of concrete includes the strength, deformation, carbonization, corrosion resistance, heat resistance, watertight, and so on. This section mainly describes the strength and deformation properties of concrete.

2.1.1 混凝土的强度
Strength of concrete

混凝土的强度是指它抵抗外力产生的某种应力的能力,即混凝土材料达到破坏时所能承受的应力。混凝土强度不仅与其材料组成等因素有关,还与其受力状态有关。混凝土的强度可分为立方体抗压强度、轴心抗压强度、轴心抗拉强度和复合应力状态下的强度等,前三个强度是单轴向受力状态下的混凝土强度。虽然实际工程中的混凝土结构和构件一般处于复合应力状态,但是单轴向受力状态下的混凝土强度是复合应力状态下强度的基础和重要参数。混凝土在结构中主要用作受压材料,单轴受压状态下的破坏过程最具有代表性。

The strength of concrete refers to its ability to resist some kind of stress produced by external force, that is, the stress that concrete material can sustain when it reaches failure. The strength of concrete is not only related to its material composition, but also to its stress state. The strength of concrete includes cubic compressive strength, axial compressive strength, axial tensile strength and strength under composite stress state. The first three strengths are concrete strength under uniaxial stress state. Although the concrete structures and members in practical engineering are generally in the state of composite stress, the concrete strength under uniaxial stress is the basis and important parameter of the strength under combined stress state. Concrete is mainly used as compression material in structure and the failure process under uniaxial compression is the most representative.

1. 混凝土立方体抗压强度和强度等级
Cubic compressive strength and strength grade of concrete

1) 混凝土立方体抗压强度的测定

1) Determination of cubic compressive strength of concrete

试验研究表明,混凝土试件在压力机上受压时,纵向压缩,横向膨胀。混凝土受压破坏是内部的微裂缝逐渐发展的结果,是其横向拉裂造成的。测定混凝土抗压强度的试件有立方体和圆柱体两种,《混凝土结构设计标准》(GB/T 50010—2010)规定采用立方体试件的抗压强度作为混凝土强度的基本指标,立方体试件的制作和试验都比较方便,强度也比较稳定,混凝土在其他各种受力情况下的强度都可与立方体抗压强度间建立起相应的换算关系。有些国家,如美国、日本等,采用直径150mm、高300mm圆柱体试件的抗压强度作为混凝土的强度指标。

Experimental research shows that when the concrete specimen is compressed on the press, it compresses longitudinally and expands laterally. Concrete compression failure is the result of the gradual development of internal micro-cracks, which are caused by its transverse tension splitting. There are two types of specimens for determining the compressive strength of concrete: cube and cylinder. "Standard for design of concrete structures" (GB/T 50010—2010) stipulates that the compressive strength of cubic specimens is used as the basic index of concrete strength. The production and testing of cube specimens are more convenient. The strength is also relatively stable, and the strength of concrete under various other stress conditions can establish a corresponding conversion relationship with the cubic compressive strength. Some countries, such as the United States, Japan, etc. use the compressive strength of a cylindrical specimen with a diameter of 150mm and a height of 300mm as the strength index of concrete.

混凝土试件的立方体抗压强度试验应根据《混凝土物理力学性能试验方法标准》(GB/T 50081—2019)的规定执行。立方体抗压强度是指按照标准方法制作的边长为150mm的立方体试件,在标准条件下养护28d或设计规定龄期,以标准试验方法测得的破坏时的平均压应力。混凝土立方体抗压强度试验的装置和试件的破坏情况如图2-1所示。

The cubic compressive strength test of concrete specimens should be carried out in accordance with the provisions of the "Standard for test methods of concrete physical and mechanical properties" (GB/T 50081—2019). The cubic compressive strength refers to the average compressive stress measured by standard test method when the cube specimen with side length of 150mm is made according to the standard method and cured for 28 days or design specified age under standard conditions. The apparatus of concrete cubic compressive strength test and the failure condition of the specimen are shown in Fig. 2-1.

(a) (b)

图 2-1 混凝土立方体抗压强度试验

(a) 试验装置;(b) 破坏情况

Figure 2-1 Cube compressive strength test of concrete

(a) Test device;(b) Failure condition

《混凝土结构设计标准》(GB/T 50010—2010)4.1.1 条指出：混凝土强度等级应按立方体抗压强度标准值确定。材料强度标准值是材料的一种特征值，其取值原则是：在符合规定质量的材料强度实测总体中，强度标准值应具有不少于 95% 的保证率，强度低于该值的概率为 5%。立方体抗压强度标准值以 $f_{cu,k}$ 表示，单位为 N/mm^2（也可记作 MPa）。

Article 4.1.1 of the "Standard for design of concrete structures"(GB/T 50010—2010) points out that the concrete strength grade should be determined according to the cubic compressive strength standard value. The characteristic value of material strength is a kind of characteristic value of material, and the principle of selecting its value is: the standard value of strength should have a guarantee rate of no less than 95% and the probability that the strength is lower than this value is 5%. The cubic compressive strength standard value is expressed in $f_{cu,k}$, and the unit is N/mm^2(also can be expressed as MPa).

特别提示 2-1

2）混凝土试件强度代表值的确定

2) Determination of strength representative value of concrete specimen

混凝土强度标准值是钢筋混凝土结构按极限状态设计时采用的材料强度的基本代表值。每组混凝土试件强度代表值的确定应符合下列规定：

（1）取 3 个试件强度的算术平均值作为每组试件的强度代表值；

（2）当一组试件中强度的最大值或最小值与中间值之差超过中间值的 15% 时，把最大值与最小值一起舍弃，取中间值作为该组试件的强度代表值；

（3）当一组试件中强度的最大值和最小值与中间值之差均超过中间值的 15% 时，该组试件的强度不应作为评定的依据。

试验表明，混凝土立方体试块尺寸越大，实测破坏强度越低，反之越高，这种现象称为尺寸效应。一般认为这是由混凝土内部缺陷引起的。

《混凝土强度检验评定标准》(GB/T 50107—2010)4.3.2 条规定：当采用非标准尺寸时，应将其抗压强度乘以尺寸折算系数，折算成边长为 150mm 的标准尺寸试件的抗压强度。尺寸折算系数 μ 按下列规定采用：

（1）当混凝土强度等级低于 C60 时，对边长为 100mm 的立方体试件取 $\mu=0.95$，对边长为 200mm 的立方体试件取 $\mu=1.05$。

（2）当混凝土强度等级不低于 C60 时，宜采用标准尺寸试件；使用非标准尺寸试件时，尺寸折算系数应由试验确定，其试件数量不应少于 30 对组。

拓展知识 2-1

任务 2-1：某工程压顶梁采用边长为 150mm 的立方体试件进行混凝土强度的检测，经过标准养护 28d 后测得受压破坏荷载分别为 757.4kN、733.3kN、734.2kN，计算该组混凝土的立方体抗压强度代表值。

Task 2-1：The concrete strength of the top beam of a project was tested by using cube specimens with side length of 150mm. After 28 days of standard curing, the compression failure loads were 757.4kN, 733.3kN and 734.2kN respectively, calculate the representative value of cubic compressive strength of this group of concrete.

3）混凝土强度等级

3) Concrete strength grade

在工程实际中，不同类型的结构和构件对混凝土强度的要求是不同的。为了应用方便，将混凝土强度按照其立方体抗压强度标准值 $f_{cu,k}$ 的大小划分为 13 个强度等级，即 C20、C25、C30、C35、C40、C45、C50、C55、C60、C65、C70、C75、C80。混凝土强度等级用符号 C 和混凝土立方体抗压强度标准值表示，例如 C30 表示立方体抗压强度标准值 $f_{cu,k}$ 为 30MPa（30N/mm^2）。

In engineering practice, different types of structures and members have different requirements for concrete strength. For the convenience of application, the concrete strength is divided into 13 strength grades according to the cubic compressive strength standard value $f_{cu,k}$, that is C20, C25, C30, C35, C40, C45, C50, C55, C60, C65, C70, C75, C80. The concrete strength grade is indicated by the symbol C and the cubic compressive strength standard value. For example, C30 represents that the standard value of cubic compressive strength $f_{cu,k}$ is 30MPa (30N/mm^2).

《混凝土结构通用规范》(GB 55008—2021)2.0.2 条关于混凝土强度等级的有关规定为：素混凝土结构构件的混凝土强度等级不应低于 C20；钢筋混凝土结构构件的混凝土强度等级不应低于 C25；预应力混凝土楼板结构的混凝土强度等级不应低于 C30，其他预应力混凝土结构构件的混凝土强度等级不应低于 C40；钢-混凝土组合结构构件的混凝土强度等级不应低于 C30；承受重复荷载作用的钢筋混凝土结构构件，混凝土强度等级不应低于 C30；抗震等级不低于二级的钢筋混凝土结构构件，混凝土强度等级不应低于 C30；采用 500MPa 及以上等级钢筋的钢筋混凝土结构构件，混凝土强度等级不应低于 C30。

The relevant provisions of Article 2.0.2 of the "General code for concrete structures" (GB 55008—2021) on concrete strength grade are as follows: the concrete strength grade of plain concrete structure members should not be lower than C20; the concrete strength grade of reinforced concrete structure members should not be lower than C25; the concrete strength grade of the prestressed concrete floor structure should not be lower than C30; the concrete strength grade of other prestressed concrete structure should not be lower than C40; the concrete strength grade of steel-concrete composite structural members should not be lower than C30; the concrete strength grade of reinforced concrete members sustaining repeated load should not be lower than C30. For reinforced concrete structural members whose seismic grade is not lower than Grade Ⅱ, the concrete strength grade should not be lower than C30; for reinforced concrete structural members using steel bars of 500MPa and above, the concrete strength grade should not be lower than C30.

 拓展知识 2-2

思考 2-1：按 C50 配置的混凝土，它的实测立方体抗压强度与 50MPa 是什么关系？

Thinking 2-1: What is the relationship between the measured cubic compressive strength and 50MPa of concrete configured according to C50?

2. 混凝土轴心抗压强度

Axial compressive strength of concrete

在实际工程中,钢筋混凝土受压时的影响区域呈棱柱体,所以采用棱柱体试件(即高度大于横截面边长的试件)比立方体试件能更好地反映混凝土的实际工作状态。采用棱柱体试件(100mm×100mm×300mm 或 150mm×150mm×450mm),按照测定立方体抗压强度的条件和方法测得的破坏时的平均压应力称为棱柱体抗压强度或轴心抗压强度。《混凝土物理力学性能试验方法标准》(GB/T 50081—2019)规定以 150mm×150mm×300mm 的棱柱体作为混凝土轴心抗压强度试验的标准试件,轴心抗压强度试验的受力分析和试件的尺寸效应如图 2-2 所示。

In practical engineering, the affected area under compression in reinforced concrete is prism, so the prism specimen (i.e. the specimen whose height is greater than the side length of the cross section) can better reflect the actual working conditions of concrete than the cube specimen. Using prism specimen (100mm×100mm×300mm or 150mm×150mm×450mm), the average stress at failure measured according to the conditions and methods for determining the cubic compressive strength is called prism compressive strength or axial compressive strength. "Standard for test methods of concrete physical and mechanical properties" (GB/T 50081—2019) stipulates that 150mm×150mm×300mm prism should be used as the standard specimen for axial compressive strength test of concrete. The force analysis and size effect of specimen in axial compression test are shown in Fig. 2-2.

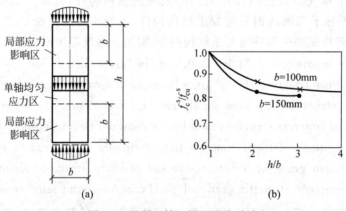

图 2-2 混凝土轴心抗压强度试验

(a) 受力分析;(b) 尺寸效应

Figure 2-2 Axial compressive strength test of concrete

(a) Stress analysis; (b) Size effect

棱柱体试件是在与立方体试件相同条件下制作的,经测试,棱柱体的抗压强度因受端部摩擦力的横向约束影响较小,试件强度不受端部摩擦力和附加偏心距的影响,中间处于均匀受压状态,比立方体抗压强度小。考虑到实际结构构件制作、养护和受力情况,实际构件强度与试件强度之间存在差异,《混凝土结构设计标准》(GB/T 50010—2010)基于安全取偏低

值,轴心抗压强度标准值 f_{ck} 与立方体抗压强度标准值 $f_{cu,k}$ 的关系按式(2-1)确定:

The prism specimen is made under the same conditions as the cube specimen. The test results show that the prism compressive strength is less affected by the lateral restraint of friction, and the strength of the specimen is not affected by the end friction and additional eccentricity, and the middle is under uniform compression, so it is smaller than the cubic compressive strength. In consideration of fabrication, maintenance and force conditions of the actual structural member, there are differences between the actual member strength and the specimen strength. "Standard for design of concrete structures" (GB/T 50010—2010) takes a low value based on safety. The relationship between the prism compressive strength standard value f_{ck} and the cubic compressive strength standard value $f_{cu,k}$ is determined according to Eq. (2-1):

$$f_{ck} = 0.88\alpha_{c1}\alpha_{c2}f_{cu,k} \tag{2-1}$$

式中,0.88 为考虑结构中混凝土的实体强度与立方体试件混凝土强度之间的差异而取用的修正系数;α_{c1} 为轴心抗压强度与立方体抗压强度之比,由试验分析可得,对混凝土强度等级为 C50 及以下的取 0.76,对 C80 混凝土取 0.82,对 C50~C80 混凝土在 0.76~0.82 之间按线性插值;α_{c2} 为高强混凝土的脆性折减系数,对 C40 及以下的混凝土取 1,对 C80 混凝土取 0.87,对 C40~C80 混凝土在 0.87~1 之间按线性插值。

Where, 0.88 is the correction coefficient to consider the difference between the real concrete strength in the structure and the concrete strength of the cube specimen; α_{c1} is the ratio of axial compressive strength on cubic compressive strength, which can be obtained from experimental analysis that for concrete strength grade lower than C50, it is 0.76, for C80 concrete, it is 0.82, and linear interpolation is used between 0.76 and 0.82 for C50-C80 concrete; α_{c2} is the brittleness reduction coefficient of high strength concrete, which is 1 for C40 and below, 0.87 for C80 concrete, and linear interpolation between 1.0 and 0.87 for C40-C80.

3. 混凝土轴心抗拉强度
Axial tensile strength of concrete

抗拉强度是混凝土的基本力学指标之一。对于不允许出现裂缝的混凝土受拉构件,如水池的池壁、有侵蚀性介质作用的屋架下弦等,轴心抗拉强度是确定混凝土构件的抗裂度和变形等的重要力学性能指标。混凝土的抗拉强度远低于抗压强度,对于普通混凝土,抗拉强度为 1/17~1/8 的抗压强度;对于高强混凝土,抗拉强度为 1/24~1/20 的抗压强度。混凝土强度等级越高,这个比值越小。因此,在工程实践中混凝土主要用于承受压力。测定轴心抗拉强度的试验方法有轴心受拉试验(直接测试法)、劈裂试验和弯折试验(间接测试法)。

英文翻译 2-2

中国建筑科学研究院等对混凝土的抗拉强度作了测定,试件用 100mm×100mm×500mm 的钢模浇筑成形,两端各预埋一根直径 16mm 的钢筋,钢筋埋入深度为 150mm 并置于试件的中心轴线上。试验时用试验机的夹具夹紧试件两端外伸的钢筋施加拉力,破坏时试件在没有钢筋的中部截面被拉断,如图 2-3 所示,其平均拉应力即为混凝土的轴心抗拉强度。

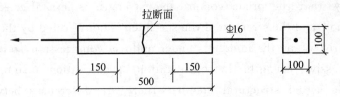

图 2-3　混凝土抗拉强度试验(直接测试法)

Figure 2-3　Tensile strength test of concrete(direct test method)

考虑到混凝土构件与试件的差别、尺寸效应、加载速度等因素的影响,《混凝土结构设计标准》(GB/T 50010—2010)考虑了从普通强度混凝土到高强混凝土的变化规律,取混凝土轴心抗拉强度标准值 f_{tk} 与立方体抗压强度标准值 $f_{cu,k}$ 之间的关系为

$$f_{tk}=0.88\times 0.395 f_{cu,k}^{0.55}(1-1.645\delta)^{0.45}\times \alpha_{c2} \tag{2-2}$$

式中,0.88、α_{c2} 的意义与式(2-1)中的相同;δ 为变异系数,对高强混凝土取 0.1;0.395、0.55 分别为轴心抗拉强度与立方体抗压强度间的折减系数。

由于混凝土内部的不均匀性、预埋的钢筋不易对中、安装试件的偏差等,混凝土轴心抗拉强度试验有时难以保持试件轴心受拉,准确测定抗拉强度是很困难的。因此,常用劈裂试验来间接测试混凝土的轴心抗拉强度,可以克服轴心受拉试验中的对中问题。劈裂试验的试件可做成圆柱体或立方体。劈裂试验用压力机通过垫条对试件中心面施加均匀线分布荷载,除垫条附近外,中心截面上将产生均匀的拉应力,当拉应力达到混凝土的抗拉强度时,试件会沿中间垂直截面劈裂成两半,如图 2-4 所示。

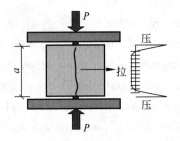

图 2-4　用劈裂试验测定混凝土抗拉强度

Figure 2-4　Determination of tensile strength of concrete by splitting test

根据弹性理论,混凝土轴心抗拉强度的试验值 f_t^0(上标 0 表示试验值)可按式(2-3)计算:

$$f_t^0=\frac{2P}{\pi la} \tag{2-3}$$

式中,P 为破坏荷载,kN;l 为圆柱体长度或立方体边长,mm;a 为圆柱体直径或立方体边长,mm。

考虑到实际构件和试件的区别、尺寸效应、加荷速度等影响,混凝土轴心抗拉强度 f_t 和立方体抗压强度 f_{cu} 的关系为

$$f_t=0.23 f_{cu}^{2/3} \tag{2-4}$$

混凝土轴心抗拉强度随混凝土强度等级的提高而增长,当混凝土强度等级为 C80 时,其轴心抗拉强度设计值约 C40 时的 1.3 倍。混凝土抗拉强度主要取决于硬化水泥浆和骨料之间的黏结强度。

> 拓展知识 2-3

4. 复合应力状态下的混凝土强度

Strength of concrete under combined stress state

混凝土结构构件很少处于理想的单向受力状态，一般受到轴力、弯矩、剪力及扭矩等共同作用，更多的是处于双向或者三向受力状态，例如框架梁承受弯矩和剪力的作用，框架柱承受弯矩、剪力和轴力的作用。框架节点区混凝土的受力状态更复杂。研究复合应力状态下混凝土材料的强度，对于更好地了解混凝土结构构件的性能，提高混凝土结构的设计及研究水平具有重要意义。

英文翻译 2-3

1）混凝土的双向受力强度

1) Strength of concrete under biaxial stress

双向应力状态下混凝土强度变化曲线如图 2-5 所示，图中 f_c 为单轴受力状态下的混凝土抗压强度，一旦超出包络线说明材料发生破坏。微分体在两个方向受到法向应力的作用，另一方向的法向应力为 0。由图 2-5 可知，在双向受拉应力状态下（第一象限），两个方向的应力 σ_1、σ_2 相互影响不大，双向受拉混凝土的强度与单轴受拉强度接近；在一向受压、一向受拉应力状态下（第二、四象限），混凝土强度低于单轴受压或者单轴受拉的混凝土强度；在双向受压应力状态下（第三象限），一个方向的混凝土强度随另一方向压应力的增大而提高，与单向受压混凝土强度相比，双向受压混凝土强度最多可提高 20%。

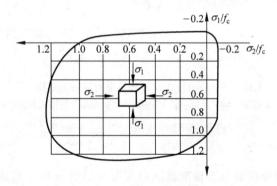

图 2-5 双向应力状态下的混凝土强度

Figure 2-5 Concrete strength under biaxial stress state

2）法向应力和剪应力组合状态下的混凝土强度

2) Strength of concrete under combined state of normal stress and shear stress

在法向应力与剪应力组合状态下的混凝土强度变化曲线如图 2-6 所示。当压应力较低时，混凝土的抗剪强度随压应力的增大而提高；但是当压应力超过 $0.5f_c$ 后，抗剪强度随应力的增大而减小；而由于剪应力的存在，混凝土的抗压强度要比单向抗压强度低，混凝土的抗剪强度随拉应力的增加而减小，混凝土的抗拉强度比单向抗拉强度低。

3）三向应力状态下的混凝土强度

3) Strength of concrete under triaxial stress state

三向受压下混凝土圆柱体的轴向应力-应变曲线可以由周围用液体压力加以约束的圆

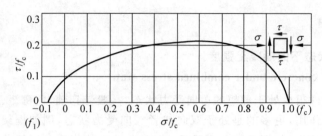

图 2-6 法向应力和剪应力组合状态下的混凝土强度

Figure 2-6 Concrete strength under the combined state of normal stress and shear stress state

柱体进行加压试验得到。在加压过程中保持液压为常值,逐渐增加轴向压力直至破坏,并量测其轴向应变的变化。如图 2-7 所示,随着侧向压力的增加,试件的强度和应变都有显著提高。混凝土在三向受压的情况下,由于受到侧向压力的约束作用,最大主压应力轴的抗压强度有较大程度的增长。

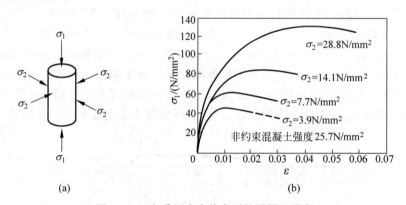

图 2-7 三向受压应力状态下的混凝土强度

(a) 混凝土三向受压试验;(b) 侧向约束对混凝土抗压强度的影响

Figure 2-7 Strength of concrete under triaxial compression stress state

(a) Triaxial compression test of concrete; (b) The influence of lateral restraint on the compressive strength of concrete

混凝土强度设计值应按其强度标准值除以混凝土材料分项系数确定,且材料分项系数取值不应小于 1.4。

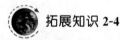 拓展知识 2-4

2.1.2 混凝土的变形

Deformation of concrete

混凝土的变形是其重要的物理力学性能之一。混凝土的变形可分为两类:一类是在荷载作用下的受力变形,如一次短期加载、长期加载和多次重复加载作用下产生的变形;另一类变形称为体积变形,如混凝土的收缩、膨胀以及温度、湿度变化等产生的变形。

The deformation of concrete is one of its important physical and mechanical properties. The deformation of concrete can be divided into two categories: one is the deformation under load, such as one short-term loading, long-term loading and multiple repeated loading; the other is called volume deformation, such as the deformation caused by shrinkage, expansion of concrete, temperature and humidity changes.

1. 混凝土在一次短期加载作用下的变形性能
Deformation properties of concrete under one short-term loading

1) 混凝土单轴受压时的应力应变关系

1) Stress-strain relationship of concrete under uniaxial compression

荷载作用下结构的性能很大程度上取决于材料的应力-应变关系。因为混凝土主要用于受压的情况，所以它的受压应力-应变曲线就很重要。混凝土在单轴受压状态下的应力应变关系是混凝土材料最基本的性能，是研究钢筋混凝土构件的承载力、变形、延性和进行受力全过程分析的重要依据。一次短期加载是指荷载从零开始单调增加至试件破坏，也称单调加载。一般采用标准棱柱体或圆柱体试件测定混凝土受压时的应力-应变曲线，测试时在试件的侧面安装应变仪读取纵向应变。

Performance of structures under load depends to a large degree on the stress-strain relationship of the material. Since concrete is used mostly in compression, its compressive stress-strain curve is important. The stress-strain relationship of concrete under uniaxial compression is the most basic performance of concrete materials, and it is an important basis for studying the bearing capacity, deformation, ductility and full process analysis of stress of reinforced concrete members. One time short-term loading means that the load increases monotonically from zero to the failure of the specimen, which is also called monotonic loading. Generally, the standard prism or cylinder specimen is used to measure the stress-strain curve of concrete under compression, and strain gauges are installed on the sides of the specimen to read the longitudinal strain.

混凝土棱柱体试件轴心受压时典型的应力-应变曲线如图 2-8(a)所示，整个曲线分为上升段与下降段两个部分。上升段又分为三段，从加载至应力为 $(0.3 \sim 0.4) f_c^0$ 的 A 点为第 1 阶段，应力-应变关系接近直线，称 A 点为比例极限。超过 A 点，进入裂缝稳定扩展的第 2 阶段，至临界点 B，临界点的应力可作为确定长期抗压强度的依据。此后，试件中所积蓄的弹性应变能保持大于裂缝发展所需要的能量，从而形成裂缝快速发展的不稳定状态直至峰点 C，这一阶段称为第 3 阶段，这时的峰值应力通常作为混凝土棱柱体抗压强度的试验值，相应的应变称为峰值应变，其值在 0.0015～0.0025 之间波动，通常取 0.002。

The typical stress-strain curve of the concrete prism specimen under axial compression is shown in Fig. 2-8(a). The entire curve is divided into two branches: the ascending branch and the descending branch. The ascending branch is further divided into three sections. The first stage is from loading to point A where the stress is about $(0.3-0.4) f_c^0$. The stress-strain relationship is close to a straight line, and point A is called the proportional limit. Beyond point A, it enters the second stage of stable crack propagation, and reaches critical point B, where the stress at the critical point can be used as the basis for long-term compressive strength. After that, the elastic strain energy accumulated in the specimen remains greater than the energy required for crack development, thereby forming an unstable state of rapid crack development until the peak point C. This stage is called the third stage, and the peak stress at this time is usually taken as the test value of the compressive strength of the concrete prism, the corresponding strain is called the peak strain, and its value fluctuates between 0.0015 and 0.0025, usually taken as 0.002.

混凝土应力-应变曲线的形状和特征是混凝土内部结构发生变化的力学标志,不同强度混凝土的应力-应变曲线有相似的形状,但也有实质性的区别。由图 2-8(a)可知,混凝土棱柱体受压时的应力-应变图形是曲线,说明混凝土是一种弹塑性材料,只有在压应力很小时才可视为弹性材料。由图 2-8(b)可知,随着混凝土强度的提高,尽管上升段和峰值应变变化不显著,但是混凝土强度对应力-应变曲线下降段有较大影响,混凝土强度越高,应力下降越快,延性越差;混凝土强度越低,下降段越平缓,延性越好。另外,混凝土受压应力-应变曲线的形状与加载速度也有密切的关系。

The shape and characteristics of the stress-strain curve of concrete are the mechanical signs of the change of the internal structure of concrete. The stress-strain curves of concrete with different strength have similar shapes, but there are substantial differences. It can be seen from Fig. 2-8(a) that the stress-strain diagram of concrete prism under compression is a curve, which indicates that concrete is an elastic-plastic material, and only when the compressive stress is very small it can be regarded as elastic material. It can be seen from Fig. 2-8(b) that with the increase of concrete strength, although the change of rising section and peak strain is not significant, the concrete strength has a great influence on the descending section of stress-strain curve. The higher the concrete strength is, the

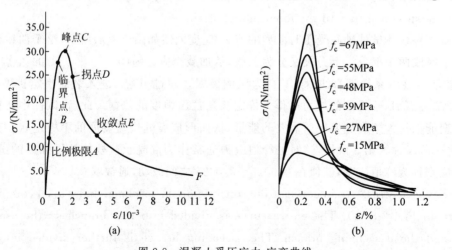

图 2-8 混凝土受压应力-应变曲线
(a)混凝土棱柱体受压应力-应变曲线;(b)不同强度的混凝土受压应力-应变曲线
Figure 2-8 Compressive stress-strain curves of concrete
(a) Compressive stress-strain curve of concrete prism;
(b) Compressive stress-strain curves of concrete with different strength

faster the stress decreases, and the worse the ductility is; the lower the concrete strength is, the gentler the descending section is, the better the ductility is. In addition, the shape of the stress-strain curve of concrete is closely related to the loading speed.

2) 混凝土的变形模量

2) Deformation modulus of concrete

与线弹性材料不同,混凝土受压应力-应变关系是曲线,在不同的应力阶段,应力与应变之比是变数,称为变形模量。混凝土的变形模量有三种表示方法:弹性模量、割线模量、切线模量。对钢筋混凝土结构,无论是进行超静定结构的内力分析,还是计算构件的变形、温度变化和支座沉降对结构构件产生的内力以及分析预应力混凝土构件等都要用到混凝土的变形模量。

英文翻译 2-4

混凝土棱柱体受压时,在应力-应变曲线的原点作一切线,其斜率为混凝土的原点模量,称为弹性模量。目前,各国对弹性模量的试验方法尚无统一标准。通常的做法是对标准尺寸的棱柱体试件,先加载至 $\sigma=0.5f_c$,然后卸载至零,再重复加载、卸载 5~10 次。由于混凝土不是弹性材料,每次卸载至零时存在残余变形,随着加载次数增加,应力-应变曲线渐趋稳定,基本上趋于直线,该直线的斜率即定为混凝土的弹性模量。《混凝土结构设计标准》(GB/T 50010—2010)通过试验统计分析得出弹性模量 E_c 与立方体抗压强度标准值 $f_{cu,k}$ 之间的关系,弹性模量计算的经验公式为

$$E_c = \frac{10^5}{2.2 + \frac{34.7}{f_{cu,k}}} \tag{2-5}$$

式中,E_c 和 $f_{cu,k}$ 的单位为 N/mm^2。

《混凝土结构设计标准》(GB/T 50010—2010)规定的混凝土受压或受拉的弹性模量见表 2-1。当有可靠试验数据时,弹性模量可根据实测数据确定;当混凝土中掺有大量矿物掺合料时,弹性模量可按规定龄期根据实测数据确定。

表 2-1 混凝土的弹性模量 E_c

Table 2-1 Elastic modulus of concrete E_c 单位:$10^4 N/mm^2$

混凝土强度等级	C20	C25	C30	C35	C40	C45	C50	C55	C60	C65	C70	C75	C80
E_c	2.55	2.80	3.00	3.15	3.25	3.35	3.45	3.55	3.60	3.65	3.70	3.75	3.80

 特别提示 2-2

混凝土的割线模量是指混凝土棱柱体受压时,连接混凝土应力-应变曲线的原点至曲线上任一点的割线的斜率。在混凝土应力-应变曲线上任一点处作一切线,则该切线处应力增量与应变增量之比称为该应力时混凝土的切线模量,即该切线的斜率。切线模量是一个变值,它随混凝土应力的增大而减小。

3) 混凝土的泊松比和剪变模量

3) Poisson's ratio and shear modulus of concrete

混凝土试件在一次短期加载(受压)时,纵向受到压缩,产生压应变,横向发生膨胀,产生

拉应变,则横向应变 ε_h 与纵向应变 ε_l 之比称为泊松比,在压应力较小时泊松比为 0.15~0.18,接近破坏时可达 0.5 以上,《混凝土结构设计标准》(GB/T 50010—2010)取 0.2。《混凝土结构设计标准》(GB/T 50010—2010)规定混凝土的剪变模量为 0.4 倍弹性模量。

When a short-term loading (compression) occurs, the concrete specimen is compressed longitudinally and produces compressive strain, while transverse expansion produces tensile strain. The ratio of the transverse strain ε_h to the longitudinal strain ε_l is called Poisson's ratio. When the compressive stress is small, Poisson's ratio is 0.15-0.18. It can reach more than 0.5 when approaching damage, "Standard for design of concrete structures" (GB/T 50010—2010) takes 0.2. According to the "Standard for design of concrete structures" (GB/T 50010—2010), the shear modulus of concrete is 0.4 times of elastic modulus.

4) 混凝土单轴受压应力-应变本构关系曲线(本构模型)

4) Stress-strain constitutive relationship curve of concrete under uniaxial compression (constitutive model)

应力-应变本构关系曲线的数学模型是应力-应变曲线的数学表达式,可根据某一应变值求出相应的应力值,应用于承载力计算和混凝土结构非线性分析。目前国内外已提出较多应力-应变本构关系曲线的数学模型,其中较常用的是美国 E. Hognestad 建议的模型,一般用于结构的非线性分析。其应力-应变曲线的上升段为二次抛物线,下降段为斜直线,如图 2-9 所示。表达式为

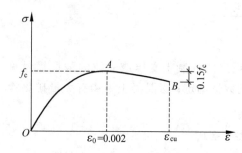

图 2-9 美国 E. Hognestad 建议的应力-应变关系

Figure 2-9 The stress-strain relationship suggested by E. Hognestad(USA)

The mathematical model of stress-strain constitutive relationship curve is a mathematical expression of stress-strain curve, which can be used to calculate the bearing capacity and nonlinear analysis of concrete structure. At present, many mathematical models of stress-strain constitutive relationship curve have been put forward at home and abroad, among which the model recommended by E. Hognestad(USA) is more commonly used, which is generally used for nonlinear analysis of structures. The ascending segment of the stress-strain curve is a quadratic parabola, and the descending segment is an diagonal straight line, as shown in Fig. 2-9. The expression is

上升段:

Ascending segment:

$$\varepsilon \leqslant \varepsilon_0, \quad \sigma = f_c \left[2\frac{\varepsilon}{\varepsilon_0} - \left(\frac{\varepsilon}{\varepsilon_0}\right)^2 \right] \tag{2-6}$$

下降段:

Descending segment:

第 2 章 混凝土和钢筋的材料性能
Material properties of concrete and reinforcement

$$\varepsilon_0 \leqslant \varepsilon \leqslant \varepsilon_{cu}, \quad \sigma = f_c\left(1 - 0.15\frac{\varepsilon - \varepsilon_0}{\varepsilon_{cu} - \varepsilon_0}\right) \tag{2-7}$$

式中，f_c 为峰值应力（混凝土轴心抗压强度）；ε_0 为相应于峰值应力时的应变，取 0.002；ε_{cu} 为极限压应变，取 0.0038。

Where, f_c is the peak stress (the axial compressive strength of concrete); ε_0 is the strain corresponding to the peak stress, taking the value as 0.002; ε_{cu} is the ultimate compressive strain, taking the value as 0.0038.

德国 Rüsch 建议的单轴受压应力-应变关系模型如图 2-10 所示。该模型形式较简单，上升段也采用二次抛物线，下降段则采用水平直线，表达式为

The stress-strain relationship model proposed by German Rüsch under uniaxial compression is shown in Fig. 2-10. The form of the model is simple. The ascending segment also adopts quadratic parabola, and the descending segment adopts horizontal straight line. The expression is

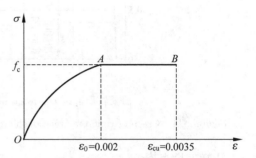

图 2-10 德国 Rüsch 建议的应力-应变关系
Figure 2-10 The stress-strain relationship suggested by Rüsch (Germany)

$$\varepsilon \leqslant \varepsilon_0, \quad \sigma = f_c\left[2\frac{\varepsilon}{\varepsilon_0} - \left(\frac{\varepsilon}{\varepsilon_0}\right)^2\right] \tag{2-8}$$

$$\varepsilon_0 \leqslant \varepsilon \leqslant \varepsilon_{cu}, \quad \sigma = f_c \tag{2-9}$$

式中，取 $\varepsilon_0 = 0.002$，$\varepsilon_{cu} = 0.0035$。
Where, take $\varepsilon_0 = 0.002$ and $\varepsilon_{cu} = 0.0035$.

2. 混凝土在长期荷载作用下的变形性能
Deformation properties under long-term load of concrete

1) 徐变的概念和对结构的影响
1) Concept of creep and its influence on structures

在荷载的长期持续作用下，应力不变，混凝土的变形随时间徐徐增长的现象称为徐变。徐变是混凝土在荷载长期作用下的重要变形性能，对于结构构件的变形、承载能力以及预应力钢筋中的应力都将产生重要影响。徐变会使钢筋与混凝土之间产生应力重分布，使混凝土应力减小，钢筋应力增大；徐变使受弯和偏心受压构件的受压区变形加大，使受弯构件挠度增加，使偏心受压构件的附加偏心距增大而导致构件承载力降低；使预应力构件产生预应力损失等。因此，设计中应考虑徐变对构件的影响。

典型的混凝土试件徐变随时间变化的关系曲线如图 2-11 所示。混凝土徐变在开始的前 4 个月增长很快，通常半年内可完成最终徐变量的 70%~80%，以后逐渐减慢，第一年内可完成 90% 左右，其余徐变在以后几年内逐渐完成，经过 2~5 年可以认为徐变基本结束，徐变量为加荷瞬时变形的 2~4 倍。

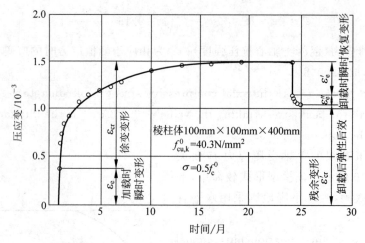

图 2-11 混凝土的徐变(加荷卸荷应变与时间关系曲线)

Figure 2-11　Creep of concrete (relationship curve of loading and unloading strain versus time)

2) 徐变的原因

2) Causes of creep

关于徐变产生的原因,目前尚无一致的解释。通常的理解为:①当应力不大时,混凝土中未晶体化的水泥凝胶体在持续的荷载作用下产生黏性流动,并把它承受的压力逐渐转移给骨料,骨料压应力增大,试件变形也随之增大;②当应力较大时,在荷载的长期作用下,混凝土内部的微裂缝不断发展和增加,也使徐变增大。

3) 影响徐变的因素

3) Factors affecting creep

影响徐变的因素很多,如应力大小、材料组成、外部环境等。混凝土徐变变形的大小和持续作用应力的大小成正比,持续作用的时间越长,徐变也越大;在保持应力不变的情况下,混凝土的加载龄期越长,徐变增长越小。

在混凝土的组成成分中,水灰比越大,水泥水化后残余的游离水越多,徐变也越大;在常用水灰比为 0.4～0.6 的情况下,徐变与水灰比呈线性关系;水灰比不变的情况下,水泥用量越多,凝胶体在混凝土中所占比重也越大,徐变也越大;水泥品种不同对徐变也有影响,用普通硅酸盐水泥制成的混凝土,其徐变要比火山灰质水泥或矿渣水泥制成的混凝土徐变大;混凝土中的骨料越坚硬,弹性模量越大,级配越好,骨料所占体积比越大,则由凝胶体流动后传给骨料的压力引起的变形也越小,徐变就越小。试验表明,当骨料所占体积比由 60% 增加到 75% 时,徐变量将减少 50%。此外,由于混凝土中水分的挥发逸散与构件的体积与表面积之比有关,因而构件尺寸越大,表面积相对越小,徐变就越小。

3. 混凝土在重复荷载作用下的变形性能
Deformation properties of concrete under repeated load

英文翻译 2-6

1) 疲劳破坏

1) Fatigue failure

对混凝土棱柱体试件加载,使其压应力达到某一数值,然后卸载到零,如此重复循环,称为多次重复荷载,如图 2-12 所示,混凝土在重复荷载下的变形性能明显不同于一次单调加载时的性能。当重复荷载达到某一循环次数时,在混凝土微裂缝、孔隙、弱骨料等内部缺陷处产生局部应力集中,引起骨料与砂浆间的黏结破坏,混凝土试件将因严重开裂或变形过大而破坏,这一现象称为疲劳破坏。混凝土发生疲劳破坏时无明显预兆,属于脆性破坏,开裂不多,但变形很大。

2) 疲劳强度

2) Fatigue strength

在实际工程中,工业厂房中的吊车梁在整个使用期限内吊车荷载作用重复次数可达 200 万～600 万次,因此在疲劳试验机上用脉冲千斤顶对 100mm×100mm×300mm 或 150mm×150mm×450mm 的棱柱体试件快速加荷、卸载的重复次数也不宜低于 200 万次。通常将加载应力为 $0.5f_c$,试件承受 200 万次(或更多次数)重复荷载,发生疲劳破坏时所能承受的最大压应力值称为混凝土的疲劳强度,如图 2-12 所示。

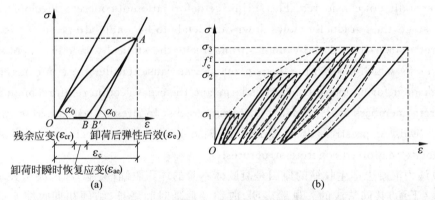

图 2-12　重复荷载作用下混凝土的应力-应变曲线
(a) 混凝土一次加荷卸载的应力-应变曲线;(b) 混凝土多次重复加荷卸载的应力-应变曲线

Figure 2-12　Stress-strain curve of concrete under repeated load
(a) The stress-strain curve of concrete under one loading and unloading;
(b) The stress-strain curve of concrete under repeated loading and unloading

试验表明,当混凝土承受变化的荷载而不是一直不变的荷载时,疲劳强度比静止时的强度低得多。混凝土的疲劳强度除了与静荷载、荷载重复次数和混凝土强度有关外,还与重复作用应力变化的幅度、湿度、龄期和加载速率有关。《混凝土结构设计标准》(GB/T 50010—2010)规定,混凝土的疲劳强度设计值(f_c^f、f_t^f)按混凝土强度设计值(f_c、f_t)乘以相应的疲劳强度修正系数确定,疲劳强度修正系数应根据不同的疲劳应力比查表确定;当混凝土受拉、压疲劳应力作用时,受拉或受压疲劳强度修正系数均为 0.6。混凝土的强度等级越高,疲劳强度也越高;荷载重复次数越多,疲劳强度越低;疲劳应力比值越小,疲劳强度越低。

4. 混凝土的收缩、膨胀和温度变形

Shrinkage, expansion and temperature deformation of concrete

1) 混凝土的收缩、膨胀

1) Shrinkage and expansion of concrete

混凝土在空气中凝结硬化时,多余的水分蒸发,混凝土硬化变干,体积会缩小,这种现象称为混凝土的收缩；而混凝土在水中结硬时体积会膨胀,恢复了很多之前由于收缩减少的体积,称为混凝土的膨胀。通常,收缩值比膨胀值大很多,膨胀一般是有利的,所以一般不考虑。混凝土的收缩变形早期发展较快,之后逐渐减慢,较长时间后趋于稳定。当混凝土受到约束不能自由收缩时,对跨度变化比较敏感的超静定结构（如拱）,将在混凝土中产生拉应力。当混凝土收缩变形较大、构件截面配筋较多时,混凝土构件将产生收缩裂缝。在预应力混凝土构件中,收缩会引起预应力损失。收缩也会对一些超静定钢筋混凝土结构产生不利影响。

The excess water evaporates, the concrete hardens and dries, the volume will shrink when concrete sets and hardens in the air. This phenomenon is called shrinkage of concrete. Conversely, if concrete hardens in the water, it expands regaining much of the volume loss from prior shrinkage, which is called expansion of concrete. In general, the expansion value is usually larger than that of shrinkage. Expansion is generally beneficial, so it is generally not considered. The shrinkage deformation of concrete develops rapidly in the early stage, then gradually slows down, and tends to be stable after a long time. When the concrete is constrained and cannot shrink freely, the statically indeterminate structure (such as arch) which is sensitive to span change can cause tensile stress in concrete. When the shrinkage deformation of concrete is large and the cross-section reinforcement is more, the concrete members will produce shrinkage cracks. In prestressed concrete members, shrinkage leads to prestress loss. Shrinkage also has adverse effects on some statically indeterminate reinforced concrete structures.

一般认为混凝土产生收缩的原因是凝胶体本身的体积收缩（凝缩）和混凝土失水产生的体积收缩（干缩）共同造成的。试验表明,收缩变形随时间增长,结硬初期收缩变形发展很快,以后逐渐减慢。蒸汽养护时,由于高温高湿条件能加速混凝土的凝结和硬化过程,减少混凝土中水分的蒸发,混凝土的收缩值比常温养护时小。影响混凝土收缩的因素很多,如混凝土的材料成分、外部环境等,各个因素对收缩的影响与对徐变的影响类似。为减小混凝土的收缩,应采取增加混凝土的密实度、掺加膨胀剂、加强对混凝土的早期养护或设置施工缝等措施。

The cause of shrinkage of concrete is generally attributed to the volume shrinkage (condensation shrinkage) of the gel itself and the volume shrinkage (dry shrinkage) caused by concrete loss of water. The test shows that the shrinkage deformation increases over time, develops rapidly at the initial stage of hardening, and then gradually slows down. In steam curing, the shrinkage value of concrete is smaller than that in normal temperature curing because the high temperature and high humidity can accelerate the setting and hardening process of concrete and reduce the evaporation of water in concrete. There are many factors that affect the shrinkage of concrete, such as the material composition of concrete, the external environment, and so on. The effects of various factors on shrinkage are similar to those on creep. In order to reduce the shrinkage of concrete, measures such as increasing the density of concrete, adding expansion agent, strengthening the early curing of concrete or setting construction joints should be taken.

2) 混凝土的温度变形

2) Temperature deformation of concrete

混凝土在温度变化时,体积热胀冷缩,称为温度变形。当混凝土的温度变形受到外界约束条件的限制而不能自由发生时,将在结构中产生大而有害的应力。钢筋对混凝土的收缩产生阻碍作用,钢筋受到压应力,混凝土受到拉应力,此拉应力可能超过混凝土的轴心抗拉强度而导致拉裂或构件损坏。混凝土的温度线膨胀系数一般可取 $1 \times 10^{-5}/℃$,钢的温度线膨胀系数为 $1.2 \times 10^{-5}/℃$,温度变化时,混凝土和钢筋之间仅引起很小的内应力,不致产生有害影响,所以可采取在结构的适当部位设置伸缩缝等措施减小温度变形的不利影响。钢筋混凝土结构伸缩缝的最大间距见《混凝土结构设计标准》(GB/T 50010—2010)。对于烟囱、水池等结构,设计中应考虑温度应力的影响。

2.2 钢筋的基本性能
Basic properties of steel bars

2.2.1 钢筋的化学成分、品种和选用
Chemical composition, types and selection of steel bars

1. 钢筋的化学成分

Chemical composition of steel bars

钢筋的物理力学性能主要取决于它的化学成分。钢筋的化学成分以铁元素为主,还含有少量的其他元素。混凝土结构中使用的钢材按化学成分可分为碳素钢和普通低合金钢。碳素钢通常可分为低碳钢、中碳钢和高碳钢。钢筋的强度随含碳量的增加而提高,但塑性和可焊性降低。在钢材中加入少量的合金元素即制成低合金钢,可以有效提高钢筋的强度,使钢筋保持较好的塑性。为了节约合金资源,近年来研制开发出细晶粒钢筋,通过控制轧钢的温度形成细晶粒的金相组织,达到与添加合金元素相同的效果。

2. 钢筋的种类

Types of steel bars

钢筋混凝土结构中所采用的钢筋有柔性钢筋和劲性钢筋。柔性钢筋即普通钢筋,是我国使用的主要钢筋形式。劲性钢筋是以角钢、槽钢、工字钢、钢轨等型钢作为结构构件的钢筋。柔性钢筋根据外形可分为光圆钢筋与变形钢筋,如图 2-13 所示。

The steel bars used in reinforced concrete structure include flexible reinforcement and stiff reinforcement. Flexible reinforcement, namely ordinary reinforcement, is the main form of reinforcement used in China. Stiff steel bars are made of angle steel, channel steel, I-shaped steel and steel rail as steel bars in structural members. Flexible steel bars can be divided into round bars and deformed bars according to the surface shape, as shown in Fig. 2-13.

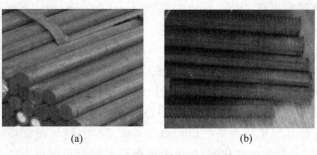

图 2-13 光圆钢筋与变形钢筋
(a) 光圆钢筋；(b) 变形钢筋
Figure 2-13　Round bar and deformed bar
(a) Round bar；(b) Deformed bar

光圆钢筋表面光滑，直径一般为 6~22mm，握裹性能差；变形钢筋表面有凸起的肋，有螺纹形、人字纹形和月牙纹形等，握裹性能好（钢筋的种类如图 2-14 所示）。变形钢筋的特点是与混凝土黏结性好，工程中常用月牙纹钢筋。变形钢筋的公称直径为 6~50mm，公称直径指相当于横截面面积相等的光圆钢筋的直径。当钢筋直径在 6~12mm 时，可采用变形钢筋，也可采用光圆钢筋；当钢筋直径大于 12mm 时，通常采用变形钢筋。直径小于 6mm 的常称为钢丝，钢丝外形多为光圆，也有的在表面上刻痕以加强钢丝与混凝土的黏结作用。

The diameter of round bar is generally 6-22mm, with smooth surface and poor binding performance; there are convex ribs on the surface of deformed bars. The ribs may be in the shape of spiral, chevron or crescent, with good binding performance (the types of steel bars are shown in Fig. 2-14). The characteristic of deformed bar is that it has good adhesion with concrete, and crescent shaped steel bars are commonly used in engineering. The nominal diameter of deformed bar is 6-50mm, and the nominal diameter refers to the diameter of round bar with equal cross-sectional area. When the steel diameter is 6-12mm, deformed bar or round bar can be used. When the diameter of steel bar is more than 12mm, the deformed bar is usually used. Steel bars with diameter less than 6mm are often

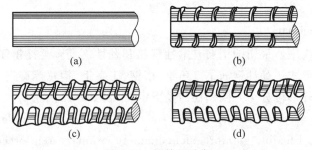

图 2-14 钢筋的种类
(a) 光圆钢筋；(b) 月牙纹钢筋；(c) 螺纹钢筋；(d) 人字纹钢筋
Figure 2-14 Types of steel bars
(a) Round bar；(b) Crescent bar；(c) Spiral bar；(d) Chevron bar

called steel wires. The surface of steel wire is generally smooth, and there are also indentations rolled into the surface to strengthen the bonding effect between steel wire and concrete.

3. 钢筋的品种

Types of steel bars

钢筋的品种较多,按生产加工工艺和力学性能的不同,用于钢筋混凝土结构和预应力混凝土结构中的钢筋或钢丝可分为热轧钢筋、中强度预应力钢丝、消除应力钢丝、钢绞线和预应力螺纹钢筋等。《混凝土结构设计标准》(GB/T 50010—2010)规定,用于钢筋混凝土结构和预应力混凝土结构中的普通钢筋可采用热轧钢筋;用于预应力混凝土结构的预应力筋,可采用中强度预应力钢丝、消除应力钢丝、钢绞线和预应力螺纹钢筋。

There are many types of steel bars. According to different production and processing technology and mechanical properties, steel bars or steel wires used in reinforced concrete structures and prestressed concrete structures can be classified as hot-rolled steel bars, medium-strength prestressed steel wires, stress-relieving steel wires, steel strands and prestressed spiral bar, etc. "Standard for design of concrete structures" (GB/T 50010—2010) stipulates that hot-rolled steel bars can be used for ordinary steel bars used in reinforced concrete structures and prestressed concrete structures. The prestressed steel bars used for prestressed concrete structures can use medium-strength prestressed steel wires, stress-relieving steel wires, steel strands and prestressed spiral steel bars.

1) 热轧钢筋

1) Hot rolled steel bar

热轧钢筋是由低碳钢、普通低合金钢或细晶粒钢在高温状态下轧制而成。热轧钢筋为软钢,其应力-应变曲线有明显的屈服点和流幅,断裂时有颈缩现象,延伸率较大。热轧钢筋根据其强度的高低分为:①HPB300,即热轧光面钢筋(hot-rolled plain steel bars)300级,工程符号为Φ;②HRB400,工程符号为Φ;③HRBF400,工程符号为Φ^F;④RRB400,即余热处理钢筋(remained heat treatment ribbed steel bars)400级,工程符号为Φ^R;⑤HRB500,工程符号为Φ,指强度标准值为500MPa的热轧带肋钢筋;⑥HRBF500,工程符号为Φ^F。其中HPB300为低碳钢,外形为光面圆形,称为光圆钢筋。HRB400、HRB500为普通低合金钢筋,HRBF400、HRBF500为细晶粒钢筋,均在表面轧有月牙肋,称为带肋钢筋或变形钢筋。RRB400为余热处理月牙纹变形钢筋。

The hot-rolled steel bar is rolled from low carbon steel, ordinary low alloy steel or fine grain steel in high temperature. The hot-rolled steel bar is flexible steel, its stress-strain curve has obvious yield point and flow plateau, and has necking phenomenon at fracture, and has large elongation. According to the strength, hot-rolled steel bars can be classified as: ① HPB300, namely hot-rolled plain steel bars 300 grade, denoted by engineering symbol Φ; ② HRB400, denoted by engineering symbol Φ; ③ HRBF400, denoted by engineering symbol Φ^F; ④RRB400, namely remained heat treatment ribbed steel bars 400 grade, denoted by engineering symbol Φ^R; ⑤HRB500, refers to hot-rolled ribbed steel bars with a standard strength value of 500MPa;

⑥HRBF500, denoted by engineering symbol Φ^F. Among them, HPB300 grade is low-carbon steel with a smooth round surface, which is called round bar. HRB400 and HRB500 are ordinary low-alloy steel bars, HRBF400 and HRBF500 are fine-grained steel bars, all with crescent ribs rolled on the surface, which are called ribbed bars or deformed bars. RRB400 grade is crescent grain deformed bar after heat treatment.

  拓展知识 2-5

2) 预应力筋

2) Prestressed tendon

中强度预应力钢丝的抗拉强度为 800~1270MPa，外形有光面（符号 ϕ^{PM}）和螺旋肋（符号 ϕ^{HM}）两种。消除应力钢丝的抗拉强度为 1570~1860MPa，外形也有光面（符号 ϕ^P）和螺旋肋（符号 ϕ^H）两种。钢绞线（符号 ϕ^S）由多根高强钢丝扭结而成，表面可以根据需要增加镀锌层、锌铝合金层、包铝层、镀铜层或涂环氧树脂等。钢绞线的抗拉强度为 1570~1960MPa。最常用的钢绞线为镀锌钢绞线和预应力钢绞线，常用预应力钢绞线的直径为 9.53~17.8mm，每根预应力钢绞线中的钢丝一般为 7 股或 3 股。预应力螺纹钢筋（符号 ϕ^T）又称精轧螺纹粗钢筋，抗拉强度为 980~1280MPa，是用于预应力混凝土的大直径高强钢筋，轧制时沿钢筋纵向全部轧有规律性的螺纹肋条，可用于螺丝套筒连接和螺帽锚固，不需要再加工螺栓，也不需要焊接。

4. 钢筋的选用

Selection of steel bars

《混凝土结构设计标准》(GB/T 50010—2010) 4.2.1 条指出混凝土结构的钢筋应按下列规定选用：

（1）纵向受力普通钢筋可采用 HRB400、HRB500、HRBF400、HRBF500 钢筋，也可采用 HPB300、RRB400 钢筋，普通钢筋是指用于钢筋混凝土结构中的钢筋和预应力混凝土结构中的非预应力钢筋。

（2）梁、柱和斜撑构件的纵向受力普通钢筋宜采用 HRB400、HRB500、HRBF400、HRBF500 钢筋。

（3）箍筋宜采用 HRB400、HRBF400、HPB300、HRB500、HRBF500 钢筋。

（4）预应力筋宜采用预应力钢丝、钢绞线和预应力螺纹钢筋。

2.2.2 钢筋的强度和变形
Strength and deformation of steel bars

钢筋的力学性能有强度和变形等，单向拉伸试验是确定钢筋性能的主要手段。根据钢筋单向拉伸应力-应变关系曲线可将钢筋分为两类：一类是有明显流幅的钢筋，如图 2-15 所示，如热轧钢筋；另一类是没有明显流幅的钢筋，如图 2-16 所示，如冷轧钢筋、预应力所用的钢丝、钢绞线及热处理钢筋等。普通钢筋应力-应变曲线都有明显的流幅，这种钢筋称软钢；没有明显流幅的钢筋和钢丝称为硬钢。钢筋拉伸试验设备如图 2-17 所示。

The mechanical properties of steel bars include strength and deformation, and uniaxial tensile test is the main method to determine the properties of steel bars. According to the uniaxial tensile stress-strain curve of steel bars, the steel bars can be classified as two categories: one is the reinforcement with obvious flow plateau, as shown in Fig. 2-15, such as hot-rolled steel bar; the other is the reinforcement without obvious flow plateau, as shown in Fig. 2-16, such as cold-rolled steel bar, steel wire for prestressing, steel strand and heat-treated steel bar. The stress-strain curves of ordinary steel bars all have obvious flow plateau, which is called flexible steel bar; the steel bars and steel wires without obvious flow plateau are called stiff steel bar. The tensile testing equipment of steel bars is shown in Fig. 2-17.

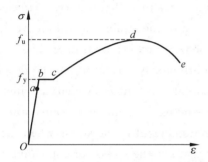

图 2-15 有明显流幅的钢筋应力-应变曲线

Figure 2-15 Stress-strain curve of steel bars with obvious flow amplitude

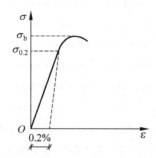

图 2-16 无明显流幅的钢筋应力-应变曲线

Figure 2-16 Stress-strain curve of steel bars without obvious flow amplitude

1. 钢筋的应力-应变曲线

Stress-strain curve of steel bars

1) 有明显流幅的钢筋

1) Steel bars with obvious flow amplitude

热轧钢筋属于有明显流幅的钢筋，其拉伸试验的典型受拉应力-应变曲线如图 2-15 所示，曲线可分为 4 个阶段：弹性阶段 Ob、屈服阶段 bc、强化阶段 cd 和破坏阶段 de。图 2-15 中，Oa 为一段斜直线，其应力与应变呈直线变化，应力与应变的比值为常数，应变在卸荷后能完全消失，称为弹性阶段，与 a 点对应的应力称为比例极限。a 点之前的应力与应变关系为 $\sigma = E_s \varepsilon$。

图 2-17 钢筋拉伸试验设备

Figure 2-17 Tensile test equipment of reinforcement

The hot-rolled steel bar belongs to the steel bar with obvious flow amplitude. Fig. 2-15 shows a typical tensile stress-strain curve of tensile test. The curve can be divided into four stages: elastic stage Ob, yield stage bc, strengthening stage cd and failure stage de. In Fig. 2-15, Oa is an diagonal straight line, the stress changes linearly with the strain, and the ratio of stress to strain is constant. The strain can completely disappear after unloading, which is called elastic stage, and the stress corresponding to point a is called proportional limit. The relationship between stress and strain before point a is $\sigma = E_s \varepsilon$.

应力超过 a 点之后，钢筋中的晶粒开始产生相互滑移错位，除弹性应变外，还有卸荷后不能消失的塑性变形，应变的增长速度略快于应力的增长速度，但若在应力达到弹性极限 b 点之前卸载，应变中的绝大部分仍能恢复。在应力到达 b 点后钢筋开始塑性流动，开始屈服，应力基本不变，应变不断增长，产生较大的塑性变形，应力-应变曲线出现水平段 bc，bc 段称为流幅或屈服台阶，相应于 b 点的应力称为钢筋的屈服强度。钢筋的屈服强度是在标准拉伸试验过程中，当荷载不增加而试件仍继续伸长时的应力或屈服台阶所对应的应力。

When the stress exceeds point a, the grains in the steel bars begin to slip and dislocation. Besides the elastic strain, there are plastic deformation which cannot disappear after unloading. The growth rate of strain is slightly faster than that of stress. However, if the stress is unloaded before the elasticity limit b, most of the strain can still recover. After the stress reaches point b, the steel bars begin plastic flow and begin yielding. The stress is basically unchanged, and the strain increases continuously, resulting in large plastic deformation. The stress-strain curve appears horizontal segment bc, which is called flow amplitude or yield plateau, and the stress corresponding to point b is called the yield strength of reinforcement. The yield strength of reinforcement is the stress when the load does not increase but the specimen continues to elongate or the stress corresponding to the yield plateau during the standard tensile test.

经过屈服阶段之后，钢筋内部晶粒经调整重新排列，抵抗荷载的能力有所提高，钢筋应力-应变关系表现为上升的曲线，cd 段称为强化阶段，表现为弹塑性性质，应力有很大的提高，变形也很大。d 点对应的钢筋应力称为钢筋的极限抗拉强度，而与 d 点应力相应的荷载是试件所能承受的最大荷载，称为极限荷载。

After the yield stage, the internal grains of the steel bars are adjusted and rearranged, and the ability to resist load is improved. The stress and strain of the steel bar show a rising curve, and the segment cd is called the strengthening stage, which shows the elastic-plastic property. There is a significant increase in stress and deformation. The steel stress corresponding to point d is called the ultimate tensile strength of the reinforcement. The load corresponding to the stress of point d is the maximum load that the specimen can bear, which is called ultimate load.

超过 d 点后，在试件某个薄弱部位的截面出现横向收缩，截面急剧缩小，塑性变形迅速增大，出现局部颈缩现象，此后应力逐渐降低，直至达到 e 点试件断裂。e 点对应的应变称为钢筋的极限应变。

After exceeding point d, the cross-section of a weak part of the specimen shrinks laterally, the section shrinks sharply, the plastic deformation increases rapidly, and the local necking phenomenon appears. After that, the stress gradually decreases until reaching point e, the specimen breaks. The strain corresponding to point e is called the ultimate strain of reinforcement.

对于有明显屈服点的钢筋，有两个强度指标：屈服强度和极限抗拉强度。屈服强度是钢筋混凝土构件承载力设计时钢筋强度取值的依据；极限抗拉强度一般作为钢筋的实际破坏强度,它是钢筋混凝土结构抗倒塌验算时钢筋强度取值的依据。

For steel bars with obvious yield point, there are two strength indexes: yield strength and ultimate tensile strength. The yield strength is the basis for determining the strength of reinforcement in the design of the bearing capacity of reinforced concrete members, and the ultimate tensile strength is generally taken as the actual failure strength of steel bars, which is the basis for determining the strength of steel bars in the check of collapse resistance of reinforced concrete structures.

另外,钢筋受压的压应力与压应变的变化曲线在屈服阶段之前与钢筋受拉基本相同。

In addition, the change curve of compressive stress and compressive strain of steel bar under compression is basically the same as that of steel bar under tension before the yield stage.

2) 无明显流幅的钢筋

2) Steel bars without obvious flow amplitude

预应力螺纹钢筋和各类钢丝属于无明显流幅的钢筋,也称为硬钢。无明显流幅钢筋的应力-应变曲线如图 2-16 所示,其应力-应变曲线上没有明显的屈服点,只有一个强度指标,即最大拉应力 σ_b 对应的极限抗拉强度,之后由于钢筋的颈缩现象曲线出现下降段,直至钢筋被拉断。此类钢筋的比例极限大约相当于其抗拉强度的 65%。对无明显流幅的钢筋,工程上一般取残余应变为 0.2% 时的应力作为屈服强度,称为条件屈服强度,作
为钢筋强度设计取值的依据。画一条应变截距为 0.2%(或 0.002)与应力-应变曲线初始弹性段平行的直线,此直线与应力-应变曲线的交点对应的应力即为屈服强度 f_y。一般情况下,含碳量高的钢筋质地较硬,没有明显的流幅,塑性变形能力较差,延伸率很小,其强度较高,下降段较短。

 特别提示 2-3, 拓展知识 2-6

3) 钢筋的弹性模量

3) Elastic modulus of steel bars

钢筋的弹性模量是根据拉伸试验中测得的弹性阶段的应力-应变曲线确定的。钢筋的强度相差较大,但其弹性模量较接近。钢筋的弹性模量 E_s 可用应力 σ_s 和应变 ε_s 表示为

The elastic modulus of steel bars is determined by the stress-strain curve of elastic stage measured in tensile test. The strength of steel bar is different from each other, but its elastic modulus is close to each other. The elastic modulus E_s of steel bars can be expressed by stress σ_s and strain ε_s as follows

$$E_s = \frac{\sigma_s}{\varepsilon_s} = 常数 \tag{2-10}$$

由于钢筋在弹性阶段的受压性能与受拉性能类同,所以同一种钢筋的受压弹性模量与受拉时相同。《混凝土结构设计标准》(GB/T 50010—2010)4.2.5 条指出普通钢筋和预应力钢筋的弹性模量 E_s 可按表 2-2 采用。

Because the compression performance and tension performance of steel bars in the elastic stage are similar, the compressive elastic modulus of the same type of reinforcement is the same as in tension. Article 4.2.5 of the "Standard for design of concrete structures" (GB/T 50010—2010) points out that the elastic modulus E_s of ordinary steel bars and prestressed steel bars should be adopted according to Table 2-2.

表 2-2 钢筋的弹性模量 E_s

Table 2-2 Elastic modulus of steel bars E_s 单位：10^5N/mm^2

牌号或种类	弹性模量
HPB300 钢筋	2.10
HRB400、HRB500 钢筋；HRBF400、HRBF500 钢筋；RRB400 钢筋；预应力螺纹钢筋	2.00
消除应力钢丝、中强度预应力钢丝	2.05
钢绞线	1.95

注：必要时可采用实测的弹性模量。

2. 钢筋的变形性能

Deformation properties of steel bars

延伸率和冷弯性能是反映钢筋塑性性能和变形能力的塑性指标。

1) 钢筋的延伸率

1) Elongation of steel bars

钢筋的延伸率是指钢筋试件断裂后标距长度（短试件取 $5d$，长试件取 $10d$，d 为钢筋试件直径）的极限伸长与原标距长度的百分比，其值大小标志钢材塑性的大小。延伸率越大，表明钢筋的塑性和变形能力越好，钢筋在拉断前有足够预兆。延伸率是选择钢筋的重要指标，国家标准规定了各种钢筋须达到的延伸率的最小值，普通热轧钢筋的最大力总延伸率应大于 2.5%。钢筋的变形能力一般用延性表示，钢筋应力-应变曲线上屈服点至极限应变点之间的应变值反映钢筋延性的大小。

延伸率仅反映钢筋拉断时残余变形的大小，而其中还包含了断口颈缩区域的局部变形，使不同量测标距长度所得的结果不一致；延伸率忽略了钢筋的弹性变形，不能反映钢筋受力时的总体变形能力；测量钢筋拉断后的标距长度时，需将拉断的两段钢筋对合后再量测，容易产生人为误差。因此，近年来国际上采用钢筋的最大力总延伸率来表示钢筋的变形能力。

钢筋的最大力总延伸率既能反映钢筋的塑性变形，又能反映钢筋的弹性变形，量测结果受原始标距的影响较小，也不产生人为误差，因此，《混凝土结构设计标准》(GB/T 50010—2010)采用钢筋的最大力总延伸率评定钢筋的塑性性能，并要求钢筋的最大力总延伸率不小于规定的数值，见附表 1-7。

2) 冷弯性能

2) Cold bending property

为了使钢筋在加工时不会断裂，使用时不会脆断，要求钢筋具有一定的冷弯性能。可对钢筋试件进行冷弯试验，如图 2-18 所示，要求常温下将钢筋围绕某个规定直径 D（规定为 $1d$、$2d$、$3d$ 等）的辊轴弯曲一定角度（90°或 180°），弯曲后的钢筋应无裂纹、鳞落或断裂现象，即认为钢筋的弯曲性能符合要求。弯转角度越大、弯心直径 D 越小，钢筋的塑性越好。

冷弯试验与受力均匀的拉伸试验相比能更有效地揭示钢筋材质的缺陷,冷弯性能是检验钢筋塑性性能的一项指标。

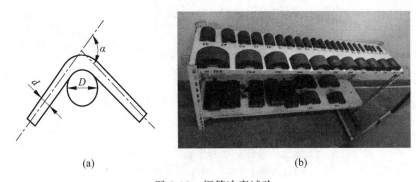

图 2-18 钢筋冷弯试验
(a) 冷弯弯曲参数;(b) 不同冷弯半径

Figure 2-18　Cold bending test of reinforcement
(a) Cold bending bending parameters; (b) Different cold bending radius

2.2.3　钢筋的本构关系
Constitutive relationship of steel bars

《混凝土结构设计标准》(GB/T 50010—2010)建议的钢筋单调加载的应力-应变本构关系曲线有以下三种。

1. 描述完全弹塑性的双直线模型

Double linear model describing complete elastoplasticity

双直线模型适用于流幅较长的低强度钢材。模型将钢筋的应力-应变曲线简化为两段直线,不计屈服强度的上限和由于应变硬化而增加的应力。图 2-19(a)中 OB 段为完全弹性阶段,B 点为屈服下限,过 C 点后即认为钢筋变形过大不能正常使用。双直线模型的数学表达式如下:

$$\varepsilon_s \leqslant \varepsilon_y, \quad \sigma_s = E_s \varepsilon_s \tag{2-11}$$

$$\varepsilon_y \leqslant \varepsilon_s \leqslant \varepsilon_{s,h}, \quad \sigma_s = f_y \tag{2-12}$$

2. 描述完全弹塑性加硬化的三折线模型

Trilinear model describing complete elastoplastic hardening

三折线模型适用于流幅较短的软钢,要求它可以描述屈服后立即发生应变硬化(应力强化)现象,并能正确地估计高出屈服应变后的应力。图 2-19(b)中 OB 及 BC 直线段分别为完全弹性和塑性阶段。C 点为硬化的起点,CD 为硬化阶段,到达 D 点时即认为钢筋破坏。

三折线模型的数学表达式如下:

$$\varepsilon_s \leqslant \varepsilon_y, \quad \sigma_s = E_s \varepsilon_s \tag{2-13}$$

$$\varepsilon_y \leqslant \varepsilon_s \leqslant \varepsilon_{s,h}, \quad \sigma_s = f_y \tag{2-14}$$

$$\varepsilon_{s,h} \leqslant \varepsilon_s \leqslant \varepsilon_{s,u}, \quad \sigma_s = f_y + (\varepsilon_s - \varepsilon_{s,h}) \tan\theta' \tag{2-15}$$

可取

$$\tan\theta' = E'_s = 0.01 E_s \tag{2-16}$$

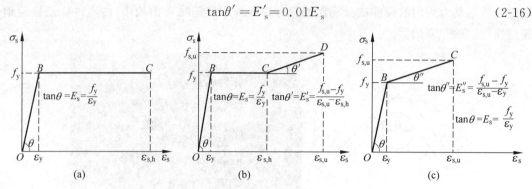

图 2-19 钢筋应力-应变曲线的数学模型
(a) 双直线;(b) 三折线;(c) 双斜线

Figure 2-19 The mathematical models of the stress-strain curve of reinforcement
(a) Double linear line; (b) Trilinear line; (c) Double diagonal line

3. 描述弹塑性的双斜线模型

Double diagonal line model describing elastoplasticity

双斜线模型可以描述没有明显流幅的高强钢筋或钢丝的应力-应变曲线。图 2-19(c)中 B 点为条件屈服点,C 点的应力达到极限值。双斜线模型的数学表达式如下:

$$\varepsilon_s \leqslant \varepsilon_y, \quad \sigma_s = E_s \varepsilon_s \tag{2-17}$$

$$\varepsilon_y \leqslant \varepsilon_s \leqslant \varepsilon_{s,u}, \quad \sigma_s = f_y + (\varepsilon_s - \varepsilon_y)\tan\theta'' \tag{2-18}$$

式中

$$\tan\theta'' = E''_s = \frac{f_{s,u} - f_y}{\varepsilon_{s,u} - \varepsilon_y} \tag{2-19}$$

2.2.4 钢筋的疲劳性能

Fatigue property of steel bars

工程结构中,吊车梁、铁路或公路桥梁、铁路轨枕、海洋采油平台等的钢筋和混凝土都承受很大的应力变化。在频繁的重复荷载作用下,构件材料抵抗破坏的情况与一次受力时有本质区别。钢筋在重复、周期动荷载作用下,在正常使用期间经过一定次数后发生脆性突然断裂,而不是单调加荷时的塑性破坏,这种破坏称为疲劳破坏,此时钢筋的最大应力低于静荷载作用下钢筋的极限强度。

Both steel bars and concrete of crane beams, railway or highway bridges, railway sleepers and offshore oil platforms in engineering structures are subjected to large numbers of stress variation. Under the action of frequent repeated loads, the failure resistance of member materials is essentially different from that under one-time loading. Under the action of repeated and periodic dynamic loads, after a certain number of times during normal service, sudden brittle fracture occurs, rather than the plastic failure under monotonic loading. This kind of failure is called fatigue failure. At this time, the maximum stress of steel bars is lower than the ultimate strength of steel bars under static load.

钢筋的疲劳强度是指在某一规定的应力变化幅度内,经过一定次数循环荷载后,发生疲劳破坏的最大应力值。钢筋的疲劳强度与一次循环应力中最大应力和最小应力的差值有关,称为疲劳应力幅。一般认为,在外力作用下钢筋发生疲劳断裂是由于钢筋内部和外表面的缺陷引起局部应力集中,钢筋中晶粒发生滑移。另一方面,由于重复荷载的作用使已产生的微裂纹时而压合,时而张开,使裂纹逐渐扩展,减小了钢筋未开裂部分的截面面积,直到面积太小导致无法抵抗外力,最后突然脆性断裂。

The fatigue strength of reinforcement refers to the maximum stress value of fatigue failure after a certain number of cyclic loading within a specified range of stress variation. The fatigue strength of reinforcement is related to the difference between the maximum stress and the minimum stress in one cycle stress, which is called fatigue stress amplitude. It is generally believed that the fatigue fracture of steel bar under external force is due to the local stress concentration caused by the defects in the internal and outer surface of steel bars, and the grain sliding in the reinforcement. On the other hand, due to the effect of repeated load, the generated microcracks are sometimes compressed and sometimes opened, so that the cracks gradually expand, this reduces the remaining uncracked cross-sectional area of steel bars until it becomes too small to resist the external force. The steel bars eventually undergo brittle fracture.

影响钢筋疲劳强度的因素很多,如疲劳应力幅、最小应力值、钢筋外表面的几何形状、钢筋直径、钢筋种类、轧制工艺和试验方法等,其中最主要的是钢筋的疲劳应力幅。《混凝土结构设计标准》(GB/T 50010—2010)规定了不同等级钢筋的疲劳应力幅度限值,并说明该值与截面同一层钢筋最小应力与最大应力的比值 ρ' 有关,ρ' 称为疲劳应力比值。对预应力钢筋,当 $\rho' \geqslant 0.9$ 时,可不进行钢筋疲劳验算。

There are many factors that affect the fatigue strength of steel bars, such as the fatigue stress amplitude, minimum stress value, the geometry of the outer surface of the steel bar, the steel diameter, the type of the steel bar, the rolling process and the test method, etc. Among them, the most important is the fatigue stress amplitude of the steel bar. "Standard for design of concrete structures" (GB/T 50010—2010) stipulates the fatigue stress amplitude limit value of different grades of steel bars, and stipulates that this value is related to the ratio ρ' of the minimum stress to the maximum stress of steel bars in the same layer of the cross-section, and ρ' is called the fatigue stress ratio. For prestressed steel bars, when $\rho' \geqslant 0.9$, it is not necessary to conduct fatigue calculation.

2.2.5 混凝土结构对钢筋性能的要求
Requirements of concrete structures for the performance of steel bars

钢筋混凝土结构对钢筋性能最主要的要求是强度高,塑性及焊接性能好,并有良好的黏结性能,所用钢筋一般应能满足下列要求。

英文翻译 2-13

1. 适当的强度和屈强比
  Proper strength and yield strength ratio

有明显流幅钢筋的强度包括屈服强度和极限强度,屈服强度是构件承载力计算的主要依据,对没有明显流幅的钢筋取它的条件屈服强度。提高强度的根本途径是改变钢筋的化学成分,生产出新的钢筋品种,使其具有良好的塑性和焊接性能,以及具有较高的强度。采用高强度钢筋可以减少构件的配筋量,节约钢材,取得较好的经济效果。可以避免配筋密集给设计、施工造成困难,减少钢筋的运输、加工、现场绑扎等工作量。但实际工程中钢筋强度并非越高越好,由于钢筋的弹性模量并不会因钢筋强度提高而增大,高强钢筋在高应力下的大变形会引起混凝土结构变形过大,造成过宽的裂缝。因此,对普通钢筋混凝土结构,不应采用高强钢丝等强度过高的钢材。对预应力混凝土结构,可以采用高强钢丝等高强度钢材。

钢筋应力-应变曲线中,屈服强度与极限抗拉强度的比值称为屈强比,它代表了钢筋的强度储备,也在一定程度上代表了结构的强度储备。屈强比小则结构的强度后备大,但屈强比太小则钢筋强度的有效利用率低,所以钢筋应具有适当的屈强比。

《混凝土结构通用规范》(GB 55008—2021)3.2.3 条规定,对按一、二、三级抗震等级设计的房屋建筑框架和斜撑构件,其纵向受力普通钢筋性能应符合下列规定:

(1) 抗拉强度实测值与屈服强度实测值的比值不应小于 1.25;
(2) 屈服强度实测值与屈服强度标准值的比值不应大于 1.30;
(3) 最大力总延伸率实测值不应小于 9%。

2. 足够的塑性变形能力

　　Sufficient plastic deformation capability

在工程设计中,要求混凝土结构承载能力极限状态为具有明显预兆的塑性破坏。若发生脆性破坏则变形很小,破坏前没有预兆,而且是突发的,很危险。工程中要避免脆性破坏,故要求钢筋断裂前要有足够的变形,使结构在破坏前有预警,以保证安全。足够大的塑性变形能力还便于钢筋施工制作,有利于提高结构构件的延性,增强结构的抗震性能。抗震结构要具有足够的延性,要求钢筋具有足够的塑性,即各种钢筋的最大力总伸长率不应小于规定值,在施工时钢筋要弯转成型,因而应具有一定的冷弯性能。

3. 可焊性

　　Weldability

要求钢筋具备良好的焊接性能,保证焊接强度,在焊接后钢筋不应产生裂纹及过大的变形,以保证焊接接头性能良好。热轧钢筋具有较好的焊接性能;细晶粒热轧带肋钢筋和直径大于 28mm 的带肋钢筋,焊接应经试验确定;热处理和冷加工钢筋在一定碳当量范围内可焊,但焊接引起的热影响区域强度降低,应采取必要的措施;余热处理钢筋焊接受热回火后强度可能降低,不宜焊接;高强钢丝、钢绞线不可焊。

4. 耐久性、耐火性和低温性能

　　Durability, fire resistance and low temperature performance

细直径钢筋特别是冷加工钢筋和预应力钢筋容易受到腐蚀而削弱截面,降低承载力和耐久性;环氧树脂涂层钢筋或镀锌钢丝均可提高钢筋的耐久性,但会降低钢筋与混凝土间的黏结性能,设计时应注意这种不利影响。热轧钢筋的耐火性能最好,冷轧钢筋其次,预应力钢筋最差。结构设计和施工时,钢筋外应有必要的混凝土保护层厚度,以满足对构件耐久性和耐火极限的要求。在寒冷地区要求钢筋具备抗低温性能,以防止钢筋低温冷脆而破坏。

5. 与混凝土具有良好的黏结

Having good bonding with concrete

黏结力是钢筋与混凝土得以共同工作的基础。变形钢筋与混凝土的黏结性能最好，设计中宜优先选用。对强度较高的钢筋或钢丝，一般在其表面轧制月牙纹横肋、螺旋肋或者刻痕等，以提高黏结强度，有助于或大大提高黏结力。钢筋表面沾染油脂、糊着泥污、长满浮锈都会削弱与混凝土的黏结作用。钢筋的锚固和有关构造要求是保证两者之间具有良好黏结力的措施。

钢筋的进场、堆放和取样见拓展知识 2-7 和图 2-20。

(a)　　　　　　　　　　　　　　(b)

图 2-20　钢筋的堆放
(a) 纵筋的堆放；(b) 箍筋的堆放
Figure 2-20　Stacking of steel bars
(a) Stacking of longitudinal bars；(b) Stacking of stirrups

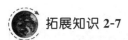 拓展知识 2-7

2.3　混凝土和钢筋的黏结

Bond of concrete and steel bars

2.3.1　混凝土与钢筋间的黏结作用

Bonding effect between concrete and steel bars

1. 黏结作用及黏结机理

Bonding effect and bonding mechanism

黏结是钢筋与周围混凝土之间一种复杂的相互作用，用以传递两者之间的应力，协调变形，它是这两种材料共同工作的保证，使之能共同承受外力、共同变形、抵抗相互间的滑移。钢筋能否可靠地与混凝土黏结直接影响到这两种材料的共同工作，关系到混凝土结构构件的安全和材料强度的充分利用。

Bond is a kind of complex interaction between steel bars and surrounding concrete, which is used to transfer the stress and coordinate deformation between them. It is the guarantee of the two materials to work together, so that they can bear external force together, deform together and resist mutual sliding. Whether steel bars can be reliably bonded with the concrete directly affects the joint work of these two materials, and relates to the safety of concrete structural members and the full utilization of material strength.

钢筋与混凝土的黏结锚固作用主要有：混凝土凝结时，由于水泥颗粒的水化作用形成了凝胶体，使钢筋和混凝土在接触面上产生胶结力；混凝土结硬时体积收缩，握裹住钢筋，在发生相互滑动时产生摩阻力；钢筋粗糙不平的表面或变形钢筋凸起的肋纹与混凝土的咬合力；当采用锚固措施后所产生的机械锚固力等。光圆钢筋的机械咬合力比变形钢筋的小。

The bonding and anchoring effects between steel bars and concrete mainly include: when concrete is condensed, the hydration of cement particles forms the gel, resulting in the bond force between steel bars and concrete on the contact surface. When the concrete is hardened, the volume shrinks and the reinforcement is wrapped, and the frictional resistance produced when mutual sliding occurs; the bite force between the rough and uneven surface or deformed rib of steel bars and the concrete, the mechanical anchoring force produced by anchoring measures. The mechanical bite force of round bar is smaller than that of deformed bar.

光圆钢筋与变形钢筋具有不同的黏结机理。

There are different bonding mechanisms between round bars and deformed bars.

(1) 光圆钢筋与混凝土的黏结力主要由三部分组成：钢筋表面与混凝土中水泥凝胶体的化学吸附作用（胶着力）；混凝土与钢筋在接触面上的摩擦力；钢筋粗糙不平的表面与混凝土之间产生的机械咬合力。光圆钢筋的黏结力主要来自胶着力和摩擦力。

(1) The bond force between round bar and concrete is mainly composed of three parts: the chemical chemisorption (glue force) between the surface of steel bars and the cement gel of concrete, the frictional force between the concrete and steel bars on the contact surface, and the mechanical biting force between the rough and even surface of steel bars and the concrete. The bonding force of round bar is mainly from the adhesive force and friction force.

(2) 变形钢筋与混凝土的黏结力仍由胶着力、摩擦力和机械咬合力组成。横肋对混凝土的挤压如同一个楔，会产生很大的机械咬合力，从而提高变形钢筋的黏结能力，如图 2-21 所示。变形钢筋的黏结力主要来自机械咬合作用。

(2) The bond force between deformed bar and concrete is still composed of adhesive force, friction force and mechanical biting force. The compression of the transverse rib on the concrete is like a wedge, which will produce a great mechanical bite force, thus improving the bonding capacity of deformed bar, as shown in Fig. 2-21. The bonding force of deformed bar is mainly from mechanical bite force.

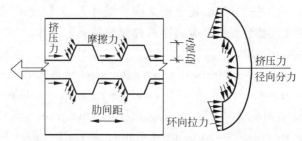

图 2-21 拔出试验（变形钢筋，肋纹的咬合作用）

Figure 2-21 Pull-out test(deformed bar, biting effect of the rib)

2. 黏结应力
Bond stress

钢筋混凝土受力后会在钢筋和混凝土的接触面上产生剪应力,通常把这种剪应力称为黏结应力,如图 2-22 所示。黏结应力按作用性质可分为两类,一是锚固黏结应力,如钢筋伸入支座或支座负弯矩钢筋在跨间截断时必须有足够的锚固长度或延伸长度,通过这段长度上黏结应力的累积,将钢筋锚固在混凝土中,使钢筋不至于在未充分发挥作用前就被拔出;二是裂缝附近的局部黏结应力,如梁截面开裂后,开裂截面的钢筋应力通过裂缝两侧的黏结应力向混凝土传递,这类黏结应力的大小反映了裂缝两侧混凝土参与受力的程度。

Reinforced concrete will produce shear stress along the contact surface between steel bars and concrete, which is usually called bond stress, as shown in Fig. 2-22. Bond stress can be divided into two types according to the property of action. One is anchoring bond stress. For example, if steel bars extend into the support or the bearing negative moment bars are cut off between spans, there must be sufficient anchorage length or development length. Through the accumulation of bond stress on this length, the reinforcement will be anchored in

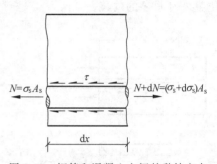

图 2-22 钢筋和混凝土之间的黏结应力
Figure 2-22 Bond stress between steel bar and concrete

the concrete, and the reinforcement will not be pulled out before it is fully used. The second is the local bond stress near cracks. For example, after cracking of the beam section, the steel stress of cracked section is transferred to the concrete through the bond stress on both sides of the crack. The magnitude of such bond stress reflects the degree to which concrete on both sides of the crack participates in the stress.

2.3.2 拔出试验
Pull-out test

钢筋与混凝土之间纵向黏结应力的分布通常用拔出试验来确定,各点的黏结应力可由相邻两点间钢筋的应力差值除以接触面积近似计算,为此需测定钢筋的应变分布。试验表明,变形钢筋的应力传递比光圆钢筋快,黏结性能也比光圆钢筋好。利用拔出试验可以计算黏结强度,并作为设计钢筋锚固长度的依据。

The distribution of longitudinal bond stress between steel bars and concrete is usually determined by pull-out test. The bond stress at each point can be approximately calculated by dividing the stress difference of reinforcement between two adjacent points by the contact area. Therefore, it is necessary to measure the strain distribution of steel bars. The test shows that the stress transfer of deformed bar is faster than that of round bar, and the bond property of deformed bar is better than that of round bar. The bond strength can be calculated by pull-out test, which can be used as the basis for the design of the anchorage length of steel bars.

拔出试验是将钢筋的一端埋置在混凝土试件中,在伸出的一端施加拉拔力,如图 2-23 所示。经测定,黏结应力的分布为曲线形,从拉拔力一边的混凝土端面开始迅速增长,在靠近端面的一定距离处达到峰值,其后逐渐衰减。而且,钢筋埋入混凝土中的长度越长,将钢筋拔出混凝土试件所需的拔出力就越大。但是埋入长度过长则过长部分的黏结力很小,甚至为零,说明过长部分的钢筋不起作用。所以,受拉钢筋在支座或节点中应有足够的埋置长度,称为锚固长度,以保证钢筋在混凝土中有可靠的锚固。光圆钢筋破坏时黏结应力分布图形接近于三角形。变形钢筋在大部分加载过程中,黏结应力的峰值均靠近加载端。随着荷载的增大,黏结应力分布长度缓慢增长,而峰值应力却显著增大,在接近破坏时,峰值应力的位置才有明显的内移。

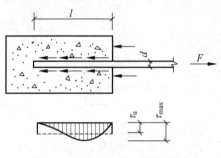

图 2-23 拔出试验(光圆钢筋)

Figure 2-23 Pull-out test (round bar)

The pull-out test is to embed one end of the steel bar in the concrete specimen, and apply pulling force on the extended end, as shown in Fig. 2-23. After measurement the distribution of bond stress is curved. The bond stress increases rapidly from the concrete end face on the side of the pull-out force, reaches the peak at a certain distance near the end face, and then decreases gradually. Moreover, the longer the length of steel bar buried in the concrete, the greater the pull-out force required to pull the steel bar out of the concrete specimen. However, if the embedded length is too long, the bond force of the too long part is very small or even zero, indicating that the steel bars in the too long part do not work. Therefore, the tensile steel bars should have adequate length in the support or node, called anchorage length, to ensure that the steel bars are reliably anchored in the concrete. The bond stress distribution pattern is close to a triangle when the round bar fails. During most of the loading process of deformed bars, the peak value of bond stress is close to the loading end. As the load increases, the bond stress distribution length grows slowly, but the stress peak increases significantly. When it is close to failure, the position of the peak stress has obvious internal shift.

根据拔出试验,沿钢筋纵向的黏结应力可取其平均值为

From the pull-out test, taking the average value of the bond stress along the longitudinal direction of the steel bar as

$$\tau_b = \frac{F}{\pi d l} \tag{2-20}$$

式中, F 为拉拔力, N; d 为钢筋直径, mm; l 为钢筋埋置长度, mm。

Where, F is the pull-out force, N; d is the bar diameter, mm; l is the embedded length of steel bar, mm.

黏结强度是黏结破坏时钢筋与混凝土界面上的最大平均黏结应力。变形钢筋的黏结强度比光面钢筋大,将光面钢筋末端做成弯钩可以大大提高拔出力。

The bond strength is the maximum average bond stress at the interface between reinforcement and concrete when bonding fails. The bond strength of deformed bars is greater than that of round bars, and the pull-out force can be greatly increased by making hooks at the end of round bars.

2.3.3 黏结破坏机理
Bond failure mechanism

各种试验结果表明,构件上裂缝的出现和裂缝的分布都与钢筋与混凝土之间的黏结力和位移量有直接关系。

1. 光圆钢筋的黏结破坏
   Bond failure of round bars

在钢筋与混凝土间出现相对滑移前,光圆钢筋的黏结作用主要取决于化学胶着力,发生滑移后黏结作用则由摩擦力和机械咬合力提供。拔出试验的破坏形态是由钢筋与混凝土相对滑移产生的,钢筋从混凝土中被拔出的剪切破坏,其破坏面就是钢筋与混凝土的接触面,黏结强度低,滑移量大。

为了提高光圆钢筋的抗滑移性能,须在光圆直钢筋的端部附加弯钩或弯转、弯折以加强锚固。附加的弯钩足以使光圆钢筋承载至屈服而不被拔出,但滑移量仍较大。

2. 变形钢筋的黏结破坏
   Bond failure of deformed bars

变形钢筋由于表面有突起的肋,能与混凝土犬牙交错紧密结合,其胶着力和摩擦力的作用有所增加,但主要还是机械咬合力发挥的作用最大,往往占黏结力的一半以上,如图 2-24 所示。根据试验,变形钢筋的黏结强度高出光圆钢筋 2~3 倍,我国生产的螺纹钢筋的黏结强度为 $25 \sim 60 \text{N/mm}^2$。

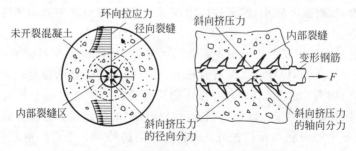

图 2-24 变形钢筋横肋处的挤压力和内部裂缝
Figure 2-24 Extrusion pressure and internal cracks at transverse ribs of deformed bars

变形钢筋的黏结强度较高,滑移量也较小,不过试验表明:如果钢筋外围混凝土很薄(如保护层厚度不足或钢筋净间距过小),且未配置环向箍筋约束,则径向裂缝很容易发展到试件表面,形成沿纵向钢筋的裂缝,使钢筋附近的混凝土沿钢筋纵向劈裂破坏。劈裂破坏不是脆性破坏,它具有一定延性特征,称之为劈裂型黏结破坏,劈裂应力约为黏结强度的 80%~85%。反之,则为沿钢筋肋外径的圆柱滑移面的剪切破坏,剪切破坏的黏结强度比劈裂破坏的大。

2.3.4 影响黏结强度的因素
Factors affecting bond strength

1. 混凝土的质量和强度
Quality and strength of concrete

混凝土的质量对黏结力的影响很大。水泥性能好、骨料强度高、配比得当、振捣密实、养护良好的混凝土与钢筋之间的黏结力大。混凝土强度越高,黏结强度也越高。

提高混凝土强度可以增大混凝土与钢筋表面的化学胶着力和机械咬合力,延迟沿钢筋纵向劈裂裂缝的出现,提高黏结强度。试验表明,黏结强度并不与混凝土立方体抗压强度成正比,而是与混凝土抗拉强度大致呈线性关系。

2. 混凝土保护层厚度和钢筋净间距
Thickness of concrete cover and net spacing of steel bars

试验结果表明,混凝土保护层厚度对光圆钢筋的黏结强度没有明显影响,而对变形钢筋的影响却很大。沿纵向钢筋的劈裂裂缝对受力和耐久性都极为不利,为了提高钢筋外围混凝土的抗劈裂能力,从而提高黏结强度,混凝土保护层不能太薄,钢筋的净间距也不能太小。混凝土保护层厚度应大于钢筋直径,以防止发生劈裂裂缝。试验表明,黏结强度随混凝土保护层厚度 c 增大而提高,当 $c \geqslant 5d$(d 为钢筋直径)时,锚固长度的取值可以减小。

3. 钢筋的外形和横向配筋
Shape of reinforcement and transverse reinforcement

钢筋的外形决定了混凝土咬合齿的形状,对黏结强度影响很大。光圆钢筋主要靠摩擦力实现黏结锚固,黏结强度较小,锚固性能很差,设计施工时要在钢筋端部做弯钩或弯折,可以增加其拔出力。

使用变形钢筋比使用光圆钢筋对提高黏结力有利得多,所以变形钢筋的末端一般无须做弯钩。变形钢筋的纹形对黏结强度有所影响,等高肋钢筋的黏结强度较大,但肋间混凝土咬合齿易被挤碎、切断,黏结延性较差;月牙肋钢筋的黏结强度略低于等高肋钢筋,锚固长度就略需加长,但黏结延性好。旋扭状钢筋(如钢绞线)的咬合齿为连续螺旋状,咬合均匀而充分,故黏结强度中等而黏结延性很好;螺旋肋钢筋的外形介于带肋钢筋和旋扭状钢筋之间,并兼有两者的优点,不仅黏结强度和刚度高,且具有很好的黏结延性。实际工程中为保证钢筋与混凝土能够共同工作,必须采取可靠的工程构造措施。

在锚固区域内配置普通箍筋或螺旋箍筋等横向钢筋可以增大混凝土的侧向约束,延缓或阻止劈裂裂缝的发展,从而提高黏结强度,提高的幅度与所配置的横向钢筋数量有关。

4. 受力状态
Stress state

试验表明,在重复荷载或反复荷载作用下,混凝土与钢筋之间的黏结强度将退化,施加的应力越大,重复或反复荷载的次数越多,黏结强度退化得越多。

上述关于混凝土与钢筋之间黏结作用的分析都是基于钢筋受拉拔出试验的结果,受压钢筋的黏结作用一般比受拉钢筋强。钢筋受压后横向膨胀，挤压周围混凝土，增加了摩擦力，黏结强度比受拉钢筋的高。

本章小结

本章内容主要包括混凝土和钢筋的强度、变形以及混凝土和钢筋之间的黏结。

(1) 混凝土的单轴向强度有立方体抗压强度、轴心抗压强度和轴心抗拉强度。混凝土强度等级是根据立方体抗压强度标准值划分的，《混凝土结构设计标准》(GB/T 50010—2010)规定的混凝土强度等级共有13个。

(2) 混凝土的变形可分为两类：一类是在荷载作用下的受力变形，如一次短期加载下的变形、重复荷载作用下的变形以及长期荷载作用下的变形；另一类与受力无关，称为体积变形，如混凝土收缩、膨胀以及由于温度变化等产生的变形。混凝土的变形模量有弹性模量、割线模量和切线模量。在计算混凝土变形时，常用到弹性模量。

(3) 钢筋按化学成分可分为碳素钢和普通低合金钢两大类。按生产加工工艺和力学性能的不同，钢筋或钢丝可分为热轧钢筋、中强度预应力钢丝、消除应力钢丝、钢绞线和预应力螺纹钢筋等。用于钢筋混凝土结构和预应力混凝土结构中的普通钢筋可采用热轧钢筋；用于预应力混凝土结构中的预应力钢筋可采用预应力钢丝、钢绞线和预应力螺纹钢筋。

(4) 热轧钢筋分为 HPB300、HRB400、HRBF400、RRB400、HRB500、HRBF500。钢筋混凝土构件中的纵向受力钢筋宜优先采用 HRB400 级钢筋。

(5) 根据钢筋拉伸试验，钢筋的应力-应变曲线可分为有明显流幅的和没有明显流幅的两大类。有明显流幅的钢筋一般以屈服强度作为钢筋强度设计取值的依据，没有明显流幅的钢筋以残余应变为 0.2% 极限抗拉强度时的应力作为条件屈服强度。

(6) 钢筋混凝土受力后会在钢筋和混凝土接触面上产生剪应力，通常把这种剪应力称为黏结应力。黏结应力主要由胶着力、摩擦力和机械咬合力三部分组成。

第 2 章拓展知识和特别提示

视频：第 2 章小结讲解

习题

2-1 选择题

1.《混凝土结构设计标准》(GB/T 50010—2010)中，混凝土各种强度指标的基本代表值是哪一个？（　　）

  A. 立方体抗压强度标准值   B. 轴心抗压强度标准值
  C. 轴心抗压强度设计值   D. 轴心抗拉强度标准值

2. 混凝土双向受力时,强度下降的情况为(　　)。
   A. 双向受拉 B. 双向受压 C. 一向受压,一向受拉

3. 对于混凝土在一次短期加荷时的受压应力-应变曲线,下列哪种叙述是正确的?(　　)
   A. 上升段是一条直线
   B. 下降段只能在刚度不大的试验机上测出
   C. 混凝土强度高时,曲线的峰部曲率较小
   D. 混凝土压应力达到最大时,并不立即破坏

4. 关于混凝土徐变,以下哪项叙述正确?(　　)
   A. 水灰比越大徐变越小   B. 水泥用量越多徐变越小
   C. 骨料越坚硬徐变越小   D. 养护环境湿度越大徐变越大

5. 对于有明显流幅的钢筋,确定其强度设计值的依据是(　　)。
   A. 屈服强度 B. 比例极限 C. 极限抗拉强度 D. 条件屈服强度

6. 对于没有明显流幅的钢筋,其强度标准值取值的依据是下列哪种?(　　)
   A. 最大应变对应的应力   B. 极限抗拉强度
   C. 0.9倍极限强度   D. 条件屈服强度

2-2 判断题

1. 混凝土立方体试块尺寸越大,量测的抗压强度越高。(　　)
2. 混凝土强度等级由轴心抗压强度标准值确定。(　　)
3. 在正常情况下,混凝土强度随时间不断增长。(　　)
4. 混凝土在不变的压力长期作用下,其应变会随时间而增长。(　　)
5. 混凝土的徐变是指在荷载作用前的养护过程中混凝土的收缩变形现象。(　　)
6. 对有明显流幅的钢筋,将其极限抗拉强度作为设计时强度取值的依据。(　　)
7. 钢筋和混凝土之间若无可靠的黏结,两者就不能共同受力,所以钢筋混凝土构件中,钢筋必须与混凝土保证可靠的黏结。(　　)

2-3 名词解释

1. 立方体抗压强度	2. 混凝土强度等级	3. 立方体抗压强度标准值	4. 混凝土轴心抗压强度
5. 混凝土强度设计值	6. 轴心抗拉强度	7. 轴心拉伸试验	8. 劈裂试验
9. 受力变形	10. 体积变形	11. 弹性模量	12. 切线模量
13. 割线模量	14. 泊松比	15. 混凝土徐变	16. 混凝土收缩
17. 混凝土膨胀	18. 重复作用	19. 疲劳破坏	20. 疲劳强度
21. 光圆钢筋	22. 变形钢筋	23. 钢丝	24. 钢绞线
25. 热轧钢筋	26. 钢筋强度等级	27. 热轧钢筋的牌号	28. 单向拉伸试验
29. 屈服强度	30. 极限抗拉强度	31. 钢筋强度设计值	32. 钢筋延伸率
33. 冷弯性能	34. 黏结应力	35. 拔出试验	36. 黏结强度

2-4 简答题

1. 混凝土立方体抗压强度是如何确定的?混凝土强度等级有哪些?
2. 什么是混凝土轴心抗压强度、轴心抗拉强度?
3. 什么是混凝土的徐变、收缩?徐变和收缩的区别是什么?如何防止?

4. 什么是软钢、硬钢？二者的应力-应变曲线的主要不同点是什么？

5. 绘制有明显流幅钢筋的应力-应变曲线，并指出各个阶段的名称以及特征点所对应的名称。

6. 工程中对有明显流幅的钢筋以屈服强度作为钢筋的设计强度，其理由是什么？

7. 钢筋的冷弯性能是如何测定的？

8. 钢筋与混凝土之间的黏结作用有哪些方面？影响黏结强度的因素是什么？

9. 钢筋为什么要进行锚固？

第 3 章 混凝土结构设计基本原则

Basic principles of concrete structure design

教学目标：

1. 了解结构的功能要求、结构安全等级、设计工作年限、作用、荷载、荷载效应、结构抗力、荷载标准值、荷载设计值、荷载代表值及结构可靠度、可靠性等概念；
2. 能根据《建筑结构荷载规范》(GB 50009—2012)分析计算实际结构构件的各种荷载；
3. 理解可靠度与失效概率的关系、失效概率与可靠指标的关系、目标可靠指标与分项系数的关系等；
4. 理解结构极限状态、极限状态设计法，重点理解和掌握分项系数表达的近似概率极限状态设计表达式；
5. 理解荷载计算对结构安全的重要性，理解土木工程师的工程责任。

导读：

结构在设计工作年限内，必须符合下列规定：应能够承受在正常施工和正常使用期间预期可能出现的各种作用，应保障结构和结构构件的预定使用要求，应保障足够的耐久性要求。结构体系应具有合理的传力路径，能够将结构可能承受的各种作用从作用点传递到抗力构件。

当发生可能遭遇的爆炸、撞击、罕遇地震等偶然事件及人为失误时，结构应保持整体稳固性，不应出现与起因不相称的破坏后果。当发生火灾时，结构应能在规定的时间内保持承载力和整体稳固性。

结构设计应包括下列基本内容：结构方案；作用的确定及作用效应分析；结构及构件的设计和验算；结构及构件的构造、连接措施；结构耐久性设计；施工可行性等。

结构应按照设计文件施工。施工过程中应采取保证施工质量和施工安全的技术措施和管理措施。结构应按设计规定的用途使用，并应定期检查结构状况，进行必要的维护和维修。

引例：

第 3 章引例

第3章 混凝土结构设计基本原则
Basic principles of concrete structure design

核心词汇：

结构的功能	functions of structure	永久荷载	permanent load
可变荷载标准值	variable load standard value	可变荷载	variable load
设计工作年限	design working life	可靠性	reliability
永久荷载设计值	permanent load design value	可靠度	reliability
设计基准期	design reference period	偶然荷载	accidental load
可变荷载设计值	variable load design value	可靠概率	reliability probability
结构安全等级	structural safety level	荷载代表值	representative value of load
荷载组合值	combination value of load	失效概率	failure probability
极限状态	limit state	荷载分项系数	partial coefficient of load
荷载准永久值	quasi-permanent value of load	可靠指标	reliability index
正常使用极限状态	serviceability limit state	荷载标准值	standard value of load
极限状态设计法	limit state design method	承载能力极限状态设计表达式	design expression of ultimate limit state
承载能力极限状态	ultimate limit state		
设计表达式	design expression	荷载设计值	design value of load
直接作用	direct action	结构重要性系数	structural importance coefficient
荷载效应	load effect	永久荷载标准值	permanent load standard value
作用效应	action effect	荷载效应组合设计值	load effect combination design value
结构抗力	structural resistance		

3.1 结构上的作用和荷载
Action and load on the structure

3.1.1 结构上的作用、效应及结构抗力
Action, effect and structural resistance on the structure

1. 结构上的作用

Structural action

结构上的作用是指施加在结构上的集中力或分布力以及引起结构内力和变形的原因。变形包括外加变形和约束变形。外加变形是结构在地震、不均匀沉降等因素作用下，边界条件发生变化而产生的位移和变形。约束变形是结构在温度变化、湿度变化及混凝土收缩等因素作用下，由于存在外部约束而产生的内部变形。按其出现的方式不同，可分为直接作用和间接作用两类。

Structural action refers to the concentrated or distributed force applied to the structure and causes of structural internal force and deformation. Deformation includes applied deformation and constrained deformation. Applied deformation is the displacement and deformation of the structure under the action of earthquake, uneven settlement and other factors, due to changes in boundary conditions. Constrained deformation is the internal deformation of the structure due to the existence of external constraints under the action of temperature changes, humidity changes and shrinkage of the concrete, etc.

According to the different ways of their appearance, they can be divided into two types: direct action and indirect action.

1) 直接作用

1) Direct action

直接作用是指直接以力(集中力或分布力)的方式施加在结构上的作用,通常也称为荷载。例如结构构件的自重、楼面和屋面上的人群及物品重量、风压力、雪压力、积水、积灰等。工程中常见的作用大部分都是直接作用。

Direct action refers to an action that is directly applied to the structure in the form of a force (concentrated or distributed), also commonly referred to as a load. For example, the self-weight of structural members, the weight of people and objects on the floor and roof, wind pressure, snow pressure, water accumulation, ash accumulation, etc. Most of the common actions in engineering are direct actions.

2) 间接作用

2) Indirect action

间接作用是指能够引起结构外加变形、约束变形或振动,产生内力效应的各种原因,例如地震、基础的差异沉降、混凝土的收缩及徐变、温度变化等。地震作用应根据《建筑抗震设计规范》(GB 50011—2010)确定。

Indirect action refers to various reasons that can cause applied deformation, constrained deformation or vibration of the structure and produce internal force effects, for example, earthquake, differential settlement of foundation, shrinkage and creep, temperature change of concrete, etc. The seismic action should be determined according to the "Code for seismic design of buildings" (GB 50011—2010).

拓展知识 3-1

2. 环境影响

Environmental influence

环境影响是指环境对结构产生的各种机械的、物理的、化学的或生物的不利影响,如温度、湿度变化、二氧化碳、氧、盐、酸等环境因素对结构的影响。环境影响会引起结构材料性能的劣化,影响结构的安全性、适用性和耐久性。

Environmental influence refers to various mechanical, physical, chemical or biological adverse effects of the environment on the structure, such as temperature, humidity changes, carbon dioxide, oxygen, salt, acid and other environmental effects on the structure.

结构的环境影响可分为无侵蚀性的室内环境影响和侵蚀性环境影响等。当把无侵蚀性的室内环境视为一个环境等级时,宜将该等级分为无高温的室内干燥环境和室内潮湿环境两个层次。根据环境侵蚀性的特点,环境影响包括生物作用、与气候等相关的物理作用、与建筑物内外人类活动相关的物理作用、介质的侵蚀作用、物理与介质的共同作用。

环境影响对结构产生的效应主要是材料性能的降低,多数情况下,环境影响的效应涉及化学的和生物的损害,其中环境湿度是最关键的因素。目前,对结构的环境影响的定量描述还比较困难,因此,目前主要根据材料特点,通过环境对结构影响程度的分级(轻微、轻度、中度、严重等)等方法进行定性描述,并在设计中采取相应的技术措施。

 特别提示 3-1

任务 3-1:分析你所在教学楼受到哪些环境因素的影响。

Task 3-1:Analyze what influence of environmental factors on your teaching building.

3. 作用效应
Action effect

结构上的作用和环境影响作用在构件上,使结构或结构构件产生内力(如弯矩、剪力、轴力、扭矩等)和变形(如挠度、转角、裂缝等)的反应称为作用效应。结构构件及其连接的作用效应通过考虑了力学平衡条件、变形协调条件、材料时变特性以及稳定性等因素的结构分析方法确定。

英文翻译 3-2

当作用为直接作用(即荷载)时,引起的效应也称为荷载效应,通常用 S 表示。荷载效应是随机变量或随机过程,是结构设计的依据之一。同时施加在结构上的各单个作用对结构的共同影响应通过作用组合来考虑,对不可能同时出现的各种作用不应考虑其组合。

任务 3-2:分析你所在教学楼的梁、板、柱在荷载下的作用效应。

Task 3-2:Analyze the action effect of beams, slabs and columns under load in your teaching building.

4. 结构抗力
Structural resistance

结构或结构构件承受作用效应(即内力和变形)和环境影响的能力称为结构抗力,用 R 表示,如构件的承载力、刚度、抗裂度、材料抗劣化能力等。影响结构抗力的因素有结构的材料性能(强度、变形模量等)、截面尺寸、配筋的数量和方式、抗力计算的基本假定和计算公式等,通常结构抗力主要取决于材料性能。这些因素都是随机变量,因此由这些因素综合而成的结构抗力也是随机变量。

任务 3-3:分析你所在教学楼的梁、板、柱在荷载作用下的结构抗力有哪些。

Task 3-3:Analyze the structural resistance under load action of beams, slabs and columns of your teaching building.

3.1.2 结构上的荷载
Loads on the structure

结构上的荷载应根据《建筑结构荷载规范》(GB 50009—2012)及相关规范、标准确定。结构上的荷载根据时间变化特性分为永久荷载、可变荷载和偶然荷载。

Structural loads should be determined according to the "Load code for the design of building structures" (GB 50009—2012) and relevant codes, standards. The loads on the structure are divided into permanent load, variable load and accidental load according to the time variation characteristics.

1. 永久荷载（或称恒荷载）

Permanent load（or constant load）

永久荷载在设计工作年限内始终存在，且其量值变化与平均值相比可以忽略不计，或其变化是单调的并趋于某个限值，例如结构构件自重、土压力、预应力等。

The permanent load exists throughout the design working life, and the magnitude change is negligible compared to the average value, or its change is monotonic and tends to a certain limit value, such as the self-weight of structural members, earth pressure, prestress, etc.

2. 可变荷载（或称活荷载）

Variable load（or live load）

可变荷载在设计工作年限内其量值和位置随时间变化，且其变化与平均值相比不可忽略不计，例如楼屋面可变荷载、路面上的行车荷载、积灰荷载、风荷载、雪荷载、吊车荷载等。

The value and position of variable load change with time during the design working life, and its variation is not negligible compared to the average value, such as the variable load on the floor and roof, the traffic load on the road, dust load, wind load, snow load, crane load, etc.

3. 偶然荷载

Accidental load

偶然荷载在设计工作年限内不一定出现，而一旦出现其值很大，且持续时间很短，例如地震、爆炸、撞击等作用。当以偶然荷载作为结构设计的主导作用时，应考虑偶然荷载发生时和偶然荷载发生后两种工况。在允许结构出现局部构件破坏的情况下，应保证结构不致因局部破坏引起连续倒塌。

Accidental load does not necessarily appear in the design working life, but once it appears, its value is large and the duration is short, such as earthquake, explosion and impact etc. When the accidental load is taken as the leading role in structural design, two working conditions should be considered when the accidental load occurs and after the accidental load occurs. Under the circumstance that local component fail of the structure is allowed, it should be ensured that the structure will not occur progressive collapse due to local damage.

任务 3-4：分析你所在教学楼楼盖上的永久荷载和可变荷载。

Task 3-4：Analyze the permanent load and variable load on the floor of your teaching building.

3.1.3 结构的荷载代表值

Load representative value of structure

荷载代表值是在建筑结构设计时，针对不同设计目的，对荷载赋予的一个规定的量值，它可以是荷载的标准值或可变荷载的伴随值。可变荷载的伴随值是在作用组合中,伴随主导作用的可变荷载值。建筑结构按不同极限状态设计时,在相应的荷载组合中对可能同时出现的各种荷载应采用不同的荷载代表值,以反映荷载在设计中的特点。

Representative value of load is a specified value given to load for different design purposes in the design of building structures, which can be a standard value of a load or a accompanying value of a variable load. The accompanying value of the variable load is the variable load value associated with the dominant action in the action combination. When the building structure is designed according to different limit states, different load representative values should be used in the corresponding load combination for various loads that may appear at the same time to reflect the characteristics of the load in the design.

1. 荷载标准值

Standard value of load

荷载标准值是指设计基准期内,结构在正常使用情况下,可能出现的最大荷载值。荷载标准值是结构设计时采用的荷载主要代表值,荷载的其他代表值都可以在荷载标准值的基础上乘以或除以适当的系数得到。荷载标准值可根据观测数据的统计、荷载的自然界限或工作经验确定,《建筑结构荷载规范》(GB 50009—2012)对荷载标准值的取值方法做了具体的规定。

The standard value of load refers to the maximum load value that may appear in the structure under normal service conditions during the design reference period. The standard value of load is the main representative value of the load used in the structural design, and any other representative value of load can be obtained by multiplying or dividing it by the standard value of load by the appropriate coefficient. The standard value of load can be determined according to the statistics of the observed data, the natural limit of the load or the working experience. Specific provisions are made in the "Load code for the design of building structures" (GB 50009—2012) for the determination of standard value of load.

1) 永久荷载标准值

1) Permanent load standard value

若结构自重变异性不大,永久荷载标准值可按结构构件的设计尺寸与材料单位体积的自重计算确定。对于自重变异性较大的材料和构件(如现场制作的保温材料、混凝土薄壁构件等),对结构不利时自重标准值取上限值,对结构有利时取下限值。位置固定的永久设备自重应采用设备铭牌重量值,当无铭牌重量时,应按实际重量计算。

If the structural self-weight has little variability, the permanent load standard value can be calculated and determined according to the design dimensions of the structural member and the weight per unit volume of the material. For materials and members with large variation in self-weight (such as on-site thermal insulation materials, thin-walled concrete members, etc.), the standard value of self-weight is taken as the upper limit when it is unfavorable to the structure, and the lower limit is taken when it is favorable to the structure. The weight of permanent equipment with fixed position should be the weight value of the equipment nameplate. If there is no nameplate weight, it should be calculated according to the actual weight.

任务 3-5：某钢筋混凝土矩形截面梁截面尺寸 $b=200\mathrm{mm}$, $h=500\mathrm{mm}$，梁两侧为 20mm 厚石灰砂浆抹面，钢筋混凝土的重度 $\gamma_1=25\mathrm{kN/m^3}$，石灰砂浆的重度 $\gamma_2=17\mathrm{kN/m^3}$，求作用在梁上沿跨度方向均匀分布的永久荷载（自重）标准值 g_k。

Task 3-5: The cross-sectional dimensions of a reinforced concrete rectangular section beam are $b=200\mathrm{mm}$, $h=500\mathrm{mm}$, both sides of the beam are plastered with 20mm thick lime mortar, the gravity of reinforced concrete is $\gamma_1=25\mathrm{kN/m^3}$, and the gravity of lime mortar is $\gamma_2=17\mathrm{kN/m^3}$. Find the standard value g_k of the permanent load (self-weight) uniformly distributed on the beam in the direction of the span.

2）可变荷载标准值

2) Variable load standard value

可变荷载标准值如楼面和屋面可变荷载、雪荷载、风荷载、积灰荷载、吊车荷载等是根据观测资料、统计分析和试验数据，并考虑工程实践经验确定的。《建筑结构荷载规范》(GB 50009—2012)和《工程结构通用规范》(GB 55001—2021)给出了各种可变荷载标准值的取值，设计时可根据各种建筑物的具体用途直接查用。当使用荷载较大、情况特殊或有专门要求时，应按实际情况采用。部分民用建筑楼面均布可变荷载标准值及其组合值、频遇值和准永久值系数见表 3-1。例如，住宅的楼面可变荷载标准值为 $2.0\mathrm{kN/m^2}$。

The variable load standard value such as floor and roof variable loads, snow loads, wind loads, ash loads, crane loads, etc. are determined based on observational data, statistical analysis and experimental data, as well as engineering practice experience. "Load code for the design of building structures" (GB 50009—2012) and "General code for engineering structures" give the values of various variable load standard values, which can be directly checked and used according to the specific purposes of various buildings during design. When the service load is large, the situation is special or there are special requirements, it should be adopted according to the actual situation. The standard value of uniformly distributed variation load and its combination value, frequent value and quasi-permanent value coefficients of floors of some civil buildings are shown in Table 3-1. For example, the variable load standard value of residential floor is $2.0\mathrm{kN/m^2}$.

任务 3-6：从表 3-1 中查出教室、实验室、宿舍、食堂、影院、商店、书库的楼面可变荷载标准值。

Task 3-6: Find out the floor variable load standard values of classrooms, laboratories, dormitories, canteens, cinemas, shops, and book stores from Table 3-1.

表 3-1 民用建筑楼面均布可变荷载标准值及其组合值、频遇值和准永久值系数

Table 3-1 Standard value and its combination value, frequency value and quasi-permanent value coefficient of uniform variable load on civil building floor

项次	类 别	标准值 /(kN/m²)	组合值系数 ψ_c	频遇值系数 ψ_f	准永久值系数 ψ_q
1	(1) 住宅、宿舍、旅馆、医院病房、托儿所、幼儿园；	2.0	0.7	0.5	0.4
	(2) 办公楼、教室、医院门诊室	2.5	0.7	0.6	0.5
2	食堂、餐厅、试验室、阅览室、会议室、一般资料档案室	2.5	0.7	0.6	0.6

续表

项次	类 别	标准值 /(kN/m²)	组合值系数 ψ_c	频遇值系数 ψ_f	准永久值系数 ψ_q
3	礼堂、剧场、影院、有固定座位的看台、公共洗衣房	3.5	0.7	0.5	0.3
4	(1) 商店、展览厅、车站、港口、机场大厅及其旅客等候室；	4.0	0.7	0.6	0.5
	(2) 无固定座位的看台	4.0	0.7	0.5	0.3
5	(1) 健身房、演出舞台；	4.5	0.7	0.6	0.5
	(2) 运动场、舞厅	4.5	0.7	0.6	0.3
6	(1) 书库、档案库、储藏室(书架高度不超过 2.5m)；	6.0	0.9	0.9	0.8
	(2) 密集柜书库(书架高度不超过 2.5m)	12.0	0.9	0.9	0.8

2. 可变荷载组合值

Variable load combination value

当两种或两种以上的可变荷载在结构上要求同时考虑时，由于所有可变荷载同时达到其标准值的可能性很小，因此，除主导荷载(产生最大作用效应的荷载)仍可用标准值作为代表值外，其他可变荷载均乘以小于 1.0 的组合值系数 ψ_c 作为荷载代表值，这种经过调整后的可变荷载代表值称为可变荷载组合值。

可变荷载组合值是组合后的荷载效应的超越概率与该荷载单独出现时其标准值荷载效应的超越概率趋于一致的荷载值，或组合后使结构具有规定可靠指标的荷载值。

可变荷载组合值可以用可变荷载的组合值系数 ψ_c 乘以相应的可变荷载标准值 Q_k 来确定，可用 $\psi_c Q_k$ 表示，$\psi_c \leqslant 1$，称为可变荷载组合值系数，相当于对荷载标准值进行了折减。表 3-1 列出了部分可变荷载组合值系数 ψ_c，以供查用。

3. 可变荷载准永久值

Quasi-permanent value of variable load

可变荷载准永久值指在设计基准期内被超越的总时间占设计基准期的比率较大的荷载值，可通过准永久值系数对作用标准值的折减来表示，对结构的影响类似于永久荷载。在正常使用极限状态的计算中要考虑荷载长期效应的影响，实际上是考虑荷载长期作用效应而对可变荷载标准值的一种折减。

可变荷载准永久值应取可变荷载准永久值系数乘以可变荷载标准值，可用 $Q_q = \psi_q Q_k$ 表示，ψ_q 为可变荷载准永久值系数，$\psi_q \leqslant 1$。表 3-1 列出了部分可变荷载组合值系数 ψ_q，以供查用，例如住宅楼面可变荷载的准永久值系数为 0.4。

4. 可变荷载频遇值

Frequent value of variable load

可变荷载频遇值是在设计基准期内被超越的总时间占设计基准期的比率较小的作用值，或被超越的频率限制在规定频率内的作用值，可通过频遇值系数对作用标准值的折减来表示，具体数值可查《建筑结构荷载规范》(GB 50009—2012)和《工程结构通用规范》(GB 55001—2021)。

特别提示 3-2

3.2 结构的功能
Functions of structure

结构的设计、施工和维护应使结构在规定的设计工作年限内以规定的可靠度满足规定的各项功能要求,而不需要大修加固,并且要经济合理。

The design, construction and maintenance of the structure should enable the structure to meet the specified functional requirements with the specified reliability in the specified design working life, no major repairs and reinforcement are needed, and be economical and reasonable.

3.2.1 结构的功能要求
Functional requirements of structure

《建筑结构可靠性设计统一标准》(GB 50068—2018)明确规定,结构在规定的设计工作年限内应满足下列功能要求。

"Uniform standard for reliability design of building structures" (GB 50068—2018) clearly stipulates that the structure should meet the following functional requirements in the specified design working life.

1. 安全性

Security

结构能够承受施工和使用期间可能出现的各种作用。当发生火灾时,在规定的时间内可保持足够的承载力。当发生爆炸、撞击、人为错误等偶然事件时,结构能保持必要的整体稳固性,不出现与起因不相称的破坏后果,防止出现结构的连续倒塌。

The structure can withstand various actions that may appear during construction and use. When a fire occurs, sufficient bearing capacity can be maintained within the specified time. In the event of accidental events such as explosion, impact, human error, etc., the structure can maintain the necessary overall stability, and no destructive consequences disproportionate to the cause occur, and prevent the continuous collapse of the structure.

2. 适用性

Applicability

结构在正常使用过程中,能保持良好的使用性能。例如,楼板不发生影响正常使用的过大挠度、过大振幅和振动或产生使用者感到不安的裂缝宽度,吊车梁不能发生过大的变形以免影响吊车的正常运行,水池不能出现过大的裂缝以免不能蓄水等。

The structure can maintain good performance during normal service. For example, the floor does not have excessive deflection, excessive amplitude and vibration that affect normal use, or create the crack width that users feel uneasy about. The crane beam should not be excessively deformed so as not to affect the normal operation of the crane, and the pool should not have excessive cracks so as not to store water.

3. 耐久性
Durability

在正常维护条件下,结构具有足够的耐久性能,如结构材料的风化、腐蚀、开裂等不超过一定限度,不能发生因混凝土保护层碳化、保护层太薄或裂缝宽度过大而导致混凝土保护层脱落、钢筋锈蚀等,以保证结构能够正常使用到预定的设计工作年限。环境对结构耐久性的影响可通过工程经验、试验研究、计算、检验或综合分析等方法进行评估。

Under normal maintenance conditions, the structure has sufficient durability, such as the weathering, corrosion, cracking, etc. of structural materials do not exceed a certain limit. No concrete protective layer will fall off and reinforcement corrosion will occur due to carbonization of concrete protective cover, too thin protective cover or excessive crack width, to ensure that the structure can be used normally until the predetermined design working life. The impact of the environment on the durability of the structure can be assessed through engineering experience, experimental research, calculation, inspection or comprehensive analysis.

结构的安全性、适用性、耐久性总称为结构的可靠性,指结构在规定时间内(设计工作年限),在规定条件下(正常设计、正常施工、正常使用和正常维护),完成预定功能(安全性、适用性、耐久性)的能力。

The safety, applicability and durability of the structure are collectively called structural reliability, which refers to within a specified time (design working life), under specified conditions (normal design, normal construction, normal service and normal maintenance), the ability of the structure to perform predetermined functions (safety, applicability, durability).

宜采取下列措施满足对结构的基本要求:采用适当的材料,采用合理的设计和构造,对结构的设计、制作、施工和使用等制定相应的控制措施。

The following measures should be taken to meet the basic structural requirements: adopt appropriate materials, adopt reasonable design and structure, and formulate corresponding control measures for the design, manufacture, construction and use of the structure.

任务 3-7:分析你所在教学楼结构的各项功能要求。

Task 3-7: Analyze the functional requirements of your teaching building structure.

3.2.2 设计工作年限、设计基准期和结构安全等级
Design working life、design reference period and structural safety level

1. 设计工作年限
Design working life

设计工作年限是设计规定的结构或结构构件不需进行大修即可按预定目的使用的年限。结构设计时,应根据工程的使用功能、建造和使用维护成本以及环境影响等因素规定设

计工作年限。《工程结构通用规范》(GB 55001—2021)规定房屋建筑结构的设计工作年限不应低于表 3-2 的规定,若建设单位提出更高的要求,经主管部门批准,也可按建设单位的要求采用。设计时应明确结构的用途,在设计工作年限内未经技术鉴定或设计许可,不得改变结构的用途和使用环境。

The design working life is the number of years that the structure or structural members specified in the design can be used for the intended purpose without major repair. When designing the structure, the design working life should be stipulated according to the factors such as the use function of the project, the construction and maintenance cost, and the environmental impact. "General code for engineering structures" (GB 55001—2021) stipulates that the design working life of building structures should not be less than the provisions in Table 3-2. If the construction unit puts forward higher requirements, it can also be adopted according to the requirements of the construction unit approved by the administrative department. The purpose of the structure should be clearly defined during the design, and the purpose and use environment of the structure should not be changed without technical appraisal or design permission during the design working life.

需注意的是,设计工作年限不等同于建筑结构的实际寿命或耐久年限,当结构的实际使用年限超过设计工作年限后,结构并不是不能继续使用,而是结构可靠度可能比设计时的预期值要小,表明完成预定功能的能力变差了,其继续使用年限需经鉴定确定。若使结构保持一定的可靠度,则设计工作年限取的越长,结构需要的截面尺寸或所需要的材料用量就越大。

It should be noted that the design working life is not equal to the actual life or durability life of the building structure. When the actual service life of the structure exceeds the design working life, the structure is not unusable, but the structural reliability may be less than the expected value at design time, indicating that the ability to complete the predetermined function has deteriorated, and its continued service life needs to be determined by appraisal. If the structure maintains a certain reliability, the longer the design working life, the larger the cross-sectional dimensions or the amount of material required for the structure.

表 3-2 房屋建筑结构的设计工作年限

Table 3-2 Design working life of building structure

序 号	类 别	设计工作年限/年
1	临时性建筑结构	5
2	普通房屋和构筑物	50
3	特别重要的建筑结构	100

必须定期涂刷防腐蚀涂层等结构的设计工作年限可为 20~30 年。预计使用时间较短的建筑物,其结构的设计工作年限不宜少于 30 年。结构的设计工作年限不同,其经济指标也不同,结构的设计工作年限越长,其工程投资越大。

The design working life of structures such as anti-corrosion coatings that must be regularly painted can be 20 to 30 years. For buildings that are expected to be used for a short period of time, the design working life of the structure should not be less than 30 years. The economic indicators vary with the design working life of the structure. The longer the design working life of the structure is, the greater the engineering investment is.

2. 设计基准期

Design reference period

结构上的可变作用、结构抗力等都与时间有关,为确定可变作用等取值而选用的时间参数称为设计基准期,它不等同于设计工作年限。《建筑结构可靠性设计统一标准》(GB 50068—2018)规定房屋建筑结构、港口工程的设计基准期为 50 年,铁路桥涵结构、公路桥涵结构等的设计基准期为 100 年。

The variable action and structural resistance of the structure are all related to time. The time parameter selected to determine the value of the variable action is called the design reference period, which is not equivalent to the design working life. "Unified standard for reliability design of building structures"(GB 50068—2018) stipulates that the design reference period of building structure and port engineering is 50 years. The design reference period of railway bridge and culvert structure, highway bridge and culvert structure, etc. is 100 years.

3. 结构安全等级

Structural safety level

由于结构物的重要性不同,一旦发生破坏对生命财产的危害程度以及社会影响也不同,因此在进行结构设计时,应根据结构破坏可能产生后果的严重性,即危及人的生命、造成经济损失、对社会或环境产生影响等的严重性,采用不同的安全等级进行设计。

Due to the different importance of the structure, once damage occurs the degree of harm to life, property and the social impact are also different. Therefore, in the structural design, different safety levels should be adopted according to the severity of possible consequences of structural damage, i. e. endangering human life, causing economic losses, affecting society or the environment, etc.

《工程结构通用规范》(GB 55001—2021)2.2.1 条规定,结构设计时,应根据结构破坏可能产生后果的严重性,采用不同的安全等级,设计时应根据具体情况进行选择。结构安全等级划分如下:一级,破坏后果很严重,例如大型的公共建筑等重要结构;二级,破坏后果严重,例如普通住宅和办公楼等一般结构;三级,破坏后果不严重,例如小型或临时性建筑等次要结构。结构及其部件的安全等级不得低于三级。

Article 2.2.1 of "General code for engineering structures" (GB 55001—2021) stipulates that in the structural design, different safety level should be adopted according to the seriousness of the possible consequences of structural damage, and the design should be selected according to the specific situation. The structural safety level is classified as follows: Level 1, the consequence of damage is very serious, important structures such as large public buildings; Level 2, the consequence of damage is serious, general structures such as ordinary residential and office buildings; Level 3, the damage is not serious, minor structures such as small or temporary buildings. The safety level of the structure and its members should not be lower than level 3.

 特别提示 3-3

任务 3-8：分析你所在教学楼结构的设计工作年限、设计基准期和结构安全等级。

Task 3-8: Analyze the design working life, design reference period and structural safety level of your teaching building structure.

3.3 结构功能的极限状态
Limit state of structural function

整个结构或结构的一部分超过某一特定状态（如承载力、变形、裂缝宽度、材料性能退化等超过某一限值）就不能满足设计规定的某一功能要求，此特定状态为该功能的极限状态。结构能够满足预定的各项功能要求时为可靠或有效状态，结构不能满足各项功能要求时为不可靠或失效状态。极限状态实质上是区分结构工作状态可靠（有效）或不可靠（失效）的界限，是结构开始失效的标志。结构要满足功能要求，就不能超过极限状态。结构的极限状态可分为三类：承载能力极限状态、正常使用极限状态和耐久性极限状态，分别规定有明确的标志和限值。

The entire structure or a part of the structure cannot meet a certain functional requirement specified in the design if it exceeds a certain state (such as bearing capacity, deformation, crack width, material performance degradation, etc. exceeds a certain limit), and this specific state is the limit state of the function. When the structure can meet the predetermined functional requirements, it is in a reliable or effective state, and when the structure cannot meet various functional requirements, it is in an unreliable or failure state. The limit state is essentially the boundary that distinguishes the reliable (effective) or unreliable (failure) working state of the structure, and it is the sign that the structure is beginning to fail. In order to meet the functional requirements, the structure must not exceed the limit state. The limit state of the structure can be divided into three categories: the ultimate limit state, the serviceability limit state and the durability limit state, which have clear signs and limits respectively.

3.3.1 承载能力极限状态
Ultimate limit state

结构或结构构件达到最大承载力或出现不适于继续承载的变形的状态称为承载能力极限状态。当结构或结构构件出现下列状态之一时，应认为超过了承载能力极限状态：①整个结构或结构的一部分作为刚体失去平衡（如倾覆、压屈等）；②结构构件或连接因所受应力超过材料强度而破坏,或因过度变形而不适于继续承载;③结构转变为机动体系;④结构或结构构件丧失稳定;⑤结构因局部破坏而发生连续倒塌;⑥地基丧失承载力而破坏(如失稳等);⑦结构或结构构件发生疲劳破坏(如因为荷载重复作用而破坏)。

The state in which the structure or structural member reaches the maximum bearing capacity or appears deformation unsuitable for continued bearing is called the ultimate limit state. When one of the following states appears in the structure or structural member, it should be considered as exceeding the ultimate limit state. ① The entire structure or part of the structure is out of balance as a rigid body (such as overturning, buckling, etc.); ②The structural member or splice is damaged due to the stress exceeding the material strength, or is not suitable for continued bearing due to excessive deformation; ③The structure is transformed into a mobile system; ④Loss of stability of a structure or structural member; ⑤The structure occurs continuous collapse due to local destruction; ⑥ The foundation is damaged due to loss of bearing capacity (such as instability, etc.); ⑦Fatigue failure of a structure or structural member(such as failure due to repeated loading).

涉及人身安全以及结构安全的极限状态应作为承载能力极限状态。承载能力极限状态是结构或结构构件达到允许的最大承载功能的状态。超过承载能力极限状态后,结构或构件就不能满足安全性的要求,可能导致人员伤亡或重大财产损失。对于任何承载的结构或构件,都需要按承载能力极限状态进行设计。

The limit state involving personal safety and structural safety should be regarded as the ultimate limit state. The ultimate limit state is the state in which the structure or structural member reaches the maximum allowable load-bearing function. After exceeding the ultimate limit state, the structure or member cannot meet the safety requirements, which may lead to casualties or significant property losses. For any bearing structure or member it is necessary to design according to the ultimate limit state.

承载能力极限状态设计的主要内容:①结构构件应进行承载力(包括失稳)计算;②直接承受重复荷载的构件应进行疲劳验算;③有抗震设防要求时,应进行抗震承载力计算;④必要时,尚应进行结构的倾覆、滑移、漂浮验算;⑤对于可能遭受偶然作用,且倒塌可能引起严重后果的重要结构,宜进行防连续倒塌设计。

The main contents of the ultimate limit state design: ① The bearing capacity (including instability) of structural members should be calculated; ②The fatigue check should be carried out for members directly subjected to repeated loads; ③When there are seismic fortification requirements, the seismic bearing capacity should be calculated; ④If necessary, overturning, sliding and floating of the structure should be checked; ⑤ For important structures that may be subjected to accidental action and whose collapse may cause serious consequences, the continuous collapse prevention design is advisable to be carried out.

任务 3-9:在网上寻找至少 3 个达到承载能力极限状态的结构或结构构件的案例或图片。

Task 3-9: Look online for at least three examples or pictures of structures or structural members that have reached the ultimate limit state.

3.3.2 正常使用极限状态
Serviceability limit state

结构或结构构件达到正常使用的某项规定限值的状态称为正常使用极限状态。当结构或结构构件出现下列状态之一时，应认为超过了正常使用极限状态：

The state in which a structure or structural member reaches a specified limit for normal service is called the serviceability limit state. When one of the following states appears in a structure or structural member, it should be considered that the serviceability limit state has been exceeded:

（1）影响外观、使用舒适性或结构使用功能的变形，如梁挠度过大影响外观、吊车梁变形过大使吊车不能平稳行驶等；

（2）造成人员不舒适或结构使用功能受限的振动，如工业厂房中设备振动导致结构的振幅超过按正常使用要求所规定的限值；

（3）影响外观、耐久性或结构使用功能的局部损坏，如梁、板裂缝过宽导致恐慌，钢筋锈蚀，水池开裂漏水影响正常使用等。

(1) Deformation that affects the appearance, comfort or structural service function, such as excessive deflection affects the appearance of the beam, deformation of the crane beam is too large so that the crane cannot run smoothly, etc.;

(2) Vibration that causes discomfort to persons or restricts the use of structures, such as the vibration of the equipment in the industrial plant causes the amplitude of the structure to exceed the limit value specified in the normal service requirements;

(3) Local damage that affects the appearance, durability or structural service function of the structure, such as excessive cracks in beams and slabs leading to panic, steel corrosion, water tank cracking and water leakage affect normal service, etc.

超过正常使用极限状态使结构不能正常工作，不能保证适用性的功能要求，会造成人们心理上的不安全感，还会影响结构的安全性。和承载能力极限状态相比，其导致的生命和财产危害较小，故出现概率允许稍高些，但仍应给予足够的重视。在结构或构件按承载能力极限状态进行计算后，还应该按正常使用极限状态进行变形、裂缝宽度或抗裂等验算。

Exceeding the serviceability limit state, the structure cannot work properly, and functional requirements for applicability cannot be guaranteed, which will cause people's psychological insecurity and also affect the safety of the structure. Compared with the ultimate limit state, it causes less harm to life and property, so the probability of occurrence is allowed to be slightly higher, but it should still be given adequate attention. After the structure or member is calculated according to the ultimate limit state, the deformation, crack width or crack resistance should also be checked according to the serviceability limit state.

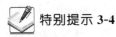

特别提示 3-4

任务 3-10：在网上寻找至少 3 个达到正常使用极限状态的结构或结构构件的案例或图片。

Task 3-10：Go online and look for at least three examples or pictures of structures or structural members that have reached the serviceability limit state.

3.3.3 耐久性极限状态
Durability limit state

结构或结构构件在环境影响下出现的劣化（材料性能随时间逐渐衰减）达到耐久性的某项规定限值或标志的状态称为耐久性极限状态。当结构或结构构件出现下列状态之一时，应认定为超过了耐久性极限状态。

（1）影响承载能力和正常使用的材料性能劣化，如钢筋、混凝土的强度降低等；

（2）影响耐久性能的裂缝、变形、缺口、外观、材料削弱等，如构件的裂缝宽度超过某一限值会引起构件内钢筋锈蚀；预应力钢筋和直径较细的受力主筋具备锈蚀条件；混凝土构件表面出现锈蚀裂缝等；

（3）影响耐久性的其他特定状态，如构件金属连接件出现锈蚀，阴极或阳极保护措施失去作用等。

When a structure or structural member has one of the following states, it should be considered that the durability limit state has been exceeded.

(1) Degradation of material properties affecting bearing capacity and normal service, such as reduction in strength of steel bars and concrete, etc. ;

(2) Cracks, deformation, gaps, appearance, material weakening, etc. that affect the durability, if the crack width of the member exceeds a certain limit, it will cause steel corrosion in the member; the prestressed reinforcement and the main stressed reinforcement with small diameter have corrosion conditions; corrosion cracks appear on the surface of concrete members. ;

(3) Other specific states that affect durability, such as corrosion of metal connectors of members, failure of cathodic or anodic protection measures, etc.

结构的耐久性极限状态设计应使结构构件出现耐久性极限状态标志或限值的年限不少于其设计工作年限。

The durability limit state design of the structure should make the number of years in which the durability limit state mark or limit value of the structural member not less than its design working life.

拓展知识 3-2

对结构的各种极限状态均应规定明确的标志或限值。结构设计时应对结构的不同极限状态分别进行计算或验算；当某一极限状态的计算或验算起控制作用时，可仅对该极限状态进行计算或验算。

For various limit states of the structure clear mark or limits should be specified. The different limit states of the structure should be calculated or checked respectively during structural design; when the calculation or check of a certain limit state plays a controlling role, only the limit state can be calculated or checked.

任务 3-11：在网上寻找至少 3 个达到耐久性极限状态的结构或结构构件的案例或图片。

Task 3-11：Look online for at least three examples or pictures of structures or structural members that have reached their durability limit state.

3.3.4 结构的设计状况
Design situations of structure

结构在建造和使用过程中所承受的作用、所处的环境条件以及经历的时间长短都是不同的，因此，设计时所采用的结构体系、可靠度水准、设计方法等也应有所区别。结构的设计状况是表征一定时段内（从设计到施工的全过程）实际情况的一组设计条件，设计时必须做到在该组条件下结构不超越有关的极限状态。进行建筑结构设计时，应根据结构在施工和使用中的环境条件和影响，区分下列四种设计状况：

（1）持久设计状况。指适用于结构正常使用的情况，在结构使用过程中一定出现，且持续期很长的设计状况，其持续期一般与设计工作年限为同一数量级，适用于建筑结构承受家具和正常人员荷载的状况。

（2）短暂设计状况。指在结构施工和使用过程中出现概率较大，而与设计工作年限相比，其持续期很短的设计状况。适用于结构施工和维修时承受施工荷载和堆料等临时情况。

（3）偶然设计状况。指在结构使用过程中出现概率很小，且持续期很短的设计状况。适用于结构遭受火灾、爆炸、非正常撞击等偶然情况。

（4）地震设计状况。适用于结构遭受地震时的设计状况。

对四种建筑结构设计状况，应分别进行下列极限状态设计：

（1）对四种设计状况均应进行承载能力极限状态设计，以保证结构的安全性。

（2）对持久设计状况，尚应进行正常使用极限状态设计，并宜进行耐久性极限状态设计，以保证结构的适用性和耐久性。

（3）对短暂设计状况和地震设计状况，根据需要进行正常使用极限状态设计，如多遇地震作用下的弹性变形验算。

（4）对偶然设计状况，允许主要承重结构因出现设计规定的偶然事件而局部破坏，但其剩余部分具有在一段时间内不发生连续倒塌的可靠度；因其持续期很短，可不进行正常使用极限状态和耐久性极限状态设计。

根据不同的设计状况，不同的极限状态均应考虑各种不同的作用组合，以确定作用控制工况和最不利的效应设计值。

任务 3-12：分析你所在教学楼结构按什么设计状况设计，需进行哪些极限状态设计。

Task 3-12：Analyze the design situation of your teaching building structure and the limit state design required.

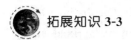
拓展知识 3-3

3.4 极限状态设计法和设计表达式
Limit state design method and design expression

极限状态设计法又称近似概率法,这个设计方法的基本思路是用概率分析方法来研究结构的可靠性。《混凝土结构设计标准》(GB/T 50010—2010)采用以概率理论为基础的极限状态设计方法,以可靠指标量度结构构件的可靠度,采用分项系数的设计表达式进行设计。

The limit state design method is also known as the approximate probability method. The basic idea of this design method is to use the probability analysis method to study the reliability of structure. "Standard for design of concrete structures" (GB/T 50010—2010) adopts the limit state design method based on probability theory, measures the reliability of structural members with the reliability index, and the design expression of partial coefficient is used for design.

3.4.1 功能函数和极限状态方程
Functional function and limit state equation

影响结构可靠度的因素有很多,如结构上的各种作用、环境影响、材料和岩土的性能、几何参数等,这些因素一般具有随机性,称为基本变量,基本变量应作为随机变量。按极限状态设计法设计建筑结构时,要求所设计的结构具有一定的预定功能,这可以用包括各有关基本变量的功能函数来表达,用 Z 表示,该函数表征一种结构完成预定功能的工作状态。当功能函数中包括作用效应 S 和结构抗力 R 这两个基本变量时,承载能力极限状态方程可以表示为功能函数 $Z=R-S$,根据 S、R 取值的不同,Z 值可能出现三种情况,按 Z 值的大小不同,可以用来描述结构所处的三种不同的工作状态。

英文翻译 3-5

当 $Z=R-S>0$ 时,结构处于可靠状态;当 $Z=R-S=0$ 时,结构达到极限状态;当 $Z=R-S<0$ 时,结构处于失效状态(破坏状态)。因此,结构可靠工作的基本条件为:$Z \geqslant 0$ 或 $R \geqslant S$。当构件的每一个截面都满足结构抗力大于作用效应即 $R \geqslant S$ 时,才认为构件是可靠的,否则认为是失效的。

3.4.2 结构的可靠度、可靠概率与失效概率
Reliability, reliability probability and failure probability of structures

结构抗力和作用效应具有随机性,安全可靠应该属于概率的范畴,应当用结构完成其预定功能的可能性(概率)的大小来衡量,而不是用一个定值来衡量。当结构完成其预定功能的概率达到一定程度时,或不能完成其预定功能的概率(失效概率)小于某一公认的、大家可以接受的程度时,就认为该结构是安全可靠的。

Structural resistance and action effect are random, safety and reliability should belong to the category of probability, and should be measured by the possibility (probability) with which the structure completes its predetermined function, rather than by a fixed value. When the probability of the structure completing its predetermined function reaches a certain degree, or the probability of failing to complete its predetermined function (failure probability) is less than a recognized and acceptable degree, the structure is considered safe and reliable.

1. 结构可靠度
Structural reliability

结构可靠度是结构在规定的时间内,在规定的条件下,完成预定功能的概率。规定的时间是指结构的设计工作年限;规定的条件是指正常设计、正常施工和正常使用的条件,不包括非正常的如人为过失的影响,人为过失应通过其他措施予以避免。结构可靠度是结构可靠性的概率量度,可靠度是用可靠概率来描述的。结构设计应保证结构既可靠又经济合理,力求以最经济的途径,使结构在设计工作年限内以适当的可靠度满足各项功能要求,这是结构设计的基本原则。

Structural reliability is the probability that the structure completes the predetermined function within the specified time and under the specified conditions. The specified time refers to the design working life of the structure, and the specified conditions refer to the normal design, normal construction and normal service conditions, excluding the abnormal effects such as human error, which should be avoided by other measures. Structural reliability is the probabilistic measure of structural reliability, and reliability is described by reliability probability. Structural design should ensure that the structure is both reliable and economical, and strive for the most economical way to make the structure meet various functional requirements with appropriate reliability in the design working life. This is the basic principle of structural design.

特别提示 3-5

2. 可靠概率
Reliability probability

结构能够完成预定功能的概率称为可靠概率,以 P_s 表示。当结构功能函数中仅有两个独立的随机变量 R 和 S,且它们都符合正态分布时,功能函数 Z 也符合正态分布,功能函数 Z 的分布曲线如图 3-1 所示,纵坐标轴以右($Z>0$)的分布曲线与横坐标 Z 轴所围成的面积即为结构的可靠概率 P_s,即 $Z>0$ 的概率。

3. 失效概率
Failure probability

结构的可靠性也可用失效概率量度,结构不能够完成预定功能的概率称为失效概率,以 P_f 表示。图 3-1 中,纵坐标轴以左($Z<0$)的阴影面积即为结构的失效概率 P_f,即 $Z<0$ 的概率。只要失效概率 P_f 足够小,则结构的可靠性必然高。结构设计的目标是对于安全等级

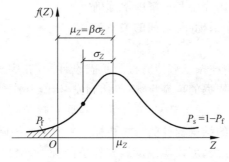

图 3-1　可靠概率与失效概率

Figure 3-1　Reliability probability and failure probability

不同的建筑物,将失效概率控制在某个可以接受的很小的数值,并使结构各部分在各种状态下的失效概率大致均衡。根据破坏形态(延性破坏、脆性破坏)的差别,失效概率的限值也应不同。

可靠概率与失效概率两者之间的关系为

$$P_s + P_f = 1 \tag{3-1}$$

3.4.3　可靠指标与目标可靠指标
Reliability index and target reliability index

1. 可靠指标

Reliability index

用失效概率量度结构可靠性具有明确的物理意义,能较好地反映问题的实质,但求失效概率 P_f 比较复杂。目前,国际上和我国都用求可靠指标 β 代替求失效概率 P_f 来量度结构的可靠性。

可靠指标 β 是量度结构可靠度的一种数值量化指标,为结构功能函数 Z 的平均值 μ_Z 与其标准差 σ_Z 之比。可靠指标与失效概率 P_f 具有数值上的一一对应关系,是失效概率负的标准正态分布函数的反函数,具有与失效概率 P_f 相对应的物理意义。可靠指标越大,则失效概率越小,结构越可靠,故称为可靠指标。

可靠指标 β 与基本变量的平均值和标准差有关,还可以考虑基本变量的概率分布类型,能反映影响结构可靠度的各主要因素的变异性,这是传统的安全系数所未能做到的。各类结构构件的安全等级每相差一级,其可靠指标的取值宜相差 0.5。

2. 目标可靠指标

Target reliability index

可靠度理论认为,安全的概念是相对的,所谓安全只是失效概率相对较小而已。要使结构或构件绝对可靠是不可能的,不存在绝对安全的结构,失效概率不可能为零,只要通过设计把失效概率控制在某一个能够接受的限值以下就可以了,这个限值称为允许失效概率。当用可靠指标表示时,限值就称为目标可靠指标[β],即设计规范所规定的、作为设计结构或结构构件时所应达到的可靠指标。结构构件设计时采用的可靠指标可根据对现有结构构件的可靠度分析,并结合使用经验和经济因素等确定。

可靠指标 β 与目标可靠指标 $[\beta]$ 之间的关系为

$$\beta \geqslant [\beta] \tag{3-2}$$

目标可靠指标是使结构在按承载能力极限状态设计时,其完成预定功能的概率不低于某一允许的水平,作为设计依据的可靠指标,以达到安全和经济上的最佳平衡。

目标可靠指标一般采用校准法确定,就是通过对原有规范可靠度的反演计算和综合分析,确定以后设计时所采用的结构构件的可靠指标。当前一些国际组织、美国、加拿大、欧洲一些国家及我国均采用此方法,可以实现可靠与经济的统一。根据校准法的确定结果,《建筑结构可靠性设计统一标准》(GB 50068—2018)规定结构构件持久设计状况承载能力极限状态设计的可靠指标不应小于表 3-3 的规定。

表 3-3 结构构件承载能力极限状态设计的可靠指标 $[\beta]$
Table 3-3 Target reliability index of ultimate limit state of structural members $[\beta]$

破 坏 类 型	结构安全等级		
	一级	二级	三级
延性破坏	3.7	3.2	2.7
脆性破坏	4.2	3.7	3.2

目标可靠指标 $[\beta]$ 主要与结构的安全等级和破坏类型有关,结构安全等级越高,其目标可靠指标越大。表 3-3 中延性破坏是指结构构件在破坏前有明显的变形或其他预兆,脆性破坏是指结构构件在破坏前没有明显的变形或其他预兆。延性破坏的危害比脆性破坏小,故目标可靠指标 $[\beta]$ 值相对低一些;脆性破坏的危害较大,所以 $[\beta]$ 值相对高一些。

拓展知识 3-4

3.4.4 承载能力极限状态设计表达式
Design expression of ultimate limit state

1. 概率极限状态设计法简介
Brief introduction of probabilistic limit state design method

英文翻译 3-8

最早的钢筋混凝土结构设计理论是以弹性理论为基础的容许应力计算法,是使结构或地基在作用标准值下产生的应力不超过规定的容许应力的设计方法,其特点是计算简单,未考虑材料的塑性性能,采用经验的安全系数进行计算。20 世纪 30 年代采用考虑塑性性能的破坏阶段计算方法,是使结构或地基的抗力标准值与作用标准值的效应之比不低于某一规定安全系数的设计方法,计算结果比较准确。20 世纪 50 年代采用极限状态计算法,近年采用基于概率理论的极限状态设计法,简称概率极限状态设计法。

结构所受的作用及结构抗力都是不确定和随机的,都是随机变量,因此传统的容许应力法、破坏阶段设计法都存在很大的不足,理论上也不完备。为克服传统设计方法的不足,20 世纪 50 年代格沃兹捷夫提出了极限状态设计方法,主要考虑各种作用、材料强度、截面尺寸等的随机性,并对这些影响结构可靠性的参数进行统计分析,以具有一定保证率的统计参数作为荷载和材料强度的标准值,然后根据可靠度理论计算出结构的可靠度。

按概率极限状态设计法设计时,一般是已知各基本变量的统计特性(如平均值和标准差),然后根据规范规定的目标可靠指标求出所需的结构抗力平均值,并转化为标准值进行截面设计。这种方法能够比较充分地考虑各有关因素的客观变异性,所设计的结构比较符合预期的可靠度要求,并且在不同结构之间,设计可靠度具有相对可比性。

 拓展知识 3-5

直接按概率极限状态设计法进行设计需进行迭代计算,计算工作量大,过于烦琐,目前只对少数十分重要的结构,如核反应堆安全壳、海上采油平台等结构直接按上述方法设计,对于大量的一般建筑结构则没有必要,需寻求简化方法。

《混凝土结构设计标准》(GB/T 50010—2010)采用的是近似概率法,将结构抗力和荷载效应作为随机变量,按给定的概率分布估算失效概率或可靠指标,在分析中采用平均值和标准差两个统计参数,采用极限状态设计表达式进行设计,且对设计表达式进行线性化处理。对于一般结构,考虑到实际工程与理论及试验的差异,直接采用荷载标准值和材料标准值进行承载力设计达不到目标可靠指标的要求,因此在承载能力设计表达式中采用分项系数进行设计计算,以荷载标准值和材料强度标准值分别与荷载分项系数和材料分项系数相联系的荷载设计值、材料强度设计值来表达。分项系数是根据结构构件基本变量的统计特性,以结构可靠度的概率分析为基础经优选确定的,起着相当于目标可靠指标$[\beta]$的作用。通过调整作用效应分项系数、材料分项系数等,使按设计表达式设计的结构或构件,其可靠指标与设计可靠指标最接近。

2. 荷载分项系数

Partial coefficient of load

荷载标准值是结构使用期间,正常情况下可能遇到的具有一定保证率的偏大荷载值。荷载是随机变量,可能超过荷载标准值。统计资料表明,不同变异性的荷载可能造成结构计算时可靠度不同,若按荷载标准值进行设计,将造成结构可靠度的严重差异,并使某些结构的实际可靠度不能达到目标可靠度的要求。因此,在承载能力极限状态设计中,将荷载标准值乘以一个分项系数进行调整,此系数称为荷载分项系数。

The standard value of load is the larger load value with a certain guarantee rate that may be encountered under normal conditions during the service of the structure. Load is random variable, it may exceed the standard value of load. Statistics show that loads with different variability may result in different reliability in structural calculation. If the design is based on the standard value of load, it will cause serious differences in structural reliability, and the actual reliability of some structures cannot meet the requirements of the target reliability. Therefore, in the ultimate limit state design, the standard value of load is multiplied by a partial coefficient for adjustment, which is called the partial coefficient of load.

结构构件极限状态设计表达式中所包含的各种分项系数宜根据有关基本变量的概率分布类型和统计参数及规定的可靠指标,通过计算分析,并结合工程经验经优化确定。当缺乏统计数据时,可根据传统的或经验的设计方法由有关标准规定各种分项系数。

Various partial coefficients contained in the limit state design expression of structural members should be determined by calculation and analysis according to the probability distribution type and statistical parameters of the relevant basic variables and the specified reliability index, and combined with engineering experience, determined by optimization. When there is a lack of statistical data, various partial coefficients may be specified by relevant standards based on traditional or empirical design methods.

考虑到荷载的统计资料尚不够完备,永久荷载标准值和可变荷载标准值的保证率不同,为了简化计算,永久荷载和可变荷载分别采用不同的荷载分项系数,用 γ_G 表示永久荷载的分项系数,γ_Q 表示可变荷载的分项系数。

Considering that the statistical data of the load is not yet complete, the guarantee rates of the permanent load standard value and the variable load standard value are different. To simplified the calculation, the permanent load and the variable load use different partial coefficient of lood respectively, and γ_G represents the partial coefficient of permanent load, and γ_Q represents the partial coefficient of variable load.

《工程结构通用规范》(GB 55001—2021)3.1.13 条规定:房屋建筑结构的作用分项系数应按下列规定取值。永久作用、预应力:当对结构不利时不应小于1.3;当对结构有利时不应大于1.0。标准值大于 $4kN/m^2$ 的工业房屋楼面活荷载,当对结构不利时不应小于1.4;当对结构有利时应取为0。除上一条之外的可变作用,当对结构不利时不应小于1.5;当对结构有利时应取为0。

Article 3.1.13 of "General code for engineering structures" (GB 55001—2021) stipulates that the partial coefficient of the action of the building structure should be specified according to the following regulations. Permanent action, prestress: when it is unfavorable to the structure, it should not be less than 1.3; when it is beneficial to the structure, it should not be greater than 1.0. When the live load on the industrial building floor with the standard value greater than $4kN/m^2$, it should not be less than 1.4 when it is unfavorable to the structure; when it is favorable to the structure, it should be taken as 0. The variable action in addition to the previous one should not be less than 1.5 when it is unfavorable to the structure; when it is beneficial to the structure, 0 should be taken.

任务 3-13:确定你所在教学楼楼盖永久荷载的分项系数和可变荷载的分项系数。

Task 3-13: Determine the permanent load partial coefficient and variable load partial coefficient of your teaching building floor.

3. 荷载设计值
Design value of load

荷载设计值是荷载分项系数与荷载标准值的乘积,如永久荷载设计值为 $\gamma_G g_k$(g_k 为永久荷载标准值),可变荷载设计值为 $\gamma_Q q_k$(q_k 为可变荷载标准值)。荷载设计值大致相当于结构在非正常使用情况下荷载的最大值,它具有比荷载标准值更大的可靠度。一般情况下,在承载能力极限状态设计中,应采用荷载设计值。当两种或两种以上可变荷载同时作用在结构上时,除主导可变荷载外,其他可变荷载标准值还应乘以组合值系数,即采用荷载组合值。

Design value of load is the product of partial coefficient of load and load standard value. For example, the design value of permanent load is $\gamma_G g_k$ (g_k is the permanent load standard value), and the design value of variable load is $\gamma_Q q_k$ (q_k is the variable load standard value). Load design value is roughly equivalent to the maximum value of load under abnormal service of the structure, and it has greater reliability than the standard value of load. In general, the load design value should be adopted in the ultimate limit state design. When two or more variable loads act on the structure at the same time, in addition to the dominant variable loads, the other variable load standard value should also be multiplied by the combination value coefficient, that is, the load combination value is used.

任务 3-14:确定你所在教学楼楼盖的永久荷载设计值和可变荷载设计值。

Task 3-14: Determine the permanent load design value and variable load design value of your teaching building floor.

4. 荷载组合值
Combination value of load

当结构上作用几个可变荷载时,各可变荷载的最大值在同一时刻出现的概率较小,如果设计中仍采用各荷载效应设计值叠加,则可能造成结构可靠度不一致,因此,必须对可变荷载设计值再乘以荷载组合值系数 ψ_{ci} 进行调整,$\psi_{ci} Q_i$ 称为可变荷载的组合值,是按极限状态设计表达式设计所得的各类结构构件所具有的可靠指标,与仅有一种可变荷载参与组合时的可靠指标有最佳的一致性。

When several variable loads act on the structure, the probability of the maximum value of each variable load appearing at the same time is small. If the design value of each load effect is still superimposed in the design, the structural reliability may be inconsistent. Therefore, the variable load design value must be multiplied by the load combination value coefficient ψ_{ci} to adjust, $\psi_{ci} Q_i$ is called the combination value of the variable load. The reliability index of various structural members designed according to the limit state design expression has the best consistency with the reliability index when only one variable load is involved in the combination.

《建筑结构荷载规范》(GB 50009—2012)给出了各类可变荷载的组合值系数,当按基本组合计算荷载组合的效应设计值时,除风荷载取 $\psi_{ci}=0.6$ 外,大部分荷载取 $\psi_{ci}=0.7$,个别可变荷载取 $\psi_{ci}=0.9\sim0.95$(如书库、储藏室的楼面可变荷载,$\psi_{ci}=0.9$)。

"Load code for the design of building structures" (GB 50009—2012) gives the coefficients for combination value of various variable loads. When calculating the effect design value of the load combination according to the basic combination, except for the wind load, $\psi_{ci}=0.6$, major load takes $\psi_{ci}=0.7$. Individual variable load is taken as $\psi_{ci}=0.9$-0.95 (such as the floor variable load of book room and storage room, $\psi_{ci}=0.9$).

5. 结构重要性系数
Structural importance coefficient

考虑到结构安全等级的差异,其目标可靠指标应相应地提高或降低。《工程结构通用规范》(GB 55001—2021)3.1.12 条规定了结构重要性系数 γ_0 的取值:对持久设计状况和短暂设计状况,安全等级为一级的结构构件 γ_0 不应小于 1.1,安全等级为二级的结构构件 γ_0 不应小于 1.0,安全等级为三级的结构构件 γ_0 不应小于 0.9;对偶然设计状况和地震设计状况,γ_0 不应小于 1.0。

Considering the difference of structural safety level, its target reliability index should be increased or decreased accordingly. Article 3.1.12 of the "General code for engineering structures"(GB 55001—2021) specifies the value of the structural importance coefficient γ_0: for persistent design situation and transient design situation, the γ_0 of structural members with safety level 1 should not be less than 1.1, the γ_0 of structural members with safety level 2 should not be less than 1.0, and the γ_0 of structural members with safety level 3 should not be less than 0.9; for accidental design situation and seismic design situation, γ_0 should not be less than 1.0.

任务 3-15:确定你所在教学楼中梁、板、柱的结构重要性系数。

Task 3-15: Determine the structural importance coefficients of beams, slabs, and columns in your teaching building.

6. 承载能力极限状态设计表达式
Ultimate limit state design expression

结构构件宜根据规定的可靠指标,采用由作用的代表值、材料性能的标准值、几何参数的标准值和各相应的分项系数构成的极限状态设计表达式进行设计,有条件时也可直接采用基于可靠指标的方法进行设计。

The structural members should be designed according to the specified reliability index by using the limit state design expression composed of the representative value of the action, the standard value of the material performance, the standard value of the geometric parameters and the corresponding partial coefficients. The method based on the reliability index can also be used directly when conditions permit.

《工程结构通用规范》(GB 55001—2021)3.1.10 条规定,对于结构或结构构件的破坏或过度变形的承载能力极限状态设计,当用内力的形式表达时,结构构件应采用下列承载能力极限状态设计表达式:

$$\gamma_0 S_d \leqslant R_d \tag{3-3}$$

式中,γ_0 为结构重要性系数;S_d 为作用组合的效应设计值,对持久设计状况和短暂设计状况应按作用的基本组合计算,对地震设计状况应按作用的地震组合计算;R_d 为结构或结构构件的抗力设计值。

Article 3.1.10 of the "General code for engineering structures" (GB 55001—2021) stipulates that for the design of ultimate limit state due to failure or excessive deformation of the structure or structural member, when expressed in the form of internal forces, the structural members should adopt the following ultimate limit state design expression:

$$\gamma_0 S_d \leqslant R_d \quad (3-3)$$

Where, γ_0 is the structural importance coefficient; S_d is the effect design value of action combination, it should be calculated according to the basic combination of actions for the persistent design situation and transient design situation, and the seismic design situation should be calculated according to the seismic combination of action; R_d is the design value of resistance of structure or structural member.

拓展知识 3-6

任务 3-16：说明承载能力极限状态设计表达式(3-3)的含义。

Task 3-16: Explain the meaning of the ultimate limit state design expression Eq. (3-3).

7. 作用组合的效应设计值

Effect design value of action combination

作用组合是在不同作用的同时影响下,为验证某一极限状态的结构可靠度而采用的一组作用设计值。当结构上作用有多种可变荷载时,由于结构上的可变荷载不止一个,对结构的影响有大有小,多个可变荷载也不一定会同时出现,因此要考虑荷载效应的组合问题。荷载效应组合是指在所有可能同时出现的各种荷载组合下,确定结构或构件内产生的总效应。

结构设计时,应根据所考虑的设计状况选用不同的组合。对持久和短暂设计状况,应采用作用的基本组合;对偶然设计状况,应采用作用的偶然组合;对地震设计状况,应采用作用的地震组合,地震组合的效应设计值应符合《建筑抗震设计规范》(GB 50011—2010)的规定。

基本组合的效应设计值按式(3-4)中的最不利值确定:

$$S_d = \sum_{i \geqslant 1} \gamma_{G_i} S_{G_{ik}} + \gamma_P S_P + \gamma_{Q_1} \gamma_{L_1} S_{Q_{1k}} + \sum_{j>1} \gamma_{Q_j} \psi_{c_j} \gamma_{L_j} S_{Q_{jk}} \quad (3-4)$$

式中, S_d 为作用组合的效应函数; γ_{G_i} 为第 i 个永久作用的分项系数; $S_{G_{ik}}$ 为第 i 个永久作用标准值的效应; γ_P 为预应力作用的分项系数; S_P 为预应力作用有关代表值的效应; γ_{Q_1}、γ_{Q_j} 分别为第 1 个可变作用(主导可变作用)和第 j 个可变作用的分项系数; γ_{L_1}、γ_{L_j} 分别为第 1 个和第 j 个考虑结构设计工作年限的荷载调整系数; $S_{Q_{1k}}$ 为第 1 个可变作用标准值的效应; ψ_{c_j} 为第 j 个可变作用的组合值系数; $S_{Q_{jk}}$ 为第 j 个可变作用标准值的效应; G_i 为第 i 个永久作用的标准值; P 为预应力作用的有关代表值; Q_{1k} 为第 1 个可变作用的标准值; Q_{jk} 为第 j 个可变作用的标准值。

当对 Q_{1k} 无法明显判断时,轮次以各可变荷载作为 Q_{1k},选其中最不利的荷载效应组合。组合值系数的确定应使按分项系数表达式设计的结构或结构构件的可靠指标 β 与目标可靠指标 $[\beta]$ 具有最佳的一致性。对每一种作用组合,建筑结构的设计均应采用其最不利的效应设计值进行。

式(3-4)中, $\gamma_G S_{G_{ik}}$ 和 $\gamma_{Q_i} S_{Q_{ik}}$ 分别称为永久荷载效应设计值和可变荷载效应设计值。

承载能力极限状态设计表达式既考虑了结构设计的传统方式,又可避免设计时直接进行概率方面的计算。

承载能力极限状态设计表达式中的作用组合应符合下列规定：

(1) 作用组合应为可能同时出现的作用的组合；

(2) 每个作用组合中应包括一个主导可变作用或一个偶然作用或一个地震作用；

(3) 当静力平衡等极限状态设计对永久作用的位置和大小很敏感时，该永久作用的有利部分和不利部分应分别作为单个作用分别考虑；

(4) 当一种作用产生的几种效应非全相关时，对产生有利效应的作用，其分项系数的取值应予以降低。

任务 3-17：说明承载能力极限状态基本组合的效应设计值式(3-4)中各参数的含义。

Task 3-17：Explain the meaning of each parameter in the effect design value Eq.(3-4) of the basic combination of ultimate limit state.

【例题 3-1】 钢筋混凝土矩形截面简支梁的结构安全等级为二级，计算跨度 $l_0 = 5\mathrm{m}$，截面尺寸为 $b \times h = 250\mathrm{mm} \times 600\mathrm{mm}$，钢筋混凝土的重度 $\gamma = 25\mathrm{kN/m^3}$，梁上可变荷载标准值 $q_k = 5\mathrm{kN/m}$，要求：按承载能力极限状态设计基本组合计算梁跨中弯矩设计值。

【Example 3-1】 The structural safety level of reinforced concrete simply supported rectangular section beam is Level 2. The calculated span $l_0 = 5\mathrm{m}$, the cross-sectional dimensions are $b \times h = 250\mathrm{mm} \times 600\mathrm{mm}$, the reinforced concrete gravity is $\gamma = 25\mathrm{kN/m^3}$. The variable load standard value on the beam is $q_k = 5\mathrm{kN/m}$. Requirement: the design value of mid-span bending moment of the beam is calculated according to the basic combination of ultimate limit state.

解：简支梁的结构安全等级为二级，永久荷载分项系数取 $\gamma_G = 1.3$，可变荷载分项系数取 $\gamma_Q = 1.5$。

(1) 确定该梁沿跨度方向均匀分布的永久荷载标准值
$$g_k = \gamma b h = 25 \times 250 \times 600 \times 10^{-6} \mathrm{kN/m} = 3.75 \mathrm{kN/m}$$

(2) 按承载能力极限状态基本组合的计算公式(3-4)计算梁跨中弯矩设计值
$$M = \gamma_G \left(\frac{1}{8} g_k l_0^2\right) + \gamma_Q \left(\frac{1}{8} q_k l_0^2\right) = \frac{1}{8}(\gamma_G g_k + \gamma_Q q_k) l_0^2$$
$$= \frac{1}{8}(1.3 \times 3.75 + 1.5 \times 5) \times 5^2 \mathrm{kN \cdot m} = 38.67 \mathrm{kN \cdot m}$$

本 章 小 结

(1) 结构上的作用分直接作用和间接作用两种，其中直接作用也称荷载，指的是施加在结构上的集中力或分布力。作用按其随时间的变异，分为永久作用、可变作用、偶然作用三种。永久作用中的直接作用亦称永久荷载，如结构重力；可变作用中的直接作用亦称可变荷载。可变荷载有标准值、组合值、准永久值三种代表值，各用于极限状态设计中的不同场合，其中标准值是荷载的基本代表值。

(2) 混凝土结构必须满足安全性、适用性、耐久性三项功能要求。我国规定房屋建筑结构的设计基准期为 50 年。结构物在设计基准时期内，在规定的条件下，完成预定功能的能力称为结构的可靠性。结构设计中结构的可靠性是用结构的极限状态来表达的。结构的极限状态是指某一功能的临界状态,当结构或结构构件超过极限状态时,就不能满足这一功能的要求。结构的极限状态有三类:承载能力极限状态是安全性功能方面的;正常使用极限状态是适用性功能方面的;耐久性极限状态是耐久性功能方面的。

(3) 荷载引起的结构或结构构件的内力、变形等称为荷载效应,用 S 表示。结构或构件承受荷载效应的能力称为结构抗力,用 R 表示。目前,我国采用的以近似概率理论为基础的极限状态设计法,主要是引入荷载分项系数、材料分项系数、结构重要性系数。建筑结构的安全等级分为一级、二级和三级,它们的结构重要性系数分别为不小于 1.1、1.0 和 0.9。

(4) 荷载分项系数乘以荷载标准值后得到荷载设计值;材料强度标准值除以材料分项系数后得到材料强度设计值;作用组合的效应设计值 S_d 与结构重要性系数 γ_0 相乘后得 $\gamma_0 S_d$,应该小于等于结构或结构构件的抗力设计值 R_d,结构才安全可靠。为了保证结构安全可靠,荷载取具有一定保证率的偏大值,材料强度取具有一定保证率的偏小值。没有绝对安全的结构,只有具有一定可靠度的结构。

第3章拓展知识和特别提示

视频:第3章小结讲解

3-1 填空题

1. 建筑结构设计的目的是使设计的结构能满足预定的功能要求,结构预定的功能要求是结构的_____、_____和_____。

2. 结构上的作用按其产生的原因分为_____作用和_____作用。

3. 结构上的荷载按其随时间的变异性,可分为_____荷载、_____荷载和_____荷载。

4. 荷载设计值可由荷载标准值乘以对应的_____系数计算。

5. 材料强度的设计值等于材料强度的标准值除以对应的_____。

6. 在规定_____内,在规定_____下,完成_____称为可靠性。

7. 建筑结构的极限状态可分为_____极限状态、_____极限状态和_____极限状态。

8. 按极限状态进行设计的方法称为_____。

9. 《混凝土结构设计标准》(GB/T 50010—2010)规定,按承载能力极限状态进行结构构件计算的一般公式为 $\gamma_0 S_d \leqslant R_d$,式中,$\gamma_0$ 表示_____,S_d 表示_____,R_d 表示_____。

3-2 选择题

1. 我国混凝土结构设计采用的设计方法是(   )。

    A. 以半概率理论为基础的极限状态设计法

    B. 以全概率理论为基础的极限状态设计法

C. 以近似概率理论为基础的极限状态设计法
D. 破损阶段设计法

2. 根据结构重要性及破坏可能产生后果的严重程度,将结构安全等级划分为(　　)级。
   A. 3　　　　　　　B. 5　　　　　　　C. 7　　　　　　　D. 10

3. 下列情况(　　)属于承载能力极限状态。
   A. 裂缝宽度超过《混凝土结构设计标准》(GB/T 50010—2010)的限值
   B. 挠度超过《混凝土结构设计标准》(GB/T 50010—2010)的限值
   C. 结构或构件视为刚体,失去平衡
   D. 预应力构件中混凝土的拉应力超过《混凝土结构设计标准》(GB/T 50010—2010)的限值

4. 下列各项中按正常使用极限状态考虑的是(　　)。
   A. 雨篷倾覆　　　　　　　　　　　B. 简支梁跨中产生塑性铰
   C. 细长柱失稳　　　　　　　　　　D. 楼板振动过大

5. 结构按极限状态设计时应符合的条件为(　　)。
   A. $S>R$　　　　B. $S\leqslant R$　　　　C. $S<R$　　　　D. $S\geqslant R$

3-3　名词解释

1. 结构上的作用	2. 直接作用	3. 间接作用	4. 作用效应
5. 结构抗力	6. 设计基准期	7. 设计工作年限	8. 结构的功能
9. 结构的可靠性	10. 结构的可靠度	11. 结构安全等级	12. 永久荷载
13. 可变荷载	14. 偶然荷载	15. 荷载代表值	16. 荷载标准值
17. 永久荷载标准值	18. 可变荷载标准值	19. 结构的极限状态	20. 承载能力极限状态
21. 正常使用极限状态	22. 耐久性极限状态	23. 结构的设计状况	24. 结构功能函数
25. 失效概率	26. 可靠指标	27. 目标可靠指标	28. 结构重要性系数
29. 永久荷载分项系数	30. 可变荷载分项系数	31. 荷载设计值	32. 永久荷载设计值
33. 可变荷载设计值	34. 荷载组合值	35. 可变荷载准永久值	

3-4　思考题

1. 可变荷载有哪些代表值?进行结构设计时如何选用这些代表值?
2. 结构的安全等级有哪几级?对应的结构重要性系数分别是多少?
3. 什么是结构的极限状态?
4. 什么是结构的功能函数?功能函数 $Z>0$,$Z<0$ 和 $Z=0$ 时各表示结构处于什么状态?
5. 结构的可靠概率与结构的失效概率有什么关系?
6. 为什么要引入荷载分项系数?如何选用荷载分项系数的值?
7. 混凝土结构的设计方法是如何演变的?
8. 按现行的设计方法进行混凝土结构设计时,需要进行哪些计算与验算?

第 4 章

受弯构件的正截面受弯承载力

Flexural capacity of normal section of flexural members

教学目标：
1. 熟悉受弯构件的一般构造要求，会确定梁的截面形式和尺寸；
2. 了解钢筋混凝土受弯构件的受力过程和特点；
3. 熟悉梁正截面受弯破坏形态；
4. 理解受弯构件正截面受弯承载力计算的基本假定；
5. 会设计计算单筋、双筋和 T 形截面受弯构件；
6. 熟悉梁的配筋构造要求。

导读：

受弯构件是在荷载作用下，截面上承受弯矩和剪力作用的构件。受弯构件是土木工程中应用最广泛的构件，如建筑中肋梁楼盖的梁和板（如图 4-1 所示），楼梯的梯段板、平台板、平台梁，门窗过梁，工业厂房中的屋面梁、吊车梁、连系梁等，桥梁中的行车道板、主梁和横隔梁，挡土墙板等。

梁一般指沿纵轴方向承受垂直荷载的线形构件，截面尺寸（宽度、高度）小于其跨度。板是具有较大平面尺寸，却有相对较小厚度的面形构件。

梁按其施工方法，可分为现浇梁、预制梁和预制现浇叠合梁。按其配筋类型，可分为钢筋混凝土梁和预应力混凝土梁。按其结构计算简图，可分为简支梁、连续梁、悬挑梁、主梁和次梁等。按其使用的部位不同有过梁、圈梁、楼盖梁、楼梯梁、雨篷梁等。

图 4-1 肋梁楼盖
Figure 4-1 Ribbed beam floor

受弯构件在弯矩和剪力共同作用下，可能发生两种主要的破坏：沿弯矩最大截面发生的正截面破坏和沿剪力最大或弯矩、剪力都较大的截面发生的斜截面破坏。正截面指与构件轴线相垂直的截面，斜截面指与构件轴线斜交的截面。进行受弯构件设计时，要同时保证不发生正截面破坏和斜截面破坏，正截面承载力和斜截面承载力分开计算,不考虑相互影响。受弯构件正截面承载力的计算一般分为两类问题:截面设计和截面复核。

受弯构件截面设计的基本步骤:

(1)根据截面所承担的弯矩选择混凝土强度等级和钢筋种类、级别;

(2)选用截面形式,估算梁、板的截面尺寸,如矩形截面梁的宽度 b、高度 h,板的厚度等;

(3)计算纵向受力钢筋的截面面积,验算配筋率;

(4)选配纵向受力钢筋的直径和根数,布置钢筋。

受弯构件的设计内容:

1) 按承载能力极限状态要求设计

受弯构件正截面在弯矩作用下发生破坏,称为受弯承载力极限状态,相应的极限弯矩称为正截面受弯承载力。根据承载能力极限状态设计表达式,受弯构件正截面受弯承载力计算应满足 $\gamma_0 M \leqslant M_u$,式中,M 为结构上的作用产生的弯矩设计值,为作用效应,采用力学方法计算;M_u 为受弯构件正截面受弯承载力设计值,为结构抗力,由钢筋混凝土材料的力学指标、截面尺寸等确定。

为保证受弯构件不因弯矩作用而破坏,根据构件控制截面(跨中或支座截面)的弯矩设计值,确定材料等级、截面尺寸和纵向受力钢筋的直径、数量等。

2) 按正常使用极限状态要求验算

受弯构件一般还需进行正常使用阶段的挠度变形和裂缝宽度验算,对使用上要求不出现裂缝的构件应进行混凝土拉应力验算。

3) 构造措施

混凝土的收缩、徐变和温度应力等对截面承载力的影响不容易计算,一般采取构造措施予以解决。所谓构造措施就是考虑施工、受力及使用等方面因素的综合影响而采取的针对性措施。构造要求是结构设计的一个重要组成部分,它是在长期工程实践经验的基础上对结构计算的必要补充,防止结构计算中没有计及或过于复杂而难以考虑的因素对结构构件的可能影响。对受弯构件除了进行计算外,还需按弯矩图、剪力图、截断、锚固、连接等要求,确定配筋构造,以保证构件的适用性和耐久性。

影响混凝土结构的因素非常多,也非常复杂,因此,在结构构件的设计中考虑构造要求十分重要,结构计算和构造措施是相互配合的。本课程涉及的构造要求比较多,也比较细,这是学习过程中需要注意和重视的问题。

引例:

第 4 章引例

第4章 受弯构件的正截面受弯承载力
Flexural capacity of normal section of flexural members

核心词汇：

受弯构件	flexural member	界限相对受压区高度	boundary relative compression zone depth
少筋梁	lightly reinforced beam	受弯承载力	flexural capacity
正截面破坏	normal section failure	截面抵抗矩系数	sectional resistance moment coefficient
界限破坏	balanced failure		
高跨比	height-span ratio		
延性破坏	ductile failure	单筋矩形截面	singly reinforced rectangular section
设计弯矩	design moment		
脆性破坏	brittle failure	内力臂系数	internal lever arm coefficient
截面设计	section design	双筋矩形截面	doubly reinforced rectangular section
开裂弯矩	cracking moment		
承载力	bearing capacity	截面有效高度	effective depth of section
极限承载力	ultimate bearing capacity	T形截面	T-shaped section
截面尺寸	cross-sectional dimension	纵向受力钢筋	longitudinally stressed reinforcement
配筋率	reinforcement ratio		
平截面假定	plane section assumption	中和轴	neutral axis
箍筋	stirrup	混凝土保护层厚度	thickness of concrete cover
等效应力	equivalent stress	翼缘	flange
弯起钢筋	bent bar	适筋梁	underreinforced beam
相对受压区高度	relative depth of compression zone	腹板	web
		超筋梁	overreinforced beam
架立钢筋	erection bar	截面复核	section review

4.1 受弯构件一般构造要求
General structural requirements of flexural members

4.1.1 受弯构件的材料选择、截面形式与尺寸
Material selection, section form and size of flexural members

在进行受弯构件正截面承载力计算之前，需要了解其有关的构造要求。

1. 受弯构件的材料选择
Material selection of flexural members

《混凝土结构通用规范》(GB 55008—2021)2.0.2条规定：钢筋混凝土结构构件的混凝土强度等级不应低于C25；预应力混凝土楼板结构的混凝土强度等级不应低于C30；钢-混凝土组合结构构件的混凝土强度等级不应低于C30。

受弯构件常用的混凝土强度等级为 C25、C30、C35、C40 等,为了防止混凝土收缩过大,一般不超过 C40;梁内纵向受力普通钢筋宜采用 HRB400、HRB500。

2. 受弯构件的截面形式

Section form of flexural members

梁常用的截面形式有矩形、T 形和 I 形,偶尔还采用环形。有时为了降低层高,还可以采用花篮形、十字形、倒 T 形截面。板的截面形式常见的有矩形、槽形和空心形等,如图 4-2 所示。

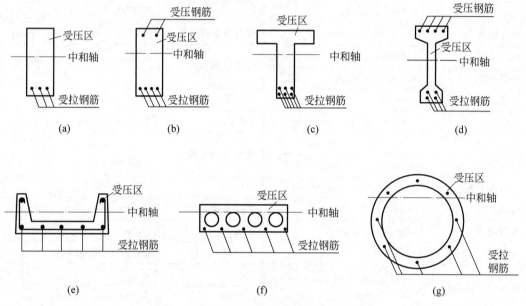

图 4-2 梁、板的截面形式

(a) 单筋矩形截面梁;(b) 双筋矩形截面梁;(c) T 形截面梁;
(d) I 形截面梁;(e) U 形截面梁;(f) 空心板;(g) 环形梁

Figure 4-2 Section form of beam and slab

(a) Singly reinforced rectangular section beam;(b) Doubly reinforced rectangular section beam;
(c) T-shaped section beam;(d) I-shaped section beam;(e) U-shaped section beam;
(f) Hollow slab;(g) Circular beam

3. 受弯构件的截面尺寸

Cross-sectional dimensions of flexural members

受弯构件的截面尺寸应满足承载力、刚度及裂缝控制的要求,还应满足使用要求、施工要求及经济要求。

1) 梁的截面尺寸

1) Cross-sectional dimensions of beams

梁的截面高度与梁的跨度及荷载大小有关,一般情况下梁的跨度、荷载越大,梁的截面高度越高。按刚度条件,梁的截面高度可根据高跨比来估计,高跨比为梁的截面高度与梁的跨度之比。根据工程经验,次梁一般可取梁高 $h=(1/18\sim1/12)l_0$(l_0 为次梁计算跨度),主梁一般可取梁高 $h=(1/12\sim1/8)l_0$(l_0 为主梁计算跨度)。

为了统一模板尺寸，便于施工，梁的截面高度一般采用 $h=250mm,300mm,350mm,\cdots,$ 750mm,800mm,900mm,1000mm 等尺寸。当梁高 $h \leqslant 800mm$ 时，梁的高度为 50mm 的倍数；当梁高 $h > 800mm$ 时，梁的高度为 100mm 的倍数。

梁的截面宽度一般采用 $b=200mm,220mm,250mm,300mm$ 等尺寸，250mm 以上取 50mm 的倍数，矩形截面框架梁的截面宽度不应小于 200mm。矩形截面梁的宽度一般取 $b=(1/3\sim1/2)h$，T 形截面梁的宽度一般取 $b=(1/4\sim1/2.5)h$（此处 b 为腹板宽度）。

2) 板的截面尺寸

2) Cross-sectional dimensions of slabs

板的厚度应由设计计算确定，由于板的混凝土用量占整个楼盖的 50%~70%，因此应使板厚尽可能接近构造要求的最小板厚，《混凝土结构设计标准》(GB/T 50010—2010) 表 9.1.2 规定了现浇钢筋混凝土板的最小厚度。《混凝土结构通用规范》(GB 55008—2021) 4.4.4 条规定，现浇钢筋混凝土实心楼板的厚度不应小于 80mm，现浇空心楼板的顶板、底板厚度均不应小于 50mm。预制钢筋混凝土实心叠合楼板的预制底板及后浇混凝土厚度均不应小于 50mm。当荷载很大及安装设备时有可能有撞击作用，板厚应增加。

《混凝土结构设计标准》(GB/T 50010—2010) 9.1.2 条规定，钢筋混凝土单向板的厚度不小于跨度的 1/30（即单向板跨厚比不大于 30），双向板的厚度不小于板跨度的 1/40（即双向板跨厚比不大于 40）。最后，应取构造要求的最小板厚和不需进行变形验算的跨厚比条件确定的板厚二者中较大者作为板的设计厚度。板的厚度一般为 10mm 的倍数，板的跨度大于 4m 时，板应适当加厚。

思考 4-1：什么是单向板、双向板？什么是板的跨厚比？《混凝土结构设计标准》(GB/T 50010—2010) 对板的最小厚度的规定有哪些？

Thinking 4-1: What are one-way and two-way slabs? What is the span-thickness ratio of the slab? What is the minimum thickness of the slab specified in the "Standard for design of concrete structures"(GB/T 50010—2010)?

混凝土结构构件的最小截面尺寸应满足承载能力极限状态、正常使用极限状态的计算要求，并应满足结构耐久性、防水、防火、配筋构造及混凝土浇筑等施工要求。

The minimum cross-sectional dimensions of concrete structural members should meet the calculated requirements of the ultimate limit state and the serviceability limit state, and should meet the construction requirements of structural durability, waterproofing, fire protection, reinforcement construction and concrete placement etc.

 拓展知识 4-1

4.1.2 混凝土保护层厚度
Thickness of concrete cover

混凝土保护层厚度是构件最外层钢筋（包括箍筋、构造筋、分布筋）的外表面到最近混凝土面的垂直距离，用 c 表示。

The thickness of concrete cover is the vertical distance from the outer surface of the outermost bars (including stirrups, structural bars and distributed bars) to the nearest concrete face, which is represented by c.

混凝土保护层有三个作用：①防止纵向钢筋锈蚀，减小混凝土的碳化；②在火灾等情况下，使钢筋的温度上升缓慢，提高混凝土结构的耐火性和耐久性；③使纵向钢筋与混凝土之间有较好的黏结。

The concrete cover has three functions: ①Prevent corrosion of longitudinal steel bars and reduce the concrete carbonization; ②In the event of fire, etc., the temperature of steel bars rises slowly and the fire resistance and durability of the concrete structure are improved; ③ There is a good bond between the longitudinal reinforcement and the concrete.

为了保证结构耐久性、耐火性以及钢筋与混凝土的黏结性能，考虑环境类别、构件种类和混凝土强度等级等因素，构件中普通钢筋及预应力钢筋的混凝土保护层厚度应满足《混凝土结构设计标准》(GB/T 50010—2010)8.2.1条的要求。

In order to ensure the structural durability, fire resistance and bonding properties between steel and concrete, the thickness of concrete cover of ordinary reinforcement and prestressed reinforcement in the member should meet the requirements of Article 8.2.1 of "Standard for design of concrete structures"(GB/T 50010—2010) considering factors such as environmental category, member type and concrete strength grade.

(1) 构件中受力钢筋的混凝土保护层厚度不应小于受力钢筋的公称直径 d。

(2) 设计工作年限为 50 年的混凝土结构，最外层钢筋的混凝土保护层厚度应符合附表 3-2 的规定；设计工作年限为 100 年的混凝土结构，最外层钢筋的混凝土保护层厚度不应小于附表 3-2 中数值的 1.4 倍。

(1) The thickness of concrete cover of the stressed reinforcement in the member should not be less than the nominal diameter d of the stressed reinforcement.

(2) For concrete structure with design working life of 50 years, the thickness of concrete cover of the outermost bars should comply with the provisions in Attached table 3-2; for concrete structure with design working life of 100 years, the thickness of concrete cover of the outermost steel bars should not less than 1.4 times the numerical value in Attached table 3-2.

《混凝土结构通用规范》(GB 55008—2021)2.0.10 条规定，混凝土保护层厚度应符合下列规定。

Article 2.0.10 of "General code for concrete structures"(GB 55008—2021) stipulates that the thickness of concrete cover should meet the following requirements.

(1) 满足普通钢筋、有黏结预应力钢筋与混凝土共同工作的性能要求；

(2) 满足混凝土构件的耐久性能及防火性能要求；

(3) 不应小于普通钢筋的公称直径，且不应小于15mm。

(1) It meets the common working performance requirements of ordinary steel bars, bonded prestressed steel bars and concrete;

(2) It meets the durability and fire performance requirements of concrete members;

(3) It should not be smaller than the nominal diameter of ordinary steel bars, and

should not be smaller than 15mm.

拓展知识 4-2

任务 4-1：估计你所在教室中梁的跨度、截面高度、截面宽度及混凝土保护层厚度。

Task 4-1: Estimate the span, section depth, section width, and thickness of concrete cover of the beam in your classroom.

4.2 受弯构件正截面受弯性能的试验研究
Experimental research on flexural performance of normal section of flexural members

钢筋混凝土材料与材料力学中的弹性、匀质、各向同性材料有很大的不同，影响钢筋混凝土正截面受弯性能的因素较多，问题也很复杂，所以受弯构件的计算理论是建立在试验研究基础上的，在试验的基础上进行理论分析，建立正截面受弯承载力的计算理论和方法。受弯构件的破坏包括正截面破坏和斜截面破坏，正截面破坏是在弯矩作用下发生的与构件轴线垂直的正截面的破坏，如图 4-3 所示。斜截面破坏是在弯矩和剪力共同作用下发生的与构件轴线倾斜的斜截面的破坏，如图 4-4 所示。

Reinforced concrete materials are very different from elastic, homogeneous and isotropic materials in material mechanics. There are many factors affecting the flexural performance of reinforced concrete in normal section, and the problems are more complex. Therefore, the calculation theory of flexural members is established on the basis of experimental research. Theoretical analysis is carried out on the basis of experimental research, and the calculation theory and method of flexural capacity of normal section are established. The failure of flexural member includes normal section failure and inclined section failure. The normal section failure is the failure of the normal section perpendicular to the axis of the member under the action of bending moment, as shown in Fig. 4-3. The inclined section failure is the failure of the inclined section inclined to the axis of the member under the combined action of bending moment and shear force, as shown in Fig. 4-4.

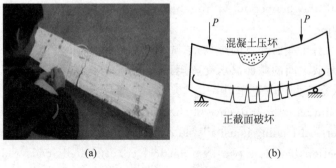

图 4-3 梁受弯正截面破坏
（a）试验梁正截面破坏；（b）梁正截面破坏示意图
Figure 4-3　Failure of normal section due to bending of the beam
(a) Failure of normal section of the test beam; (b) Schematic diagram of the failure of normal section of the beam

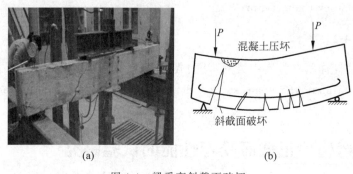

图 4-4 梁受弯斜截面破坏
（a）试验梁斜截面破坏；（b）梁斜截面破坏示意图
Figure 4-4　Failure of inclined section due to bending of the beam
（a）Failure of inclined section of the test beam；（b）Schematic diagram of the failure of inclined section of the beam

4.2.1　简支试验梁受弯承载力试验
Flexural capacity test of simply supported test beams

为了研究受弯构件的破坏过程，分析其受力特点、变形过程和破坏形态，用钢筋混凝土矩形截面简支梁进行试验研究。如图 4-5、图 4-6 所示为钢筋混凝土矩形截面简支试验梁受弯承载力试验，采用适量配筋的钢筋混凝土试验梁，截面尺寸为 $b \times h = 150\text{mm} \times 350\text{mm}$，混凝土强度等级 C25，纵向受拉钢筋为 3⌀14。试验采用荷载从小到大的逐级加载方式，每级加载后观测并记录裂缝的出现和发展情况，并记录梁的挠度，直至正截面受弯破坏。

In order to study the failure process of the flexural member, analyze its stress characteristics, deformation process and destruction form, a reinforced concrete simply supported rectangular beam is used for experimental research. Fig. 4-5 and Fig. 4-6 show the flexural capacity test of a reinforced concrete rectangular section simply supported test beam. Reinforced concrete test beam is with appropriate reinforcement, cross-sectional dimensions are $b \times h = 150\text{mm} \times 350\text{mm}$, concrete strength grade is C25, longitudinal tensile reinforcement is 3⌀14. The test adopts the loading method step by step from small to large. After each stage loading, the occurrence and development of cracks are observed and recorded, and the deflection of the beam is recorded until the normal section is damaged by bending.

为了消除剪力对正截面受弯的影响，通常采用两点对称加载。在忽略梁自重的情况下，使两个对称集中荷载之间的截面只承受弯矩而无剪力，称为纯弯段。纯弯段外的两个区域为剪弯段，同时承受剪力和弯矩作用。

In order to eliminate the influence of shear force on the bending of the normal section, two-point symmetrical loading is usually adopted. In the case of ignoring the self-weight of the beam, the section between two symmetrical concentrated loads that only bears the bending moment without the shear force is called pure bending segment. The two areas outside the pure bending segment are shear bending segments simultaneously bearing shear force and bending moment.

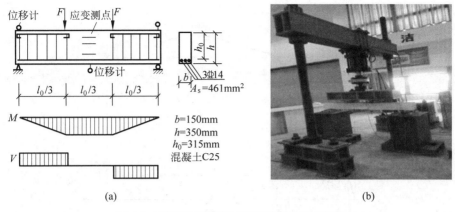

图 4-5 矩形截面简支试验梁受弯承载力试验
（a）正截面受弯承载力试验示意图；(b) 试验设备

Figure 4-5 Flexural capacity test of simply supported test beam with rectangular section
(a) Schematic diagram of normal section flexural capacity test; (b) Test equipment

图 4-6 试验梁中的钢筋骨架和试验过程观测
（a）试验梁中的钢筋骨架；(b) 试验过程的观测

Figure 4-6 Steel skeleton in the test beam and observation of the test process
(a) Steel skeleton in the test beam; (b) Observation of the test process

4.2.2 钢筋混凝土梁正截面的三个工作阶段
Three working stages of normal section of reinforced concrete beams

经试验，测得典型单筋矩形截面梁的弯矩 M^0 与截面曲率 φ^0 关系曲线如图 4-7 所示，图中纵坐标 M^0 为各级荷载下梁跨中截面的实际弯矩，横坐标 φ^0 为跨中截面曲率的实测值，这里的上标 0 表示实测值。在纯弯曲段，弯矩将使正截面转动，此弯曲变形可用截面曲率来量度，即单位长度的梁上正截面的转角。当梁上荷载逐渐从零增加到使梁破坏时，M^0-φ^0 关系曲线上有两个明显的转折点 c 和 y，它们把适筋梁的受力过程分为三个不同的阶段：弹性阶段Ⅰ、带裂缝工作阶段Ⅱ、破坏阶段Ⅲ。

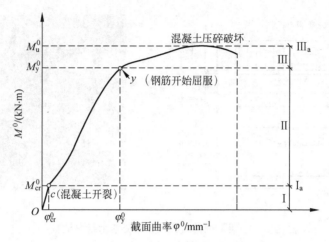

图 4-7　梁的弯矩 M^0 与截面曲率 φ^0 关系曲线

Figure 4-7　Relationship curve between bending moment M^0 and sectional curvature φ^0 of the beam

1. 第Ⅰ阶段(弹性阶段)
Stage Ⅰ (elastic stage)

第Ⅰ阶段表示从开始加载到梁截面受拉区混凝土即将开裂,此阶段截面弯矩 M 较小,梁截面受拉区未出现裂缝,构件基本处于弹性受力阶段,M^0 和 φ^0 的关系接近直线变化。截面应变为直线分布,受压区混凝土应力接近于直线分布,受拉区混凝土的应力在此阶段前期为直线分布,后期为曲线分布,混凝土中的应力很小且和应变成正比,受拉区塑性变形有所发展,沿截面高度混凝土的应力分布如图 4-8(a)、(b)所示。当 M 达到开裂弯矩 M_{cr} 时,梁即将出现裂缝,第Ⅰ阶段结束。

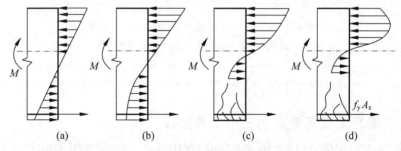

图 4-8　梁受力各阶段正截面应力分布

(a) 第Ⅰ阶段前期;(b) 第Ⅰ阶段后期;(c) 第Ⅱ阶段;(d) 第Ⅲ阶段

Figure 4-8　The stress distribution of the normal section of the beam at each stage of stress

(a) Early stage Ⅰ; (b) Late stage Ⅰ; (c) Stage Ⅱ; (d) Stage Ⅲ

2. 第Ⅱ阶段(带裂缝工作阶段)
Stage Ⅱ (working stage with cracks)

第Ⅱ阶段表示从梁截面受拉区混凝土开裂到受拉钢筋屈服。在这个阶段,当荷载持续增加,应力和应变相应增大且不再成比例。截面弯矩 M 超过开裂弯矩 M_{cr},弯矩与梁中挠度呈曲线关系,截面曲率增长较快。很快达到混凝土的抗拉强度,梁截面受拉区混凝土出现受拉裂缝,随着荷载继续增加,裂缝不断开展,挠度不断增长。梁带裂缝工作,裂缝处受拉区混凝土大部分退出工作,不传递拉应力,纵向受拉钢筋承担全部的拉应力,但受拉钢筋没有屈服。受压区混凝土有塑性变形,但发展尚不充分,压应力图形为曲线,如图 4-8(c)所示。

梁的正常使用阶段处于第Ⅱ阶段,该阶段的应力状态是建立正常使用极限状态的变形和裂缝宽度计算模型的依据。

3. 第Ⅲ阶段(破坏阶段)
Stage Ⅲ (failure stage)

第Ⅲ阶段表示从受拉钢筋屈服后到截面受弯破坏。此阶段截面弯矩 M 增加不多,弯矩与挠度关系接近水平直线,在后期有所上升。梁截面受拉区混凝土裂缝急剧变宽并向上发展,同时梁的挠度急剧增加。当钢筋突然屈服并发生很大变形,弯矩 M 增加到梁的极限弯矩 M_u,受压区混凝土达到极限压应变被压碎,最后达到梁的受弯承载力而破坏。

受拉区大部分混凝土退出工作,受压区混凝土压应力曲线图形比较饱满,如图 4-8(d)所示。受压区边缘混凝土的应变超过峰值应变进入应力-应变曲线的下降段,直至边缘应变达到极限压应变,混凝土被压碎,截面达到承载能力极限状态,而靠近中和轴附近的混凝土仍处于应力-应变曲线的上升段,即应变还没有达到峰值应变,受压区有明显的应力重分布。

混凝土开裂引起了钢筋应力的突变,使钢筋应力与弯矩的增长不再呈线性变化。钢筋屈服后的力学性能集中反映了钢筋和混凝土的塑性性能。混凝土的开裂和受压塑性性能使截面的应力分布图形发生变化,要保持截面受力平衡,中和轴的位置必然发生变化。

4.2.3 受弯构件正截面的破坏形态
Destruction form of normal section of flexural members

试验表明,梁正截面的破坏形态主要和纵向受拉钢筋的配筋量有关,纵向受拉钢筋的配筋量一般用配筋率表示。试验结果表明,当梁的截面尺寸和材料强度一定时,根据受弯构件纵向受力钢筋配筋率的不同,达到承载能力极限状态时的破坏特征不同,梁正截面受弯破坏时有三种破坏形态:适筋破坏、超筋破坏、少筋破坏,如图 4-9 所示。三种破坏形态对应的梁分别称为适筋梁、超筋梁、少筋梁。

The tests indicate that the destruction form of the normal section of the beam is mainly related to the reinforcement amount of longitudinal tensile steel bars, and the reinforcement amount of longitudinal tensile steel bars is generally expressed by the reinforcement ratio. The experimental results show that when the cross-sectional dimensions and material strength of the beam are constant, according to the different reinforcement ratio of the longitudinally stressed bars of the flexural member, the failure characteristics are different when the ultimate limit state is reached. There are three destruction forms when the failure of normal section is flexural failure: underreinforced failure, overreinforced failure, and lightly reinforced failure, as shown in Fig. 4-9, the beams corresponding to the three destruction forms are called underreinforced beam, overreinforced beam, and lightly reinforced beam respectively.

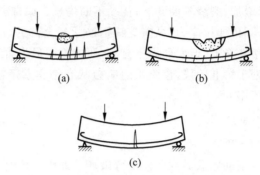

图 4-9　钢筋混凝土梁的正截面受弯破坏形态
(a) 适筋破坏；(b) 超筋破坏；(c) 少筋破坏

Figure 4-9　Bending destruction form of normal section of reinforced concrete beams
(a) Underreinforced failure；(b) Overreinforced failure；(c) Lightly reinforced failure

1. 适筋破坏
Underreinforced failure

配有适量纵向受拉钢筋，配筋率适中的梁称为适筋梁，适筋梁发生的破坏即为适筋破坏，如图 4-9(a)所示。在整个加载过程中，梁经历比较明显的三个受力阶段。适筋破坏的主要特征为：受拉区混凝土开裂，裂缝充分开展，纵向受拉钢筋的应力首先达到屈服强度，继续增加荷载后，受压区边缘混凝土的塑性变形充分发展，达到极限压应变，混凝土的压应力达到轴心抗压强度设计值，最后混凝土被压碎。

Beams with appropriate amount of longitudinal tensile reinforcement and moderate reinforcement ratio are called underreinforced beams, and the failure of underreinforced beam is underreinforced failure, as shown in Fig. 4-9 (a). During the whole loading process, the beam undergoes three obvious stress stages. The main characteristics of underreinforced failure are: concrete in tension zone cracks, the cracks are fully developed, the stress of the longitudinal tensile reinforcement first reaches the yield strength, and after continuing to increase the load, the plastic deformation of the concrete at the edge of the compression zone is fully developed, reaching the ultimate compressive strain, the compressive stress of the concrete reaches the design value of axial compressive strength, finally the concrete is crushed.

适筋破坏是延性破坏，即破坏不是突然发生的，从受拉钢筋屈服到承载能力极限状态有一个较长的塑性变形过程，破坏前裂缝和挠度有明显增长，有明显的破坏预兆，如图 4-10 所示。由于适筋梁破坏前有明显的塑性变形，且钢筋和混凝土的材料性能都能得到充分利用，因此实际工程中的梁都应设计成适筋梁。受弯构件正截面承载力计算是建立在适筋破坏基础上的。

The underreinforced failure is ductile failure, that is, the failure does not occur suddenly. There is a long plastic deformation process from the yield of tensile reinforcement to the ultimate limit state. Before failure, cracks and deflections increase significantly, and there are obvious warning of failure, as shown in Fig. 4-10. Due to the underreinforced beam has obvious plastic deformation before failure, and the material properties of steel and concrete can be fully utilized, the beams in practical engineering should be designed as underreinforced beams. The calculation of the bearing capacity of normal section of flexural member is based on the underreinforced failure.

图 4-10　适筋梁破坏前裂缝的增长

Figure 4-10　Crack growth of underreinforced beam before failure

2. 超筋破坏
Overreinforced failure

如果梁内纵向受拉钢筋配置过多，配筋率很大，在荷载作用下构件达到承载能力极限状态时，受压区边缘混凝土的压应变先达到其极限压应变，在纵向受拉钢筋开始屈服前，混凝土已经达到抗压强度，混凝土因压碎而破坏。截面破坏前纵向受拉钢筋的应力没有达到屈服强度，还处于弹性阶段，所以裂缝延伸不高，宽度也不大，不能形成一条宽度较大的主裂缝，梁的挠度也不大，如图 4-9(b) 所示。

If the longitudinal tensile reinforcement in the beam is too much and the reinforcement ratio is very large, when the member reaches ultimate limit state under the load, the compressive strain of concrete at the edge of the compression zone first reaches its ultimate compressive strain. Concrete has reached compressive strength before the longitudinal tensile bars begin to yield. Concrete fails by crushing. Before the failure of the section, the stress of the longitudinal tensile reinforcement does not reach the yield strength, and is still in the elastic stage, so the crack extension is not high and the width is not large, cannot form a major crack with large width, and the deflection of the beam is not large, as shown in Fig. 4-9(b).

超筋破坏是脆性破坏，破坏前构件的挠度变形很小，破坏发生前预兆不明显，混凝土压碎而导致的受压破坏比较突然，破坏过程短暂。由于破坏突然，用钢量大，纵向受拉钢筋的强度没有充分利用，不经济，设计中应避免采用超筋梁。梁的破坏最好先源于钢筋屈服而不是混凝土被压碎。

Overreinforced failure is brittle failure. Before failure, the deflection deformation of the member is very small, and the warning is not obvious before the failure. The compression failure caused by the crushing of concrete is relatively sudden and the failure process is short. Due to the sudden failure and large amount of steel used, the strength of longitudinal tensile reinforcement is not fully utilized, and it is not economical, so the use of overreinforced beams should be avoided in the design. The failure of the beam is best caused by the yielding of steel bar, rather than the crushing of concrete.

特别提示 4-1

3. 少筋破坏
Lightly reinforced failure

如果纵向受拉钢筋配置过少,配筋率很小,则受拉区混凝土一旦开裂,在裂缝截面处原来由受拉区混凝土承担的拉力几乎全部移交给钢筋承担,导致钢筋应力显著增大,很快达到其屈服强度,甚至经历整个流幅进入强化阶段,个别情况下钢筋甚至可能被拉断。受拉区的裂缝也会迅速开展,而且一般集中开展一条宽度很宽的裂缝,沿梁的高度延伸较长,但受压区混凝土并不会被压碎,但裂缝宽度太大,已标志着梁的破坏,如图 4-9(c)所示。

If the configuration of longitudinal tensile reinforcement is too little, the reinforcement ratio is very small, once the concrete in the tension zone cracks, almost all the tensile force originally borne by the concrete in the tension zone is transferred to the reinforcement at the cracked section, resulting in the significant increase of the steel stress, quickly reaches its yield strength, and even enters the strengthening stage through the entire flow amplitude, and the steel bar may even break in some cases. Cracks in the tension zone can also develop rapidly, and generally focus on a wide crack, which extends longer along the depth of the beam, but the concrete in the compression zone will not be crushed, but the crack width is too large, which has marked the failure of the beam, as shown in Fig. 4-9(c).

少筋破坏承载力低,破坏时的极限弯矩小,具有一裂就坏的显著特点,破坏突然,无预兆,是脆性破坏,工程结构中要严格避免,不允许采用少筋梁。

The bearing capacity of lightly reinforced failure is low, the ultimate bending moment at failure is small, and it has the remarkable characteristic of failure at the first crack. The failure is sudden without warning, which is brittle failure. In engineering structure, it should be strictly avoided and it is not allowed to use lightly reinforced beams.

4. 三种破坏形态的比较
Comparison of three destruction forms

适筋破坏和超筋破坏的受拉区混凝土裂缝开展都比较充分,破坏时受压区混凝土都会被压碎,但是超筋破坏的变形比适筋破坏小,受压区的范围比适筋破坏大,受拉区混凝土裂缝的发展高度比适筋破坏短,裂缝宽度也较小。少筋破坏梁一裂就坏,梁的挠曲变形很小。

由三种破坏形态的荷载-挠度曲线(如图 4-11 所示)可知,少筋破坏的变形性能很差,承载力低;超筋破坏的变形性能差,但承载力较高;适筋破坏的变形性能非常好,承载力也比较高。结构构件的设计既要保证足够的承载力,又要有良好的变形能力,适筋梁能满足这些要求。

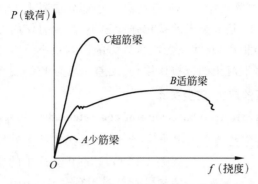

图 4-11　适筋、超筋、少筋梁的荷载-挠度曲线

Figure 4-11　Load-deflection curves of underreinforced, overreinforced and lightly reinforced beams

4.3 正截面受弯承载力计算原理
Calculation principle of flexural capacity of normal section

4.3.1 受弯构件正截面承载力计算的基本假定
Basic assumptions for the calculation of normal section bearing capacity of flexural members

受弯构件正截面承载力计算应采用符合工程需求的混凝土应力-应变本构关系,并应满足变形协调条件和静力平衡条件。受弯构件在荷载作用下,截面开裂后继续带裂缝工作,混凝土受压区的塑性变形不断发展直至破坏。由于截面应力-应变分布的复杂性,为了便于简化分析和工程应用,《混凝土结构设计标准》(GB/T 50010—2010)6.2.1 条规定:受弯构件正截面承载力应按下列基本假定进行计算。

1. 平截面假定

Plane section assumption

假定受弯构件加载前截面为平面,加载后正截面受力发生弯曲变形,截面继续保持平面,截面应变分布也保持平面,截面应变分布符合平截面假定。

国内外大量试验研究结果表明,受弯构件在混凝土开裂前的第Ⅰ阶段中,纵向受拉钢筋屈服前,截面各点的混凝土和钢筋的纵向应变沿截面高度呈直线变化,即正截面内任意点的应变与该点到中和轴的距离成正比,符合平截面假定。在开裂后的第Ⅱ、Ⅲ阶段,在跨越一定标距范围内,跨越若干条裂缝后,截面各点的混凝土和钢筋的平均应变仍能符合平截面假定。分析表明,采用平截面假定引起的计算值与实际情况的误差不大,完全能符合工程计算精度的要求。国际上一些主要国家的有关规范均采用了平截面假定。

2. 不考虑混凝土的抗拉强度

Without considering the tensile strength of concrete

受拉区混凝土开裂后,裂缝截面处,受拉区混凝土大部分退出工作,截面受拉区的拉力全部由钢筋承担。虽然在靠近中和轴附近尚有一小部分混凝土承担着拉应力,但与钢筋承担的拉应力相比数值很小,且混凝土的抗拉强度很小,其合力作用点离中和轴较近,拉力的内力臂不大,提供的弯矩很小,对截面受弯承载力的影响很小。因此,在正截面承载力计算中,一般可不考虑受拉区混凝土的抗拉作用,其误差一般为 1%~2%。

3. 假定混凝土受压的应力-应变关系

Assume the stress-strain relationship of concrete under compression

混凝土受压的应力-应变关系采用曲线-水平线,由抛物线上升段和水平段两部分组成,如图 4-12 所示,图中公式中的参数取值具体见《混凝土结构设计标准》(GB/T 50010—2010)。随着混凝土强度等级的提高,混凝土的峰值应变 ε_0 不断增大,而极限压应变 ε_{cu} 却逐渐减小,材料的脆性加大。

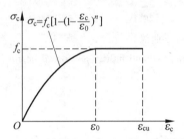

图 4-12 设计采用的混凝土受压应力-应变曲线

Figure 4-12 Compressive stress-strain curve of concrete used in the design

实际上,混凝土的应力-应变关系与混凝土强度等材料性能及轴向压力的偏心程度有关,要进行准确描述是非常复杂的。正截面受压区的应变不均匀,应变速率也不一样,实际的应力-应变关系曲线与按轴心受压确定的应力-应变关系曲线不同。按近似公式确定的受压区混凝土应力图形必然存在一定的误差,但能满足工程设计的精度要求。

4. 纵向受拉钢筋的极限拉应变取为 0.01

The ultimate tensile strain of longitudinal tensile steel bars is taken as 0.01

纵向受拉钢筋的极限拉应变取 0.01,对于软钢,相当于钢筋进入了屈服阶段,将其作为构件达到承载能力极限状态的标志之一,即只要达到混凝土极限压应变或纵向受拉钢筋极限拉应变的其中一个,则标志着构件达到了承载能力极限状态。

5. 假定纵向钢筋的应力取值

Assume the stress value of longitudinal steel bars

纵向钢筋的应力取钢筋应变与其弹性模量的乘积,且钢筋应力不应超过钢筋抗压、抗拉强度设计值。对于轴心受压构件,钢筋的抗压强度设计值取值不应超过 $400\text{N}/\text{mm}^2$。

钢筋混凝土构件的纵向受力钢筋一般采用热轧钢筋,热轧钢筋的应力-应变关系采用折线,不考虑钢筋屈服后的强化阶段,纵向受拉钢筋具有理想的弹塑性关系,如图 4-13 所示。在钢筋屈服前(上升段),钢筋应力 σ_s 与应变 ε_s 成正比,$\sigma_s = \varepsilon_s E_s$($E_s$ 为钢筋的弹性模量);在钢筋屈服后(水平段),钢筋应力 σ_s 保持不变,$\sigma_s = f_y$(f_y 为钢筋的抗拉或抗压强度设计值)。

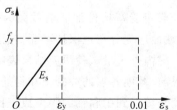

图 4-13 设计采用的钢筋应力-应变关系

Figure 4-13 Stress-strain relationship of reinforcement used in the design

4.3.2 受压区等效矩形应力图形
Equivalent rectangular stress block of compression zone

以单筋矩形截面适筋梁受弯为例，按照上述基本假定由适筋梁第Ⅲ阶段正截面应力-应变图形可知，截面在承载能力极限状态时，受压区混凝土实际应力图形为曲线形，受压边缘混凝土的应变达到了极限压应变 ε_{cu}。假定此时截面受压区高度为 x_c，可以利用非线性的应力-应变关系进行积分计算，求出任一应变情况下的弯矩以及截面的弯矩-曲率关系，但比较麻烦，不便于工程应用。

在实际应用中为简化计算，《混凝土结构设计标准》(GB/T 50010—2010)采用等效矩形应力图形来代换实际的曲线形应力图形，即假定混凝土的压应力在受压区是均匀分布的，不需要进行积分计算，这样既简化了计算，又能比较容易地根据平衡条件计算截面极限承载力。等效代换的原则是：保证受压区混凝土压应力的合力大小相等（等效矩形应力图形的面积等于曲线应力图形的面积）、作用点不变（等效矩形应力图形的形心位置应与曲线应力图形的形心位置相同），这样才能保证极限承载能力不变，如图 4-14 所示。等效矩形应力图形并非真实的应力分布，但能满足承载能力不变的条件，所以称为等效矩形应力图形。

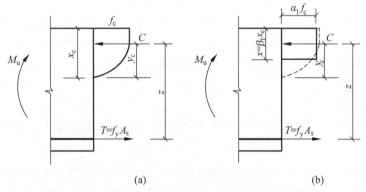

图 4-14　等效矩形应力图形代换曲线应力图形
(a) 曲线应力分布；(b) 等效矩形应力分布

Figure 4-14　Replacing the curve stress block with the equivalent rectangular stress block
(a) Curve stress distribution; (b) Equivalent rectangular stress distribution

为简化计算，《混凝土结构设计标准》(GB/T 50010—2010)将等效矩形应力图形的压应力取为 $\alpha_1 f_c$，等效矩形应力图形的高度取为 x，混凝土实际受压区高度为 x_c，令

$$x = \beta_1 x_c \tag{4-1}$$

《混凝土结构设计标准》(GB/T 50010—2010) 6.2.6 条建议采用的受压区混凝土等效矩形应力图系数 α_1、β_1 见表 4-1，当混凝土强度等级大于 C50 时，α_1 和 β_1 值随混凝土强度等级的提高而减小。

表 4-1 受压区混凝土等效矩形应力图系数 α_1、β_1 值

Table 4-1 Value of equivalent rectangular stress block coefficient α_1, β_1 of concrete in compression zone

混凝土强度等级	≤C50	C55	C60	C65	C70	C75	C80
α_1	1.00	0.99	0.98	0.97	0.96	0.95	0.94
β_1	0.80	0.79	0.78	0.77	0.76	0.75	0.74

4.3.3 相对受压区高度和界限相对受压区高度
Relative depth of compression zone and boundary relative compression zone depth

1. 相对受压区高度 ξ

Relative depth of compression zone ξ

等效矩形应力图形的混凝土受压区高度 x 与截面有效高度 h_0 的比值称为相对受压区高度 ξ，即

The ratio of the concrete compression zone depth x to the effective depth of section h_0 in the equivalent rectangular stress block is called the relative depth of compression zone, namely

$$\xi = \frac{x}{h_0} \tag{4-2}$$

式中，x 为等效矩形应力图形的混凝土受压区高度，mm；h_0 为截面有效高度，mm。

Where, x is the depth of concrete compression zone of the equivalent rectangular stress block, mm; h_0 is the effective depth of section, mm.

 特别提示 4-2

截面有效高度 h_0 是从纵向受拉钢筋的合力作用点到受压混凝土边缘的垂直距离，表达式为

Effective depth of section h_0 is the vertical distance from the action point of the resultant force of the longitudinal tensile steel bars to the edge of the compression concrete, and the expression is

$$h_0 = h - a_s \tag{4-3}$$

式中，h_0 为截面有效高度，mm；h 为截面高度，mm；a_s 为纵向受拉钢筋的合力作用点到受拉混凝土边缘的垂直距离，mm。

Where, h_0 is the effective depth of section, mm; h is the section depth, mm; a_s is the vertical distance from the action point of the resultant force of the longitudinal tensile steel bars to the edge of the tensile concrete, mm.

截面的抵抗弯矩主要取决于受拉钢筋的拉力与受压区混凝土的压力所形成的力矩，所以，在对受弯构件进行截面设计和复核时，只能采用截面有效高度，其取值与纵向受拉钢筋的直径及布置有关。

Resistance moment of the section mainly depends on the moment formed by the tensile force of tensile reinforcement and the compressive force of concrete in the compression zone. Therefore, when designing and reviewing the section of flexural members, only the effective depth of the section can be used, its value is related to the diameter and placement of longitudinal tensile steel bars.

在正截面受弯承载力设计中,在钢筋直径、数量和层数等还未知的情况下,纵向受拉钢筋合力作用点到截面受拉混凝土边缘的距离 a_s 往往需要预先估计。根据最外层钢筋的混凝土保护层最小厚度的规定,考虑箍筋直径和纵向受拉钢筋直径,当环境类别为一类(室内环境),梁中纵向受拉钢筋放置一层时,a_s 一般取 35mm 或 40mm,放置二层时,a_s 取 60mm 或 65mm;板中 a_s 取 20mm 或 25mm。当混凝土强度等级不大于 C25 时,a_s 应再增加 5mm。

In the design of the flexural capacity of normal section, when the diameter, quantity and number of layers of steel bars are still unknown, the distance a_s from the action point of the resultant force of the longitudinal tensile steel bars to the edge of the tensile concrete section often needs to be estimated in advance. According to the regulations for the minimum thickness of concrete cover outside of the outermost bars, considering the diameter of stirrup and the diameter of longitudinal tensile steel bar, when the environment category is Class Ⅰ (indoor environment), when the longitudinal tensile steel bars in the beam are placed in one layer, the a_s is generally taken as 35mm or 40mm, when placed in two layers, the a_s is 60mm or 65mm; the a_s in the slab is 20mm or 25mm. When the concrete strength grade is not greater than C25, the a_s should be increased by another 5mm.

2. 界限相对受压区高度 ξ_b
Boundary relative compression zone depth ξ_b

适筋、超筋、界限配筋梁破坏时的正截面平均应变如图 4-15 所示。总会有一个界限配筋率,在纵向受拉钢筋应力 σ_s 达到屈服强度 f_y 的同时,受压区边缘混凝土纤维的压应变 ε_c 恰好达到混凝土受弯时的极限压应变 ε_{cu},此破坏形式称为界限破坏。

The average strain of normal section in failure of underreinforced beams, overreinforced beams and boundary reinforced beams is shown in Fig. 4-15. There is always a balanced reinforcement ratio, when the longitudinal tensile steel bars stress σ_s reaches the yield strength f_y, the compressive strain ε_c of the concrete fiber at the edge of the compression zone is exactly the ultimate compressive strain ε_{cu} of the concrete under bending. This failure mode is called balanced failure.

界限相对受压区高度 ξ_b 是指受弯构件由适筋到超筋的界限破坏时,等效矩形应力图形中混凝土受压区高度 x_b 与截面有效高度 h_0 的比值。如图 4-16 所示,设界限破坏时实际受压区高度为 x_{cb},相应等效矩形应力图形中受压区高度为 x_b,对有屈服点的普通钢筋,取 $\varepsilon_y = f_y/E_s$,则

Boundary relative compression zone depth ξ_b refers to the ratio of the depth x_b of the concrete compression zone to the effective depth of section h_0 in the equivalent rectangular stress block when balanced failure of flexural members from underreinforced failure to

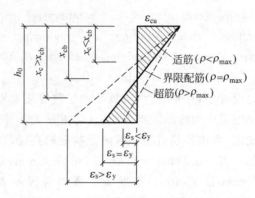

图 4-15 适筋、超筋、界限配筋梁破坏时的正截面平均应变

Figure 4-15 Average strain of normal section of beams during underreinforced failure, overreinforced failure and balanced failure

overreinforced failure. As shown in Fig. 4-16, assuming the depth of actual compression zone in case of balanced failure is x_{cb}, the compression zone depth in the corresponding equivalent rectangular stress block is x_b. For ordinary steel bars with yield points, take $\varepsilon_y = f_y/E_s$, then

$$\xi_b = \frac{x_b}{h_0} = \frac{\beta_1 x_{cb}}{h_0} = \frac{\beta_1 \varepsilon_{cu}}{\varepsilon_{cu} + \varepsilon_y} = \frac{\beta_1}{1 + \dfrac{f_y}{E_s \varepsilon_{cu}}} \tag{4-4}$$

式中,x_b 为界限破坏时等效矩形应力图形中混凝土受压区高度,mm;ε_{cu} 为非均匀受压时的混凝土极限压应变;E_s 为钢筋的弹性模量,N/mm²。

Where, x_b is the depth of the concrete compression zone in the equivalent rectangular stress block in case of balanced failure, mm; ε_{cu} is the ultimate compressive strain of concrete under nonuniform compression; E_s is the elastic modulus of steel bars, N/mm².

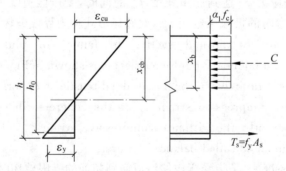

图 4-16 界限破坏时的梁截面应力-应变状态和混凝土受压区高度

Figure 4-16 Stress strain state of beam section and depth of concrete compression zone during balanced failure

为便于应用,对采用不同强度等级的混凝土和有屈服点钢筋的受弯构件,由式(4-4)算得的界限相对受压区高度 ξ_b 值见表 4-2,设计时可直接查用。由图 4-15 可知,当 $\xi < \xi_b$ 时,为适筋破坏;当 $\xi > \xi_b$ 时,为超筋破坏;当 $\xi = \xi_b$ 时,为界限破坏。

For the convenience of application, for flexural members with concrete of different strength grades and steel bars with yield points, boundary relative compression zone depth ξ_b calculated by Eq. (4-4) is shown in Table 4-2, and can be directly referred to when design. It can be seen from Fig. 4-15 that when $\xi < \xi_b$, it is underreinforced failure; when $\xi > \xi_b$, it is overreinforced failure; and when $\xi = \xi_b$, it is balanced failure.

表 4-2 有屈服点钢筋的界限相对受压区高度 ξ_b 值
Table 4-2 Value of boundary relative compression zone depth ξ_b of reinforcement with yield point

钢筋级别	≤C50	C60	C70	C80
HPB300	0.576	0.556	0.537	0.518
HRB400、HRBF400、RRB400	0.518	0.499	0.481	0.463
HRB500、HRBF500	0.482	0.464	0.447	0.429

4.3.4 纵向受拉钢筋的配筋率
Reinforcement ratio of longitudinal tensile steel bars

1. 配筋率

Reinforcement ratio

受弯构件纵向受拉钢筋的配筋率 ρ 对正截面破坏形态有较大影响,其表达式为

The reinforcement ratio of longitudinal tensile steel bars of flexural members has a great influence on the destruction form of normal section, and the expression is

$$\rho = \frac{A_s}{bh_0} \tag{4-5}$$

式中,ρ 为纵向受拉钢筋的配筋率,一般用百分数表示;A_s 为纵向受拉钢筋的截面面积,mm^2;b 为矩形截面梁的宽度或 T 形截面梁的肋宽,mm;h_0 为截面有效高度,mm。

Where, ρ is the reinforcement ratio of longitudinal tensile steel bars, generally expressed as a percentage; A_s is the cross-sectional area of longitudinal tensile reinforcement, mm^2; b is the width of rectangular section beam or the rib width of T-shaped section beam, mm; h_0 is the effective depth of section, mm.

2. 界限配筋率

Balanced reinforcement ratio

界限破坏时纵向受拉钢筋的配筋率称为界限配筋率 ρ_b,此时钢筋应力达到钢筋屈服强度的同时,受压区边缘混凝土纤维的压应变也恰好达到混凝土极限压应变。界限配筋率是适筋破坏的最大配筋率,超筋破坏时的配筋率大于界限配筋率。

In case of balanced failure, the reinforcement ratio of longitudinal tensile steel bars is called the balanced reinforcement ratio ρ_b. At this time, when the steel stress reaches the yield strength of the steel bars, the compressive strain of concrete fiber at the edge of the compression zone just reaches the ultimate compressive strain of the concrete. The balanced reinforcement ratio is the maximum reinforcement ratio for underreinforced failure, and the reinforcement ratio in the case of overreinforced failure is greater than the balanced reinforcement ratio.

3. 最大配筋率
Maximum reinforcement ratio

界限破坏是适筋梁与超筋梁的界限,界限破坏时纵向受拉钢筋的配筋率就是适筋梁的最大配筋率,即界限配筋率 ρ_b。最大配筋率 ρ_{max} 的表达式为

The balanced failure is the boundary between an underreinforced beam and an overreinforced beam. The reinforcement ratio of longitudinal tensile steel bars in case of balanced failure is the maximum reinforcement ratio of underreinforced beam, that is, the balanced reinforcement ratio ρ_b. The expression of the maximum reinforcement ratio ρ_{max} is

$$\rho_{max} = \frac{A_{s,max}}{bh_0} = \left(\alpha_1 \xi_b \frac{f_c bh_0}{f_y}\right) \bigg/ bh_0 = \xi_b \frac{\alpha_1 f_c}{f_y} \qquad (4\text{-}6)$$

由式(4-6)可知,对于材料强度给定的截面,最大配筋率 ρ_{max} 与界限相对受压区高度 ξ_b 之间存在明确的换算关系,只要确定了 ξ_b,就相当于确定了 ρ_{max},这二者从不同方面确定了适筋梁的上限值。在实际计算中,常采用 ξ_b 进行计算,比较方便。

From Eq. (4-6), it can be known that for a given section of material strength, there is a clear conversion relationship between the maximum reinforcement ratio ρ_{max} and the boundary relative compression zone depth ξ_b. As long as ξ_b is determined, it is like confirming ρ_{max}. The two determine the upper limit of the underreinforced beam from different aspects. In actual calculation, ξ_b is often used for calculation, which is more convenient.

4. 最小配筋率
Minimum reinforcement ratio

最小配筋率 ρ_{min} 理论上是少筋梁和适筋梁的界限配筋率。最小配筋率 ρ_{min} 的确定原则是按Ⅲ$_a$ 阶段计算的钢筋混凝土受弯构件的正截面承载力 M_u 与同样条件下素混凝土受弯构件按Ⅰ$_a$ 阶段计算的开裂弯矩 M_{cr} 相等。当实际配筋率小于最小配筋率时为少筋梁,少筋梁一开裂就破坏,在建筑结构中不允许采用。

Minimum reinforcement ratio ρ_{min} is theoretically the balanced reinforcement ratio of the lightly reinforced beam and the underreinforced beam. The principle of determining the minimum reinforcement ratio ρ_{min} is that the normal section bearing capacity M_u of reinforced concrete flexural member calculated according to stage Ⅲ$_a$ is equal to the cracking moment M_{cr} calculated according to stage Ⅰ$_a$ of plain concrete flexural member under the same conditions. When the actual reinforcement ratio is less than the minimum reinforcement ratio, they are lightly reinforced beams and they will be destroyed as soon as they crack, and they are not allowed to be used in building structures.

最小配筋率的影响因素较多,考虑混凝土收缩和温度应力的不利影响等因素,并参考以往的经验,《混凝土结构通用规范》(GB 55008—2021)4.4.6 条规定受弯构件一侧的受拉钢

筋的最小配筋率为

There are many factors affecting the minimum reinforcement ratio. Taking into account the adverse effects of shrinkage of concrete and temperature stress, and referring to previous experience, Article 4.4.6 of "General code for concrete structures" (GB 55008—2021) stipulates that the minimum reinforcement ratio of the tensile reinforcement on one side of the flexural member is

$$\rho_{min} = \max\left\{0.45\frac{f_t}{f_y}, 0.2\%\right\} \quad (4\text{-}7)$$

最小配筋率随混凝土强度的提高而增大,随钢筋抗拉强度的提高而降低。

The minimum reinforcement ratio increases with the increase of concrete strength and decreases with the increase of tensile strength of steel bars.

对于矩形截面,最小配筋率的限值是对全截面面积而言;对于T形或I形截面,受压区翼缘挑出部分面积的影响很小,可以忽略不计;对I形或倒T形截面,最小受拉钢筋面积计算应考虑受拉区翼缘挑出部分的面积$(b_f-b)h_f$。

For rectangular section, the limit value of the minimum reinforcement ratio is for the full section area; for T-shaped or I-shaped section, the effect of the overhanging part of the flange in the compression zone is very small and can be ignored; for I-shaped section or inverted T-shaped section, the calculation of minimum tensile steel area should take into account the area of the overhanging part of the flange in the tension zone $(b_f-b)h_f$.

对矩形或T形截面钢筋混凝土构件,纵向受拉钢筋的最小面积为

For rectangular or T-shaped section reinforced concrete members, the minimum area of longitudinal tensile reinforcement is

$$A_{s,min} = \rho_{min} bh \quad (4\text{-}8)$$

式(4-8)中计算截面所应满足的纵向受拉钢筋最小面积$A_{s,min}$时,混凝土采用全截面面积bh,而不采用bh_0,这是因为混凝土开裂退出工作的部分包括受拉钢筋以下部分的混凝土,而进行承载力计算时,截面的有效面积只有bh_0。

In Eq. (4-8), when calculating the longitudinal tensile reinforcement minimum area $A_{s,min}$, concrete uses full cross-sectional area bh instead of using bh_0. This is because the part of the concrete that cracks and exits the work includes the part of the concrete below the tensile reinforcement and when the bearing capacity is calculated, the effective area of the section is only bh_0.

任务4-2:某钢筋混凝土梁的混凝土强度等级为C25,配置纵向受拉钢筋为HRB400,确定此梁的最小配筋率。

Task 4-2: The concrete strength grade of the reinforced concrete beam is C25, and the longitudinal tensile reinforcement is HRB400, determine the minimum reinforcement ratio of the beam.

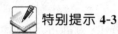 特别提示 4-3

5. 适筋梁的配筋率要求

Reinforcement ratio requirements of underreinforced beams

适筋梁的配筋率应在最小配筋率和最大配筋率之间,见式(4-9),小于最小配筋率是少筋破坏,大于最大配筋率是超筋破坏。

The reinforcement ratio of underreinforced beam should be between the minimum reinforcement ratio and the maximum reinforcement ratio, as shown in Eq. (4-9), less than the minimum reinforcement ratio is the lightly reinforced failure, and greater than the maximum reinforcement ratio is the overreinforced failure.

$$\rho_{\min} \leqslant \rho \leqslant \rho_{\max} \tag{4-9}$$

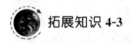 拓展知识 4-3

4.4 单筋矩形截面受弯承载力计算
Calculation of flexural capacity of singly reinforced rectangular section

单筋截面梁是只在梁的截面受拉区配置纵向受拉钢筋(承受由弯矩产生的拉应力,其直径和根数通过计算确定),在梁的截面受压区只布置固定箍筋所需要的纵向架立钢筋,架立钢筋不计其受力作用,如图 4-17 所示。

Longitudinal tensile steel bars are only arranged in the tension zone of the beam section in singly reinforced section beam (to bear the tensile stress generated by the bending moment, whose diameter and number are determined by calculation), and only the longitudinal erection bars required for fixing stirrups are arranged in the compression zone of the beam section, regardless of the stress of erection bars, as shown in Fig. 4-17.

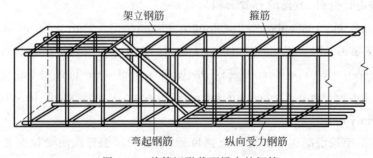

图 4-17 单筋矩形截面梁中的钢筋

Figure 4-17 Reinforcement in the singly reinforced rectangular section beam

4.4.1 基本计算公式及适用条件
Basic calculation formula and applicable conditions

1. 基本计算公式
Basic calculation formula

单筋矩形截面受弯构件正截面受弯承载力计算简图如图 4-18 所示，根据截面力的平衡条件和力矩平衡条件，可得基本计算公式。

The calculation diagram of normal section flexural capacity of the singly reinforced rectangular section flexural member is shown in Fig. 4-18. According to the force and the moment equilibrium conditions of the section, the basic calculation formula can be obtained.

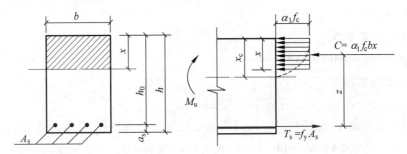

图 4-18 单筋矩形截面受弯构件正截面受弯承载力计算简图

Figure 4-18 Calculation diagram of normal section flexural capacity of singly reinforced rectangular section flexural members

由水平力的平衡条件，水平方向的力大小相等，得

From the horizontal force equilibrium condition, forces equal in horizontal direction, one obtains

$$\alpha_1 f_c b x = f_y A_s \tag{4-10}$$

由力矩平衡条件，对纵向受拉钢筋合力作用点取力矩，得

From the moment equilibrium condition, taking the moment about the action point of the resultant force of the longitudinal tensile steel bars gives

$$M \leqslant M_u = \alpha_1 f_c b x \left(h_0 - \frac{x}{2}\right) \tag{4-11}$$

或对受压区混凝土压应力的合力作用点取力矩，得

Or taking the moment about the action point of the resultant force of the compressive stress of the concrete in the compression zone gives

$$M \leqslant M_u = f_y A_s \left(h_0 - \frac{x}{2}\right) \tag{4-12}$$

式中，α_1 为混凝土受压区等效矩形应力图系数，取值同前所述，可查表 4-1；f_c 为混凝土轴心抗压强度设计值，N/mm²；b 为梁截面宽度，mm；若是现浇板，通常取 1m 宽板带进行计算，即 $b=1000$mm；x 为混凝土受压区高度或受压区计算高度，mm；f_y 为纵向受拉钢筋的

抗拉强度设计值，N/mm²；A_s 为纵向受拉钢筋截面面积，mm²；h_0 为截面有效高度，mm；M 为弯矩设计值，N·mm；M_u 为极限弯矩，即正截面受弯承载力设计值，N·mm。

Where, α_1 is the equivalent rectangular stress block coefficient of the concrete compression zone, the value is the same as above, and can be found in Table 4-1; f_c is the design value of concrete axial compressive strength, N/mm²; b is the beam section width, mm, in case of cast-in-place slab, 1m wide slab strip is usually taken for calculation, that is, $b = 1000$mm; x is the depth of concrete compression zone or the calculated depth of compression zone, mm; f_y is the design value of tensile strength of longitudinal tensile steel bars, N/mm²; A_s is the cross-sectional area of longitudinal tensile steel bars, mm²; h_0 is the effective depth of section, mm; M is the design value of bending moment, N·m; M_u is the ultimate bending moment, the design value of flexural capacity of normal section, N·m.

2. 公式的适用条件

Applicable conditions of formula

1) 最大配筋率要求

1) Maximum reinforcement ratio requirement

为了防止超筋破坏，保证构件破坏时纵向受拉钢筋首先屈服，基本计算公式应满足

In order to prevent overreinforced failure and ensure that the longitudinal tensile steel bars first yield when the member is damaged, the basic calculation formula should satisfy

$$\xi \leqslant \xi_b \text{ 或 } x \leqslant \xi_b h_0 \quad \text{或} \quad \rho \leqslant \rho_{\max} \tag{4-13}$$

或

$$M \leqslant M_u = \alpha_1 f_c b h_0^2 \xi_b (1 - 0.5\xi_b) \tag{4-14}$$

受弯构件的正截面受弯承载力随配筋率的增加而逐渐增大，达最大配筋率时，其承载能力达到最大值 M_u。其中，M_u 为适筋截面所能承担的最大弯矩，在截面尺寸、材料种类等因素确定的情况下，M_u 是一个定值，与钢筋数量无关。

The normal section flexural capacity of the flexural member increases gradually with the increase of the reinforcement ratio. When the maximum reinforcement ratio is reached, the bearing capacity reaches the maximum value M_u, where M_u is the maximum moment that the underreinforced section can bear. When the cross-sectional dimensions, material types and other factors are determined, M_u is a fixed value and has nothing to do with the number of steel bars.

2) 最小配筋率要求

2) Minimum reinforcement ratio requirement

为了防止少筋破坏，基本计算公式应满足

In order to prevent lightly reinforced failure, the basic calculation formula should satisfy

$$\rho = \frac{A_s}{bh_0} \geqslant \rho_{\min} \frac{h}{h_0} \quad \text{或} \quad A_s \geqslant A_{s,\min} = \rho_{\min} bh \tag{4-15}$$

满足以上适用条件则可以保证受弯构件破坏时是适筋破坏。

If the above applicable conditions are met, the failure of the flexural member can be ensured to be underreinforced failure.

4.4.2 截面设计
Section design

结构设计计算中，受弯构件正截面受弯承载力计算有截面设计和截面复核两种情况。

In structural design calculation, there are two cases in the calculation of normal section flexural capacity of flexural members, the section design and the section review.

进行单筋矩形截面受弯构件的截面设计时，已知截面弯矩设计值 M，需要选择混凝土强度等级和钢筋级别，确定截面尺寸等，计算纵向受拉钢筋截面面积并选用钢筋。设计时应满足 $M \leqslant M_u$，从经济性考虑，一般按 $M = M_u$ 进行计算。在实际工程中，要根据力学的有关理论计算出结构内力，并进行内力组合才能求出截面设计弯矩 M，结构的整体设计要综合几门课程的相关知识才能完成。

When designing the section of singly reinforced rectangular section flexural member, the design value of bending moment M of section is known, it is necessary to select concrete strength grade and reinforcement grade, determine the cross-sectional dimensions, etc., calculate the cross-sectional area of the longitudinal tensile steel bars and select the reinforcement. $M \leqslant M_u$ should be satisfied in the design. From an economic perspective, it is generally calculated according to $M = M_u$. In practical engineering, according to the relevant theory of mechanics the internal force of the structure should be calculated and the internal force combination should be carried out to obtain the design bending moment M of the section. The overall design of structure can only be completed by integrating the relevant knowledge of several courses.

截面设计的方法主要有解方程法和计算系数法。按基本计算公式求解一元二次方程，计算过程比较复杂，为简化计算，根据基本计算公式给出一些计算系数，可以简化计算过程。下面主要介绍计算系数法，计算系数法又有两种具体计算方法。

The methods of section design mainly include equation solving method and calculation coefficient method. Solving the quadratic equation with one variable according to the basic calculation formula, the calculation process is quite complex. In order to simplify the calculation, some calculation coefficients are given according to the basic calculation formula, which can simplify the calculation process. The following mainly introduces the calculation coefficient method, and the calculation coefficient method has two specific calculation methods.

1. 计算系数法
Calculation coefficient method

1) 方法一：纵向受拉钢筋截面面积 A_s 以相对受压区高度 ξ 表达

1) Method 1: the cross-sectional area A_s of longitudinal tensile steel bars is expressed by relative depth of compression zone ξ

由基本计算公式

From the basic calculation formula

$$M = \alpha_1 f_c bx \left(h_0 - \frac{x}{2}\right) \tag{4-16}$$

因为相对受压区高度 $\xi = x/h_0$,所以混凝土受压区高度 $x = \xi h_0$,代入式(4-16)中,得

Because the relative depth of compression zone $\xi = x/h_0$, the depth of concrete compression zone $x = \xi h_0$, and substituting it into Eq. (4-16), and get

$$M = \alpha_1 f_c b h_0^2 \xi(1 - 0.5\xi) \tag{4-17}$$

令 $\xi(1 - 0.5\xi) = \alpha_s$,$\alpha_s$ 称为截面抵抗矩系数,式(4-17)变为

Let $\xi(1 - 0.5\xi) = \alpha_s$, α_s is called the sectional resistance moment coefficient, Eq. (4-17) becomes

$$M = \alpha_s \alpha_1 f_c b h_0^2 \tag{4-18}$$

因此,由式(4-18)得

So, from Eq. (4-18), and get

$$\alpha_s = \frac{M}{\alpha_1 f_c b h_0^2} \tag{4-19}$$

将混凝土受压区高度 $x = \xi h_0$ 代入力的平衡方程 $\alpha_1 f_c bx = f_y A_s$,得纵向受拉钢筋截面面积 A_s:

Substituting the depth of concrete compression zone $x = \xi h_0$ into the force equilibrium equation, and get the cross-sectional area of longitudinal tensile steel bars A_s:

$$A_s = \frac{\alpha_1 f_c bx}{f_y} = \frac{\alpha_1 f_c b \xi h_0}{f_y} \tag{4-20}$$

2) 方法二:纵向受拉钢筋截面面积 A_s 以内力臂系数 γ_s 表达

2) Method 2: the cross-sectional area A_s of longitudinal tensile steel bars is expressed by internal lever arm coefficient γ_s

将混凝土受压区高度 $x = \xi h_0$ 代入力矩平衡方程 $M = f_y A_s \left(h_0 - \frac{x}{2}\right)$,得

Substituting the depth of concrete compression zone $x = \xi h_0$ into the moment equilibrium equation $M = f_y A_s \left(h_0 - \frac{x}{2}\right)$, and get

$$M = f_y A_s h_0 (1 - 0.5\xi) \tag{4-21}$$

令 $\gamma_s = 1 - 0.5\xi$,γ_s 称为内力臂系数,则

Let $\gamma_s = 1 - 0.5\xi$, γ_s is called the internal lever arm coefficient, and then

$$M = f_y A_s h_0 \gamma_s \tag{4-22}$$

得纵向受拉钢筋截面面积 A_s:

The cross-sectional area A_s of the longitudinal tensile steel bars is obtained:

$$A_s = \frac{M}{f_y \gamma_s h_0} \tag{4-23}$$

3) 有关系数及相互关系

3) Related coefficients and their relationship

截面抵抗矩系数 α_s:

Sectional resistance moment coefficient α_s:

$$\alpha_s = \xi(1-0.5\xi) \tag{4-24}$$

相对受压区高度 ξ：

Relative depth of compression zone ξ：

$$\xi = 1 - \sqrt{1-2\alpha_s} \tag{4-25}$$

内力臂系数 γ_s：

Internal lever arm coefficient γ_s：

$$\gamma_s = 1 - 0.5\xi \tag{4-26}$$

$$\gamma_s = \frac{1+\sqrt{1-2\alpha_s}}{2} \tag{4-27}$$

以上三个系数之间存在一一对应关系，截面抵抗矩系数 α_s、内力臂系数 γ_s 都是相对受压区高度 ξ 的函数，ξ 值增大，截面抵抗矩系数 α_s 呈非线性增大，内力臂系数呈非线性减小。利用 ξ、α_s、γ_s 的关系求受弯构件纵向受拉钢筋截面面积 A_s，可以不必求解一元二次方程。还可以将以上各系数的关系制成表格，直接查用，方便设计计算。

There is a one-to-one correspondence between the above three coefficients. The sectional resistance moment coefficient α_s and the internal lever arm coefficient γ_s are both functions of the relative depth of compression zone ξ. When the value of ξ increases, the sectional resistance moment coefficient α_s increases nonlinearly, and the internal force arm coefficient decreases nonlinearly. Using the relationship of ξ、α_s、γ_s to obtain the cross-sectional area A_s of longitudinal tensile steel bars of flexural member, it is not necessary to solve the quadratic equation with one variable. The relationship between the above coefficients can also be made into a table, which can be used directly to facilitate the design and calculation.

4）截面抵抗矩系数最大值 $\alpha_{s,max}$

4) Maximum value of sectional resistance moment coefficient $\alpha_{s,max}$

当 $\xi = \xi_b$ 时，$\alpha_s = \alpha_{s,max} = \xi_b(1-0.5\xi_b)$，界限破坏时的截面抵抗矩系数 $\alpha_{s,max}$ 计算结果见表4-3。

When $\xi = \xi_b$, $\alpha_s = \alpha_{s,max} = \xi_b(1-0.5\xi_b)$, the calculation results of the sectional resistance moment coefficient $\alpha_{s,max}$ at balanced failure are shown in Table 4-3.

表4-3　界限破坏时的截面抵抗矩系数 $\alpha_{s,max}$

Table 4-3　Sectional resistance moment coefficient $\alpha_{s,max}$ at balanced failure

钢筋种类	≤C50	C60	C70	C80
HPB300	0.410	0.402	0.393	0.384
HRB400、HRBF400、RRB400	0.384	0.374	0.365	0.356
HRB500、HRBF500	0.366	0.357	0.347	0.337

2. 单筋矩形截面受弯构件设计计算步骤

Design and calculation steps of singly reinforced rectangular section flexural members

1）计算截面有效高度 h_0

根据环境类别、混凝土强度等级，按《混凝土结构设计标准》(GB/T 50010—2010)的要求确定混凝土保护层最小厚度 c(查附表 3-2),再根据截面设计弯矩 M 的大小,假定纵向受拉钢筋配置一层还是二层,初步估计 a_s (a_s 为纵向受拉钢筋合力作用点至受拉混凝土边缘的距离),再计算截面有效高度,即 $h_0 = h - a_s$。

1) Calculate the effective depth of section h_0

According to the environmental category and concrete strength grade, determine the minimum thickness c of concrete cover (checking Attached table 3-2) according to the requirements of "Standard for design of concrete structures" (GB/T 50010—2010), and then according to the design moment M of section, assume the longitudinal tensile steel bars in one or two layers. Preliminary estimate a_s (a_s is the distance from the action point of longitudinal tensile steel bars to the edge of tensile concrete), and then calculate the effective depth of section $h_0 = h - a_s$.

2) 计算截面抵抗矩系数 α_s

2) Calculate the sectional resistance moment coefficient α_s

$$\alpha_s = \frac{M}{\alpha_1 f_c b h_0^2} \tag{4-28}$$

3) 计算相对受压区高度 ξ 或内力臂系数 γ_s

3) Calculate the relative depth of compression zone ξ or internal lever arm coefficient γ_s

$$\xi = 1 - \sqrt{1 - 2\alpha_s} \tag{4-29}$$

或

$$\gamma_s = 1 - 0.5\xi \tag{4-30}$$

按基本计算公式的适用条件,验算是否超筋,若 $\xi \leqslant \xi_b$,则截面未超筋,可以继续计算纵向受拉钢筋截面面积 A_s;若 $\xi > \xi_b$,则为超筋破坏,需加大截面尺寸或提高混凝土强度等级,重新计算。

According to the applicable conditions of the basic calculation formula, check whether the reinforcement is overreinforced. If $\xi \leqslant \xi_b$, it is not overreinforced, and the cross-sectional area A_s of longitudinal tensile steel bars can be further calculated; if $\xi > \xi_b$, it is overreinforced failure, and it is necessary to increase the cross-sectional dimensions or increase the concrete strength grade and recalculate.

4) 计算纵向受拉钢筋截面面积 A_s

4) Calculate the cross-sectional area A_s of longitudinal tensile steel bars

若 $\xi \leqslant \xi_b$,则截面未超筋,按式(4-31)计算纵向受拉钢筋截面面积 A_s:

If $\xi \leqslant \xi_b$, the section is not overreinforced, calculate the cross-sectional area A_s of longitudinal tensile steel bars according to Eq. (4-31):

$$A_s = \frac{\alpha_1 f_c b \xi h_0}{f_y} \tag{4-31}$$

或按式(4-32)计算纵向受拉钢筋截面面积 A_s:

Or calculate the longitudinal tensile steel bars A_s according to Eq. (4-32):

$$A_s = \frac{M}{f_y \gamma_s h_0} \tag{4-32}$$

特别提示 4-4

5) 设置梁、板中的钢筋

5) Set steel bars in beams and slabs

梁中一般配置有纵向受力钢筋、架立钢筋、弯起钢筋、箍筋、拉筋和梁侧纵向构造钢筋等。此处主要介绍纵向受力钢筋、架立钢筋的配置,弯起钢筋、箍筋、拉筋和梁侧纵向构造钢筋的配置在下一章介绍。单向板中主要配置有受力钢筋和分布钢筋,双向板中双向配置有受力钢筋。

Generally the beam is equipped with longitudinally stressed reinforcement, erection bar, bent bar, stirrup, tie bar and longitudinal structural reinforcement on the side of the beam. The configuration of longitudinally stressed bar and erection bar is mainly introduced here. The configuration of bent bar, stirrup, tie bar and longitudinal structural reinforcement on the side of the beam is introduced in the next chapter. The one-way slab is mainly equipped with stressed and distributed reinforcement, and the two-way slab is equipped with stressed reinforcement in both directions.

(1) 设置纵向受力钢筋。根据计算出来的纵向受力钢筋截面面积 A_s 查附表 2-1(梁)或附表 2-2(板),选配钢筋直径和根数(梁)或钢筋直径和间距(板),实际选用的钢筋截面面积应与计算值接近,一般不应小于计算值,不宜超过或小于计算值的 5%。

(1) Set longitudinally stressed reinforcement. Refer to the Attached table 2-1(beam) or Attached table 2-2 (slab) according to the calculated cross-sectional area A_s of the longitudinally stressed reinforcement, and select the diameter and number of reinforcement (beam) or the diameter and spacing of reinforcement (slab). The cross-sectional area of reinforcement actually selected should be close to the calculated value, generally, it should not be less than the calculated value, and should not exceed or be less than 5% of the calculated value.

梁中纵向受力钢筋宜采用 HRB400、HRB500、HRBF400、HRBF500,伸入梁支座范围内的纵向受力钢筋根数不应少于 2 根,为便于施工,钢筋根数不宜过多。梁内纵向受力钢筋常用直径为 12~28mm,同一截面内钢筋的直径宜尽可能相同,也可以选用不同直径的钢筋以更好地满足钢筋截面面积的要求,直径不同时种类不宜过多,以便施工。纵向受力钢筋直径不宜相差太大,以免截面受力不均。同一构件中钢筋直径相差不宜小于 2mm,以便在施工中能用肉眼辨别。减少不同钢筋直径的数量对结构是有好处的。

The longitudinally stressed reinforcement in the beam should use HRB400, HRB500, HRBF400, HRBF500, and the number of longitudinally stressed steel bars extending into the beam support should not be less than 2. For the convenience of construction, the number of steel bars should not be too large. The commonly used diameter of longitudinally stressed steel bars in the beam is 12-28mm, and the diameter of reinforcement in the same section should be the same as possible. It is desirable to select steel bars of different diameters to better meet steel area requirements. The diameter of longitudinally stressed reinforcement should not be too different to avoid uneven section stress. The diameter difference of steel bars in the same member should not be less than 2mm, in order to be able to distinguish with naked eye during construction. There is some advantage to the structure to minimize the number of different steel bar diameters.

选择的梁宽应能提供足够的空间,足以放得下所选的钢筋,并保证足够的混凝土保护层厚度和钢筋间距。给定梁宽的梁一层所能放置的最大钢筋数量取决于钢筋直径和间距的要求。当一层放不下所有纵向受拉或受压钢筋时,可以放置在多层。不同的钢筋应相对梁截面的竖中线对称布置。

The selected beam width should provide adequate space to accommodate the selected steel bars, and ensure adequate thickness of concrete cover and bar spacing. The maximum number of steel bars that can be placed on one single layer in the beam with a given beam width depends on the steel bar diameter and spacing requirements. When all the longitudinal tensile or compressive steel bars can not be placed on one single layer in the beam, they can be placed on multiple layers. Different bars should be placed symmetrically about the vertical centerline of the beam section.

梁中纵向钢筋的根数、钢筋级别和直径的标注:如 3⌀16 表示梁中纵向钢筋的根数为 3 根,钢筋级别为 HRB400,钢筋直径为 16mm。

Marking of the number, grade and diameter of longitudinal steel bars in the beam: for example, 3⌀16 means that the number of longitudinal steel bars in the beam is 3, the reinforcement grade is HRB400, and the bar diameter is 16mm.

(2) 设置梁中架立钢筋。当梁混凝土受压区按计算不需要配置纵向受压钢筋时,在梁上部没有纵向受力钢筋的区段,梁顶面箍筋角点处应设置与梁底部纵向受拉钢筋平行的架立钢筋,主要作用是沿梁跨连续布置上部纵筋以支撑箍筋,与纵向受拉钢筋、箍筋形成钢筋骨架,同时承受混凝土收缩和温度变化产生的拉应力,防止发生裂缝。若混凝土受压区配有计算需要的纵向受压钢筋时,受压钢筋可兼作架立钢筋。

架立钢筋的直径与梁的计算跨度 l_0 有关。当 $l_0<4m$ 时,架立钢筋的直径不宜小于 8mm;当 $4m \leqslant l_0 \leqslant 6m$ 时,直径不应小于 10mm;当 $l_0>6m$ 时,直径不宜小于 12mm。

(3) 设置板中受力钢筋。板中受力钢筋常采用 HPB300、HRB400、HRBF400、RRB400,板中纵向受力钢筋常用直径为 6、8、10、12、14mm,当板厚较大时,钢筋直径可用 14~18mm。为了防止施工时板面钢筋被踩下,现浇板的板面钢筋直径不宜小于 8mm。板内受力钢筋间距一般为 70~200mm,当板厚 $h \leqslant 150mm$ 时,钢筋间距不宜大于 200mm;当板厚 $h>150mm$ 时,钢筋间距不宜大于 $1.5h$,且不宜大于 250mm。

板中受力钢筋常按每米板宽所需钢筋截面面积 A_s 选用钢筋的直径和间距,如经公式计算得每米板宽内所需钢筋截面面积 A_s 为 $230mm^2$,查附表 2-2,选用 A_s 为 $251mm^2$,⌀8@200,表示板中钢筋级别为 HRB400,钢筋直径为 8mm,钢筋间距即相邻钢筋中心到中心的距离为 200mm。

(4) 设置板中分布钢筋。当按单向板设计时,应在垂直于受力钢筋沿板的长跨方向布置分布钢筋。分布钢筋的主要作用是浇筑混凝土时固定受力钢筋的位置,抵抗混凝土收缩或温度变化所产生的内力和变形,承担并分布板上局部荷载引起的内力。对四边支承的单向板,可承担在长跨方向板内实际存在的一些弯矩。

英文翻译4-9

板中分布钢筋应放置在跨中受力钢筋及支座处负弯矩钢筋的内侧,如图 4-19 所示,宜采用 HPB300 钢筋,直径不应小于 6mm,常用直径为 6mm 和 8mm,间距不宜大于 250mm。单位长度上分布钢筋的截面面积不应小于单位宽度上受力钢筋截面面积的 15%,且配筋率不宜小于 0.15%。当集中荷载较大时,分布钢筋的配筋面积尚应增加,且间距不宜大于 200mm。在温度、收缩应力较大的现浇板区域,应在板的表面双向配置防裂构造钢筋,其配筋率不宜小于 0.10%,间距不宜大于 200mm。

6) 验算最小配筋率

6) Check the minimum reinforcement ratio

$$\rho = \frac{A_s}{bh_0} \geq \rho_{\min} \frac{h}{h_0} \quad 或 \quad A_s \geq \rho_{\min} bh \quad (4-33)$$

若不满足上式,则发生少筋破坏,按 $A_s = \rho_{\min} bh$ 配筋。配筋率计算有关参数如图 4-20 所示。注意,验算最小配筋率时,要采用查表后实际选用的纵向受拉钢筋截面面积 A_s 进行计算。

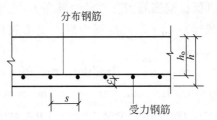

图 4-19 板中受力钢筋和分布钢筋

Figure 4-19 Stressed and distributed steel bars in slab

If the above formula is not satisfied, there will be lightly reinforced failure, and reinforced according to $A_s = \rho_{\min} bh$. The relevant parameters of reinforcement ratio calculation are shown in Fig. 4-20. Note that when checking the minimum reinforcement ratio, the cross-sectional area A_s of the longitudinal tensile steel bars actually selected after referring to the table should be used for calculation.

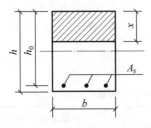

图 4-20 配筋率计算参数示意图

Figure 4-20 Schematic diagram of calculation parameters of reinforcement ratio

7) 在梁截面内布置纵向钢筋

7) Arrange longitudinal steel bars in the beam section

梁中纵向受拉钢筋应配置在梁截面的受拉区,纵向受力钢筋尽量布置在一层,当一层钢筋根数过多或直径过大,放不下所有钢筋时,也可以布置二层,但上层钢筋应直接放置在下层钢筋对应位置之上,不得错缝排列。纵向受力钢筋由放置一层改成二层时,a_s 发生变化,需要重新计算与 a_s 有关的截面有效高度 h_0,重新计算纵向受拉钢筋截面面积 A_s,并重新选配钢筋。箍筋的每个角落至少放置一根纵向钢筋。

The longitudinal tensile steel bars in the beam should be arranged in the tension zone of the beam section, and the longitudinally stressed reinforcement should be arranged in one single layer as far as possible. When the number of steel bars in one single layer is too large or the diameter is too large to arrange all the steel bars, they can also be arranged in two layers. However, the upper layer of bars should be placed on the corresponding position directly above those in the bottom layer and should not be staggered. When the longitudinally stressed steel bars are arranged from one layer to two layers, a_s changes, it is necessary to recalculate the effective depth of section h_0 related to a_s, recalculate the cross-sectional area A_s of the longitudinal tensile reinforcement, and reselect the reinforcement. At least one longitudinal bar must be placed in each corner of the stirrup.

为了便于浇筑混凝土,保证钢筋周围混凝土的密实性,使钢筋与混凝土之间有良好的黏结性能,梁中纵向受力钢筋的净间距不能太小,有必要使相邻钢筋间维持一个最小的净间距以保证钢筋周围有适量的混凝土,如图4-21所示。《混凝土结构设计标准》(GB/T 50010—2010)9.2.1条规定梁中纵向受力钢筋的间距要求见表4-4。

In order to facilitate the pouring of concrete, ensure the compactness of the concrete around the steel bars, and make the steel bar and concrete have a good bonding performance, the clear distance of the longitudinally stressed reinforcement in the beam should not be too small. It is necessary to maintain a minimum clear distance between adjacent steel bars to ensure that there is an appropriate amount of concrete around steel bars, as shown in Fig. 4-21. The spacing requirements of longitudinally stressed bars in the beam specified in Article 9.2.1 of "Standard for design of concrete structures" (GB/T 50010—2010) are shown in Table 4-4.

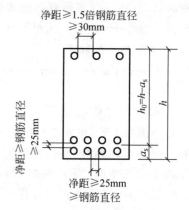

图 4-21 梁中纵向钢筋的间距要求

Figure 4-21 Spacing requirements of longitudinal steel bars in beam

表 4-4 梁纵向钢筋的最小间距

Table 4-4 Minimum spacing of longitudinal steel bars of beam

间距类型	水平净距		垂直净距(层距)
钢筋位置	截面上部	截面下部	
最小间距(取两者中大值)	30mm, 1.5d	25mm, d	25mm, d

注: d 为钢筋的最大直径。

①梁上部钢筋水平方向的净间距不应小于30mm 和 1.5d;②梁下部钢筋水平方向的净间距不应小于25mm 和 d,当下部钢筋多于2层时,2层以上钢筋水平方向的中距应比下面2层的中距增大一倍;③各层钢筋之间的竖向净间距不应小于25mm 和 d,d 为钢筋的最大直径。

①The clear distance in the horizontal direction of the reinforcement in the upper part of the beam should not be less than 30mm and 1.5d; ② The clear distance in the horizontal direction of the reinforcement in the lower part of the beam should not be less than 25mm and d. When there are more than two layers of lower reinforcement, the horizontal middle distance between the steel bars above two layers should be twice that of the two layers below; ③ The vertical clear distance between the steel bars of each layer should not be less than 25mm and d, where d is the maximum diameter of the steel bars.

8）绘制梁、板截面配筋图

8) Draw reinforcement diagram of beam and slab section

 拓展知识 4-4

【例题 4-1】 钢筋混凝土矩形截面梁的截面尺寸为 $b \times h = 250\text{mm} \times 500\text{mm}$，环境类别为一类，截面弯矩设计值 $M = 150\text{kN} \cdot \text{m}$，混凝土强度等级 C30，采用 HRB400 钢筋。求所需的纵向受拉钢筋截面面积，选配钢筋直径和根数，并画出梁横截面配筋图。

视频 4-4

【Example 4-1】 The cross-sectional dimensions of a reinforced concrete rectangular section beam are $b \times h = 250\text{mm} \times 500\text{mm}$, the environmental category is Class I, the design value of bending moment of the section is $M = 150\text{kN} \cdot \text{m}$, the concrete strength grade is C30, and HRB400 steel bars are used. Find the required cross-sectional area of the longitudinal tensile steel bars, select the diameter and number of the reinforcement, and draw the cross-section reinforcement diagram of the beam.

解：环境类别为一类，查附表 3-2 得 C30 混凝土梁的混凝土保护层最小厚度 $c = 25\text{mm}$，由于截面弯矩设计值较小，假定纵向受拉钢筋单层设置，取 $a_s = 35\text{mm}$，则截面有效高度 $h_0 = h - a_s = (500 - 35)\text{mm} = 465\text{mm}$。

由混凝土强度等级和钢筋级别，查附表 1-3、附表 1-4、附表 1-6 得：C30 混凝土轴心抗压强度设计值 $f_c = 14.3\text{N/mm}^2$，混凝土轴心抗拉强度设计值 $f_t = 1.43\text{N/mm}^2$，HRB400 钢筋抗拉强度设计值 $f_y = 360\text{N/mm}^2$。查表 4-2 得界限相对受压区高度 $\xi_b = 0.518$。

截面抵抗矩系数

$$\alpha_s = \frac{M}{\alpha_1 f_c b h_0^2} = \frac{150 \times 10^6}{1.0 \times 14.3 \times 250 \times 465^2} = 0.194$$

相对受压区高度

$$\xi = 1 - \sqrt{1 - 2\alpha_s} = 1 - \sqrt{1 - 2 \times 0.194} = 0.218 < \xi_b = 0.518, 满足要求。$$

内力臂系数 $\gamma_s = 1 - 0.5\xi = 1 - 0.5 \times 0.218 = 0.891$

纵向受拉钢筋截面面积

$$A_s = \frac{M}{f_y \gamma_s h_0} = \frac{150 \times 10^6}{360 \times 0.891 \times 465}\text{mm}^2 = 1006\text{mm}^2$$

或

$$A_s = \frac{\xi f_c b h_0}{f_y} = \frac{0.218 \times 14.3 \times 250 \times 465}{360}\text{mm}^2 = 1006\text{mm}^2$$

根据计算出的 $A_s = 1006\text{mm}^2$，查附表 2-1 中的钢筋公称截面面积表，选用 4⌀18，$A_s = 1017\text{mm}^2$（选用钢筋时应满足有关间距、直径及根数等的构造要求）。

根据混凝土保护层厚度、钢筋根数和直径、钢筋净距,假定箍筋直径为8mm,放置一层钢筋需要的最小截面宽度为 $b_{min}=[2\times(25+8)+4\times18+3\times25]mm=213mm<b=250$mm,满足要求。

验算基本计算公式的适用条件:

(1) 验算是否会发生超筋破坏

$x=\xi h_0=0.218\times465$mm$=101.37$mm$<\xi_b h_0=0.518\times465mm=240.87$mm

不会发生超筋破坏,满足要求。

(2) 验算最小配筋率

$$\rho=\frac{A_s}{bh_0}=\frac{1017}{250\times465}=0.87\%>\rho_{min}\frac{h}{h_0}=0.45\frac{f_t}{f_y}\cdot\frac{h}{h_0}$$

$$=0.45\times\frac{1.43}{360}\cdot\frac{500}{465}=0.192\%$$

且

$$\rho=0.87\%>\rho_{min}\frac{h}{h_0}=0.2\%\times\frac{h}{h_0}$$

$$=0.2\%\times\frac{500}{465}=0.215\%$$

满足要求。

注意,验算适用条件时,要采用查表实际选用的纵向受拉钢筋截面面积 $A_s=1017$mm^2 进行计算。

梁横截面配筋如图 4-22 所示。

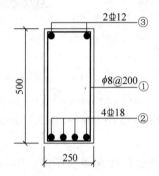

图 4-22 梁横截面配筋

Fig. 4-22 Cross section reinforcement of beam

【例题 4-2】 钢筋混凝土单向板承受均布可变荷载标准值 $q_k=3$kN/m^2,板厚 $h=100$mm,计算跨度 $l_0=3$m,环境类别为一类,结构安全等级为二级,混凝土强度等级为 C30,采用 HRB400 钢筋。求纵向受拉钢筋截面面积,选配钢筋。

【Example 4-2】 The uniformly distributed variable load standard value of reinforced concrete one-way slab is $q_k=3$kN/m^2, the slab thickness $h=100$mm, the effective span $l_0=3$m, the environmental category is Class I, the structural safety level is Level 2, the concrete strength grade is C30, using HRB400 steel bar. Calculate the cross-sectional area of the longitudinal tensile steel bars, select the steel bars.

解:环境类别为一类,假定 $a_s=25$mm,则截面有效高度 $h_0=h-a_s=(100-25)$mm$=75$mm。由混凝土强度等级和钢筋级别,查附表 1-3、附表 1-4、附表 1-6 得:混凝土轴心抗压强度设计值 $f_c=14.3$N/mm^2,混凝土轴心抗拉强度设计值 $f_t=1.43$N/mm^2,钢筋抗拉强度设计值 $f_y=360$N/mm^2。

因为是单向板,可以取板宽 $b=1000$mm 的板带作为计算单元进行计算,板的自重标准值为

$$g_k=\gamma h=25\times0.1\text{kN/m}^2=2.5\text{kN/m}^2$$

板的跨中最大弯矩设计值

$$M=\frac{1}{8}(\gamma_G g_k+\gamma_Q q_k)l_0^2=\frac{1}{8}(1.3\times2.5+1.5\times3)\times3^2\text{kN}\cdot\text{m}=8.72\text{kN}\cdot\text{m}$$

计算板中纵向受拉钢筋截面面积 A_s：

$$\alpha_s = \frac{M}{\alpha_1 f_c b h_0^2} = \frac{8.72 \times 10^6}{1.0 \times 14.3 \times 1000 \times 75^2} = 0.108$$

$$\xi = 1 - \sqrt{1 - 2\alpha_s} = 1 - \sqrt{1 - 2 \times 0.108} = 0.115 < \xi_b = 0.518, 满足要求$$

$$\gamma_s = 1 - 0.5\xi = 1 - 0.5 \times 0.115 = 0.943$$

则

$$A_s = \frac{M}{f_y \gamma_s h_0} = \frac{8.72 \times 10^6}{360 \times 0.943 \times 75} \text{mm}^2 = 343 \text{mm}^2$$

或

$$A_s = \frac{\xi f_c b h_0}{f_y} = \frac{0.115 \times 14.3 \times 1000 \times 75}{360} \text{mm}^2 = 343 \text{mm}^2$$

根据计算出的 $A_s = 343 \text{mm}^2$，查附表 2-2 每米板宽内的钢筋截面面积表，并考虑板钢筋直径和受力钢筋间距方面的构造要求，选用 $\Phi 8@140$，实配钢筋截面面积 $A_s = 359 \text{mm}^2$。

验算最小配筋率：

$$\rho_{\min} = \max\left\{0.45 \frac{f_t}{f_y}, 0.2\%\right\} = \max\left\{0.45 \times \frac{1.43}{360}, 0.2\%\right\}$$

$$= \max\{0.18\%, 0.2\%\} = 0.2\%$$

$$A_s = 359 \text{mm}^2 > A_{s,\min} = \rho_{\min} bh = 0.2\% bh$$

$$= 0.2\% \times 1000 \times 100 \text{mm}^2 = 200 \text{mm}^2$$

所以，配筋率满足要求。

4.4.3 截面复核
Section review

进行截面复核时，已知弯矩设计值、混凝土强度等级、钢筋级别、构件截面尺寸和纵向受拉钢筋截面面积，求截面所能承担的极限弯矩 M_u，验算正截面受弯承载力是否足够，判断其安全性。截面复核一般出现在结构的改造、加固等设计计算中，结构本身已经存在，需要验算结构构件中的配筋是否满足结构功能要求。受弯构件正截面受弯承载力复核的主要计算步骤如下。

英文翻译 4-10

1. 计算混凝土受压区高度 x

Calculate the depth of concrete compression zone x

由力的平衡方程 $\alpha_1 f_c b x = f_y A_s$，得

$$x = \frac{f_y A_s}{\alpha_1 f_c b} \tag{4-34}$$

2. 验算基本计算公式的适用条件，计算正截面受弯承载力 M_u

Check the applicable conditions of the basic calculation formula and calculate the flexural capacity M_u of the normal section

（1）若 $A_s < A_{s,\min} = \rho_{\min} bh$，则为少筋梁，说明纵向受拉钢筋配置太少，构件不安全，需

要修改设计或进行加固处理。

(2) 若 $x > \xi_b h_0$,则为超筋梁,纵向受拉钢筋达不到屈服,取 $x = \xi_b h_0$,代入力矩平衡方程 $M_u = \alpha_1 f_c bx(h_0 - x/2)$,得

$$M_u = \alpha_1 f_c b h_0^2 \xi_b (1 - 0.5\xi_b) \tag{4-35}$$

式中,M_u 为对应于界限破坏时的受弯承载力,kN·m。

(3) 若 $x \leqslant \xi_b h_0$,且 $A_s \geqslant \rho_{min} bh$,则为适筋梁,正截面受弯承载力(极限弯矩):

$$M_u = \alpha_1 f_c bx(h_0 - x/2) \tag{4-36}$$

3. 截面复核的步骤

Steps of section review

(1) 按上述方法计算截面的正截面受弯承载力(极限弯矩)M_u。

(2) 将 M_u 与截面所承担的弯矩设计值 M 进行比较。

(3) 判断安全性:若 $M_u \geqslant M$,则承载力足够,构件是安全的;否则承载力不够,构件是不安全的,需修改原设计(新建工程)或进行补强加固处理(既有工程)。若 M_u 远大于 M,则不经济,必要时也应修改原设计。

【例题 4-3】 钢筋混凝土矩形截面梁的截面尺寸为 $b \times h = 250\text{mm} \times 450\text{mm}$,纵向受拉钢筋 4⌀16,查附表 2-1 得纵向受拉钢筋截面面积为 $A_s = 804\text{mm}^2$,混凝土强度等级为 C40,承受的弯矩设计值 $M = 80\text{kN·m}$,环境类别为二 a 类,验算此梁截面是否安全。

【Example 4-3】 Cross-sectional dimensions of the reinforced concrete rectangular section beam are $b \times h = 250\text{mm} \times 450\text{mm}$, the longitudinal tensile reinforcement is 4⌀16, check the Attached table 2-1 the cross-sectional area of the longitudinal tensile reinforcement is $A_s = 804\text{mm}^2$, the concrete strength grade is C40, and the design value of bending moment is $M = 80\text{kN·m}$, and the environmental category is Class Ⅱ$_a$. Check whether the beam section is safe.

解:混凝土强度等级为 C40,查附表 1-3、附表 1-4 得 $f_c = 19.1\text{N/mm}^2$,$f_t = 1.71\text{N/mm}^2$,对于 HRB400 钢筋查附表 1-6 得 $f_y = 360\text{N/mm}^2$,查表 4-2 得 $\xi_b = 0.518$。

查附表 3-2 得环境类别为二 a 类的梁混凝土保护层最小厚度为 25mm,根据承受弯矩设计值的大小,设纵向受拉钢筋放置一层,取 $a_s = 35\text{mm}$,则 $h_0 = h - a_s = (450-35)\text{mm} = 415\text{mm}$。

$$x = \frac{f_y A_s}{\alpha_1 f_c b} = \frac{360 \times 804}{1 \times 19.1 \times 250}\text{mm} = 60.6\text{mm} < \xi_b h_0 = 0.518 \times 415\text{mm} = 214.97\text{mm}$$

此梁为适筋梁,满足适用条件。

正截面受弯承载力(极限弯矩)为

$$M_u = \alpha_1 f_c bx(h_0 - x/2) = 1 \times 19.1 \times 250 \times 60.6 \times (415 - 60.6/2)\text{kN·m}$$
$$= 111.3\text{kN·m} > M = 80\text{kN·m}$$

所以此梁是安全的。

4.5 双筋矩形截面受弯承载力计算
Calculation of flexural capacity of doubly reinforced rectangular section

4.5.1 双筋截面概述
Overview of doubly reinforced section

1. 双筋截面的概念
Concept of doubly reinforced section

单筋截面梁在混凝土受压区配有架立钢筋,架立钢筋虽然受压,但对正截面受弯承载力的贡献很小,仅在构造上起作用,在计算中可以忽略其抗压作用。如果在梁截面受压区配置的纵向受压钢筋数量较多,则不仅可以起到架立钢筋的作用,而且其受压作用在计算中不能忽略,可协助混凝土共同承担压力,能提高正截面受弯承载力,改善受弯构件的延性,这种截面称为双筋截面,如图 4-23 所示。双筋截面梁就是同时配有纵向受压钢筋和纵向受拉钢筋的梁。

2. 采用双筋截面的情况
Case of using a doubly reinforced section

由于在受弯构件中采用纵向受压钢筋协助混凝土承担压力一般不够经济,所以就受弯承载力而言,双筋截面主要应用于以下情况。

(1) 当梁承受的弯矩设计值 M 很大,超过了单筋矩形截面梁所能承担的最大弯矩 M_u,即 $M > M_u = \alpha_1 f_c b h_0^2 \xi_b (1-0.5\xi_b)$,截面超筋,但梁的截面尺寸、材料强度受限制不能再增大,特别是截面高度不能增大,单筋截面梁无法满足要求时,应设计成双筋截面。

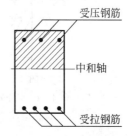

图 4-23 双筋截面梁
Figure 4-23 Doubly reinforced section beam

(2) 荷载效应组合使框架梁上承受的弯矩可能改变方向(变号弯矩),如承受不同方向的风力或地震力的框架梁,即梁的受拉区与受压区相互调换,设计成双筋截面既能满足变号弯矩的需要,又能提高截面延性,减少截面开裂和构件的徐变变形,可减少受弯构件在长期荷载作用下的挠度。

(3) 抗震结构中为提高截面延性,框架梁必须配置一定比例的纵向受压钢筋。

3. 双筋截面的特点
Characteristics of doubly reinforced section

采用双筋截面可以提高截面受弯承载力和延性,减小构件在荷载作用下的变形。但在同样配筋的情况下,配置在受拉区的钢筋比配置在受压区的更有效,因此,一般的梁尽量不采用双筋截面,而且双筋截面的用钢量大,不经济。

4.5.2 基本计算公式及适用条件
Basic calculation formulas and applicable conditions

试验表明，双筋截面受弯构件的受力特点和破坏特征与单筋截面类似，区别在于双筋截面的受压区配有纵向受压钢筋，和混凝土一起受压。双筋矩形截面受弯构件正截面承载力计算简图如图 4-24 所示，因此，可与单筋矩形截面类似建立基本计算公式。

The test shows that the stress and failure characteristics of the doubly reinforced section flexural members are similar to those of the singly reinforced section. The differente is that the compression zone of the doubly reinferced section is equipped with longitudinal compression steel bars, which are compressed together with the concrete. The calculation diagram of normal section bearing capacity of doubly reinforced rectangular section flexural members is shown in Fig. 4-24. Therefore, the basic calculation formula can be established similar to that of singly reinforced rectangular section.

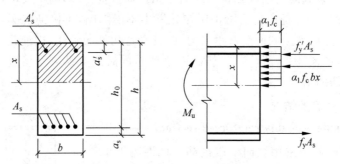

图 4-24 双筋矩形截面受弯构件正截面承载力计算简图

Figure 4-24 Calculation diagram of normal section bearing capacity of doubly reinforced rectangular section flexural members

1. 基本计算公式

Basic calculation formula

根据平衡条件，可得双筋矩形截面正截面受弯承载力的基本计算公式。

According to the equilibrium condition, the basic calculation formula of the normal section flexural capacity of doubly reinforced rectangular section can be obtained.

力的平衡条件：

Force equilibrium condition：

$$\alpha_1 f_c bx + f'_y A'_s = f_y A_s \tag{4-37}$$

力矩平衡条件：

Moment equilibrium condition：

$$M \leqslant M_u = \alpha_1 f_c bx \left(h_0 - \frac{x}{2}\right) + f'_y A'_s (h_0 - a'_s) \tag{4-38}$$

式中，f'_y 为钢筋抗压强度设计值，N/mm^2；A'_s 为纵向受压钢筋截面面积，mm^2；a'_s 为纵向受压钢筋合力作用点到截面受压区边缘之间的距离，mm；其他符号意义同前。

Where, f'_y is the design value of compressive strength of steel bar, N/mm²; A'_s is the cross-sectional area of longitudinal compression reinforcement, mm²; a'_s is the distance between the action point of the longitudinal compression reinforcement and the edge of the compression zone of the section, mm; other symbols have the same meaning as before.

2. 公式的适用条件

Applicable conditions of formula

应用上述基本计算公式时，必须满足以下条件：

When applying the above basic calculation formula, the following conditions must be met：

（1）为了防止超筋破坏，保证构件破坏时纵向受拉钢筋首先屈服，应满足以下条件：

(1) In order to prevent overreinforced failure and ensure that the longitudinal tensile reinforcement is first yielded when the member is damaged, the following conditions should be met：

$$\xi \leqslant \xi_b \quad 或 \quad x \leqslant \xi_b h_0 \tag{4-39}$$

当 $\xi > \xi_b$ 时，说明发生超筋破坏，应增加纵向受压钢筋截面面积 A'_s 或截面尺寸。

When $\xi > \xi_b$, it means that in case of overreinforced failure, the cross-sectional area A'_s of the longitudinal compression reinforcement or cross-sectional dimensions should be increased.

（2）为保证纵向受压钢筋的强度充分发挥，保证受压钢筋应力达到屈服强度，受压钢筋合力作用点不能离中和轴太近，混凝土受压区高度 x 应满足式(4-40)：

(2) In order to ensure that the strength of the longitudinal compression reinforcement is given full play and the compressive steel stress reaches the yield strength, the action point of the resultant force of the compressive steel bars cannot be too close to the neutral axis, and the depth of concrete compression zone x should meet Eq. (4-40)：

$$x \geqslant 2a'_s \tag{4-40}$$

若混凝土受压区高度满足 $x \geqslant 2a'_s$，且配置必要的封闭箍筋，则纵向受压钢筋的应力可取《混凝土结构设计标准》(GB/T 50010—2010)规定的抗压强度设计值 f'_y，抗压强度可充分发挥。

If the depth of concrete compression zone satisfies $x \geqslant 2a'_s$, and the necessary closed stirrups are configured, the stress of longitudinal compressive steel bars can take the compressive strength design value f'_y specified in "Standard for design of concrete structures"(GB/T 50010—2010) and the compressive strength can be fully developed.

当 $x < 2a'_s$ 时，表明受压钢筋的位置离中和轴太近，纵向受压钢筋面积 A'_s 较大，受压钢筋的应变 ε'_s 太小，双筋截面破坏时，受压钢筋应力 σ'_s 达不到钢筋抗压强度设计值 f'_y 而为未知。通常可近似取 $x = 2a'_s$，对受压钢筋的合力作用点取矩，如图 4-25 所示，则正截面受弯承载力为

When $x < 2a'_s$, it indicates that the position of the compressive reinforcement is too close to the neutral axis, the area of longitudinal compressive reinforcement A'_s is large, the strain ε'_s of compressive reinforcement is too small, and when the doubly reinforced section is damaged, the compressive steel stress σ'_s can't reach the compressive strength design value f'_y of the steel bar and still unknown, usually it can be approximated as $x=2a'_s$, and taking the moment about the resultant force action point of the compressive reinforcement, as shown in Fig. 4-25, then the flexural capacity of normal section is

$$M \leqslant M_u = f_y A_s (h_0 - a'_s) \tag{4-41}$$

按式(4-41)求得的钢筋截面面积 A_s 可能比不考虑受压钢筋而按单筋矩形截面计算的 A_s 还大，此时应按单筋矩形截面的计算结果配筋。

The cross-sectional area A_s of the steel bar obtained according to the Eq. (4-41) may be larger than A_s calculated according to single-reinforced rectangular section without considering the compressive reinforcement. At this time, the reinforcement should be arranged according to the calculation results of singly reinforced rectangular section.

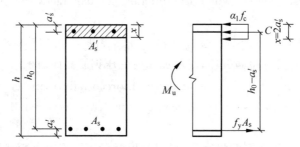

图 4-25　$x=2a'_s$ 时的双筋截面梁计算简图

Figure 4-25　Calculation diagram of doubly reinforced section beam when $x=2a'_s$

特别提示 4-5

4.5.3　截面设计
Section design

与单筋截面类似，双筋矩形截面受弯构件正截面承载力计算也包括截面设计和截面复核两类问题。

Similar to singly reinforced section, the calculation of normal section bearing capacity of doubly reinforced rectangular section flexural member also includes two types of problems: section design and section review.

1. 情形 1：已知弯矩设计值 M、截面尺寸和材料强度 f_c、f_y、f'_y，求纵向受压和受拉钢筋截面面积 A'_s、A_s

   Case 1: known design value M of bending moment, cross-sectional dimensions and material strength f_c, f_y, f'_y, find: the cross-sectional area of longitudinal compressive and tensile steel bars A'_s, A_s

两个基本计算公式式(4-37)、式(4-38)中有 A'_s、A_s 和 x 三个未知数，需补充一个条件才能求解。考虑到受压钢筋是用来协助受压区混凝土承受压力的,计算受压钢筋时,应在充分利用受压区混凝土抗压强度之后,再由受压钢筋承担受压区混凝土不能承担的压力。

The two basic calculation formulas Eq. (4-37)、Eq. (4-38) have three unknowns, A'_s、A_s and x, which can only be solved by adding a condition. Considering that the compressive steel bars are used to assist the concrete in the compression zone to bear the compressive force, when calculating the compressive steel bars, the compressive strength of the compression zone concrete should be fully utilized, and then the compressive steel bars should bear the compressive force that the compression zone concrete cannot bear.

为了减少基本计算公式中未知数的数量,充分利用混凝土受压,使总的钢筋用量 $A_s + A'_s$ 最少,取得较好的经济效果,令 $x = \xi_b h_0$,双筋截面的截面设计过程如下。

To reduce the number of unknowns in the basic calculation formula, make full use of concrete compression, minimize the total amount of reinforcement $A_s + A'_s$, and achieve better economic results, let $x = \xi_b h_0$, the section design process of doubly reinforced section is as follows.

1) 判断构件是否需要设计成双筋截面

1) Determine whether the member needs to be designed as doubly reinforced section

若构件截面弯矩设计值 M 满足式(4-42),则需要设计成双筋截面,否则按单筋截面设计。

If the design value of bending moment M of the member section satisfies Eq. (4-42), it needs to be designed as doubly reinforced section, otherwise it is designed as singly reinforced section.

$$M > M_u = \alpha_{s,max} \alpha_1 f_c b h_0^2 = \xi_b (1 - 0.5\xi_b) \alpha_1 f_c b h_0^2 \quad (4-42)$$

任务 4-3:查表确定 HRB400 钢筋的 ξ_b、$\alpha_{s,max}$。

Task 4-3: Check the table to determine ξ_b、$\alpha_{s,max}$ of HRB400 steel bar.

2) 计算纵向受压钢筋截面面积 A'_s

2) Calculate the cross-sectional area of longitudinal compressive steel bars A'_s

由双筋矩形截面的力矩平衡方程

From the moment equilibrium equation of doubly reinforced rectangular section

$$M = \alpha_1 f_c bx \left(h_0 - \frac{x}{2}\right) + f'_y A'_s (h_0 - a'_s) \quad (4-43)$$

将 $x = \xi_b h_0$ 代入式(4-43),得

Substituting $x = \xi_b h_0$ into Eq. (4-43), one obtains

$$A'_s = \frac{M - \xi_b(1 - 0.5\xi_b)\alpha_1 f_c b h_0^2}{f'_y(h_0 - a'_s)} = \frac{M - \alpha_{s,max}\alpha_1 f_c b h_0^2}{f'_y(h_0 - a'_s)} \quad (4-44)$$

3) 计算纵向受拉钢筋截面面积 A_s

3) Calculate the cross-sectional area of longitudinal tensile steel bars A_s

若受压钢筋的 $A'_s \leq 0$,则说明不需要配置受压钢筋,可按单筋截面计算受拉钢筋的 A_s;若 $A'_s > 0$,则由双筋截面力的平衡方程

If the compressive reinforcement $A'_s \leq 0$, it means that there is no need to configure the compressive reinforcement, and the tensile reinforcement A_s can be calculated according to the singly reinforced section; if $A'_s>0$, from the force equilibrium equation of doubly reinforced section

$$\alpha_1 f_c bx + f'_y A'_s = f_y A_s \quad (4-45)$$

将 $x=\xi_b h_0$ 代入式(4-45),得
Substituting $x=\xi_b h_0$ into Eq. (4-45), one obtains

$$A_s = \xi_b \frac{\alpha_1 f_c}{f_y} bh_0 + \frac{f'_y}{f_y} A'_s \quad (4-46)$$

4) 选用纵向受压、受拉钢筋,合理布置钢筋

4) Select longitudinal compression and tension reinforcement and reasonably arrange reinforcement

根据计算出的纵向受压钢筋截面面积 A'_s 和纵向受拉钢筋截面面积 A_s 查附表 2-1,分别选用受压、受拉钢筋的直径和根数,并在梁内按构造要求合理布置钢筋。

According to the calculated cross-sectional area A'_s of longitudinal compressive steel bars and cross-sectional area A_s of longitudinal tensile steel bars, select the diameter and number of compressive and tensile steel bars respectively from Attached table 2-1, and arrange the steel bars reasonably in the beam according to the structural requirements.

5) 验算最小配筋率 $\rho_{\min}$

5) Check and calculate the minimum reinforcement ratio $\rho_{\min}$

双筋截面弯矩设计值 M 较大,配筋较多,一般不会出现少筋破坏,可不必验算。

There are more reinforcement in the doubly reinforced section due to relatively large design value of the bending moment M, and generally there will be no lightly reinforced failure, so it is not necessary to check.

【例题 4-4】 钢筋混凝土矩形截面梁的截面尺寸为 $b\times h=200\text{mm}\times 500\text{mm}$,混凝土强度等级为 C40,钢筋采用 HRB400,截面弯矩设计值 $M=330\text{kN}\cdot\text{m}$,环境类别为一类,求所需纵向受压和受拉钢筋截面面积。

【Example 4-4】 The cross-sectional dimensions of a reinforced concrete rectangular section beam are $b\times h=200\text{mm}\times 500\text{mm}$, the concrete strength grade is C40, HRB400 steel bar is used, the design value of section moment is $M=330\text{kN}\cdot\text{m}$, and the environmental category is one. Calculate: the required cross-sectional area of longitudinal compressive and tensile reinforcement.

解:由混凝土和钢筋的强度等级,查附表 1-3、附表 1-6 得 $f_c=19.1\text{N/mm}^2$,$f_y=f'_y=360\text{N/mm}^2$,查表 4-1 得 $\alpha_1=1.0$,查表 4-2 得 $\xi_b=0.518$。

由于弯矩设计值 M 较大,假定纵向受拉钢筋放两层,则 $a_s=60\text{mm}$,$h_0=h-a_s=(500-60)\text{mm}=440\text{mm}$。

假定纵向受压钢筋放一层,则 $a'_s=35\text{mm}$,得

$$\begin{aligned}M_{u1}&=\alpha_1 f_c bh_0^2 \xi_b(1-0.5\xi_b)=1.0\times 19.1\times 200\times 440^2\times\\&\quad 0.518\times(1-0.5\times 0.518)\text{kN}\cdot\text{m}\\&=283.9\text{kN}\cdot\text{m}<M=330\text{kN}\cdot\text{m}\end{aligned}$$

说明如果设计成单筋截面,将会出现超筋的情况。若不能加大截面尺寸,又不能提高混凝土

强度等级,则应设计成双筋截面。

纵向受压钢筋截面面积

$$A'_s = \frac{M - M_{u1}}{f'_y(h_0 - a'_s)} = \frac{330 \times 10^6 - 283.9 \times 10^6}{360 \times (440 - 35)} \text{mm}^2 = 316.2 \text{mm}^2$$

纵向受拉钢筋截面面积

$$A_s = A'_s \frac{f'_y}{f_y} + \xi_b \frac{\alpha_1 f_c b h_0}{f_y}$$

$$= \left(316.2 \times \frac{360}{360} + 0.518 \times \frac{1.0 \times 19.1 \times 200 \times 440}{360}\right) \text{mm}^2$$

$$= 2734.7 \text{mm}^2$$

任务 4-4:根据例题 4-4 中计算出的 A'_s、A_s 查附表 2-1,分别选配纵向受压、受拉钢筋的直径、根数,并画出梁横截面配筋图。

Task 4-4: According to the calculation of A'_s、A_s in Example 4-4 check the Attached table 2-1, select the diameter and number of longitudinal compressive and tensile steel bars respectively, and draw the reinforcement diagram of the beam cross-section.

2. 情形 2:已知纵向受压钢筋截面面积 A'_s,计算纵向受拉钢筋截面面积 A_s

Case 2: known the cross-sectional area A'_s of longitudinal compressive steel bars, calculate the cross-sectional area A_s of longitudinal tensile steel bars

有时由于计算或构造要求,纵向受压钢筋截面面积 A'_s 已知,只要计算纵向受拉钢筋截面面积 A_s。此时只有 A_s 和 x 两个未知数,利用平衡方程就可以直接求解,为避免联立求解方程,也可采用计算系数法求解。

(1) 纵向受压钢筋压应力的合力对纵向受拉钢筋合力作用点取矩

$$M_2 = f'_y A'_s (h_0 - a'_s) \tag{4-47}$$

(2) 受压区混凝土压应力的合力对纵向受拉钢筋合力作用点取矩

$$M_1 = M - M_2 = M - f'_y A'_s (h_0 - a'_s) = \alpha_1 f_c b x \left(h_0 - \frac{x}{2}\right) = \alpha_s \alpha_1 f_c b h_0^2 \tag{4-48}$$

(3) 截面抵抗矩系数 α_s

$$\alpha_s = \frac{M_1}{\alpha_1 f_c b h_0^2} = \frac{M - f'_y A'_s (h_0 - a'_s)}{\alpha_1 f_c b h_0^2} \tag{4-49}$$

(4) 相对受压区高度 ξ

$$\xi = 1 - \sqrt{1 - 2\alpha_s} \tag{4-50}$$

(5) 纵向受拉钢筋截面面积 A_s 和纵向受压钢筋截面面积 A'_s

① 若 $\xi > \xi_b$,则说明纵向受压钢筋 A'_s 太小,截面超筋,应按 A'_s 未知,重新计算 A'_s 及 A_s;

② 若 $2a'_s \leqslant x \leqslant \xi_b h_0$,由双筋截面力的平衡方程

$$\alpha_1 f_c b x + f'_y A'_s = f_y A_s \tag{4-51}$$

考虑截面受压区高度 $x = \xi h_0$,求得纵向受拉钢筋截面面积 A_s:

$$A_s = \frac{\alpha_1 f_c b x + f'_y A'_s}{f_y} = \frac{\alpha_1 f_c b \xi h_0 + f'_y A'_s}{f_y} \tag{4-52}$$

③ 若 $x<2a'_s$，则表明纵向受压钢筋 A'_s 的压应力不能达到其抗压强度设计值 f'_y，令 $x=2a'_s$，对纵向受压钢筋合力作用点取矩，由式(4-47)得纵向受拉钢筋截面面积 A_s：

$$A_s = \frac{M}{f_y(h_0 - a'_s)} \tag{4-53}$$

4.5.4 截面复核
Section review

截面复核是指已知双筋截面弯矩设计值 M、截面尺寸、混凝土强度等级和钢筋级别、纵向受拉钢筋截面面积 A_s、纵向受压钢筋截面面积 A'_s，求截面所能承担的极限弯矩 M_u，并与弯矩设计值 M 比较，验算正截面受弯承载力 M_u 是否足够，构件是否安全。截面复核的步骤如下。

1. 计算混凝土受压区高度 x

Calculate the depth of concrete compression zone x

根据双筋截面力的平衡方程 $\alpha_1 f_c bx + f'_y A'_s = f_y A_s$，计算混凝土受压区高度 x：

$$x = \frac{f_y A_s - f'_y A'_s}{\alpha_1 f_c b} \tag{4-54}$$

2. 计算双筋截面受弯构件正截面受弯承载力 M_u

Calculate the flexural capacity of the normal section of the doubly reinforced section flexural member M_u

(1) 若 $2a'_s \leq x \leq x_b = \xi_b h_0$，则满足基本计算公式的适用条件，根据双筋截面的力矩平衡方程，得正截面受弯承载力 M_u：

$$M_u = \alpha_1 f_c bx \left(h_0 - \frac{x}{2}\right) + f'_y A'_s (h_0 - a'_s) \tag{4-55}$$

(2) 若 $x<2a'_s$，则说明纵向受压钢筋应力未达到屈服强度，强度不能充分发挥，取 $x=2a'_s$，如图 4-25 所示，对纵向受压钢筋合力作用点取矩确定 $M_u = f_y A_s (h_0 - a'_s)$。

(3) 若 $x>\xi_b h_0$，则说明为超筋梁，取 $x=\xi_b h_0$，即 $\xi = \xi_b$，近似按下式计算 M_u：

$$\begin{cases} M_u = \alpha_{s,\max} \alpha_1 f_c b h_0^2 + f'_y A'_s (h_0 - a'_s) \\ \alpha_{s,\max} = \xi_b (1 - 0.5\xi_b) \end{cases} \tag{4-56}$$

3. 判断构件是否安全

Determine whether the member is safe

若 $M_u \geq M$，则说明截面受弯承载力足够，构件安全；若 $M_u < M$，则说明截面受弯承载力不够，构件是不安全的，需重新设计或进行补强加固处理。

【例题 4-5】 钢筋混凝土矩形截面梁的截面尺寸为 $b \times h = 250\text{mm} \times 500\text{mm}$，截面弯矩设计值 $M=300\text{kN}\cdot\text{m}$，环境类别为一类，结构安全等级为二级。混凝土强度等级为 C25，钢筋采用 HRB400，箍筋直径为 10mm，纵向受压钢筋为 3⎬18，纵向受拉钢筋为 4⎬20，验算此梁是否安全。

【**Example 4-5**】 The cross-sectional dimensions of the reinforced concrete rectangular section beam are $b \times h = 250\text{mm} \times 500\text{mm}$, the bending moment design value of the section is $M = 300\text{kN} \cdot \text{m}$, the environmental category is Class Ⅰ, and the structural safety level is Level 2. The concrete strength grade is C25, the reinforcement is HRB400, the diameter of the stirrup is 10mm, the longitudinal compressive reinforcement is 3 ⌀ 18, and the longitudinal tensile reinforcement is 4 ⌀ 20. Check whether the beam is safe or not.

解：由混凝土强度等级和钢筋级别，查附表 1-3、附表 1-6、表 4-1、表 4-2 得 $f_c = 11.9\text{N/mm}^2$，$f_y = 360\text{N/mm}^2$，$f'_y = 360\text{N/mm}^2$，$\alpha_1 = 1.0$，$\xi_b = 0.518$。

因为环境类别为一类，查附表 3-2 得最外层钢筋（箍筋）的混凝土保护层最小厚度为 20mm，箍筋直径为 10mm，则 $a_s = c + d_{箍筋} + d_{受拉纵筋}/2 = (20 + 10 + 20/2)\text{mm} = 40\text{mm}$，$a'_s = c + d_{箍筋} + d_{受压纵筋}/2 = (20 + 10 + 18/2)\text{mm} = 39\text{mm}$，截面有效高度 $h_0 = h - a_s = (500 - 40)\text{mm} = 460\text{mm}$。

由式（4-54）可得混凝土受压区高度

$$x = \frac{f_y A_s - f'_y A'_s}{\alpha_1 f_c b} = \frac{360 \times 1256 - 360 \times 763}{1.0 \times 11.9 \times 250}\text{mm} = 60\text{mm} < x_b$$

$$= \xi_b h_0 = 0.518 \times 460\text{mm} = 238.28\text{mm}$$

满足基本计算公式的适用条件。

根据双筋截面的力矩平衡方程，求得正截面受弯承载力（极限弯矩）M_u：

$$\begin{aligned}
M_u &= \alpha_1 f_c b x \left(h_0 - \frac{x}{2}\right) + f'_y A'_s (h_0 - a'_s) \\
&= [1.0 \times 11.9 \times 250 \times 60 \times (460 - 60/2) + 360 \times 763 \times (460 - 39)]\text{N} \cdot \text{mm} \\
&= 192 \times 10^6 \text{N} \cdot \text{mm} \\
&= 192\text{kN} \cdot \text{m} < M = 300\text{kN} \cdot \text{m}
\end{aligned}$$

说明正截面受弯承载力足够，构件安全。

4.6 T 形截面受弯承载力计算
Calculation of flexural capacity of T-shaped section

4.6.1 T 形截面概述
Overview of T-shaped section

1. T 形截面的形成

Formation of T-shaped section

矩形截面受弯构件达到承载能力极限状态破坏时，大部分受拉区混凝土因开裂已退出工作，原来由受拉区混凝土承担的拉力转移到受拉钢筋上，正截面承载力计算时不考虑混凝土的抗拉作用，可挖去受拉区的部分混凝土，并把原有的纵向受拉钢筋集中布置，保持钢筋截面重心高度不变，形成 T 形截面，如图 4-26 所示。这种 T 形截面和原来的矩形截面所能承

英文翻译 4-14

受的弯矩是相同的,这样可以节约混凝土,减轻构件自重,取得较好的经济效果。

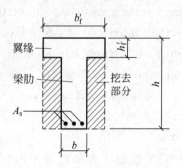

图 4-26 T 形截面的形成和组成

Figure 4-26 Formation and composition of T-shaped section

2. T 形截面的组成
Composition of T-shaped section

如图 4-26 所示,T 形截面由两部分组成:中间 $b \times h$ 部分称为腹板,腹板宽度为 b,截面高度为 h;伸出部分 $(b'_f - b) \times h'_f$ 称为翼缘,翼缘宽度为 b'_f,翼缘高度为 h'_f。

3. T 形截面的应用
Application of T-shaped section

T 形截面在土木工程中有广泛应用,如图 4-27 所示,有整体现浇肋梁楼盖中的梁、预制独立的吊车梁、屋面梁、I 形截面薄腹梁、预制槽形板、空心板以及箱形梁等,箱形梁可换算成力学性能等效的 T 形截面或 I 形截面梁进行计算。

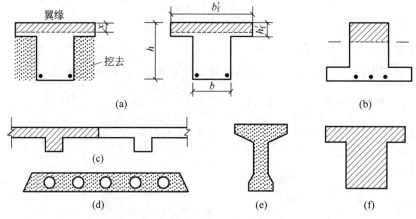

图 4-27 工程中的各类 T 形截面

(a) T 形;(b) 倒 T 形;(c) 肋梁楼盖的梁板;(d) 圆孔空心板;(e) 薄腹梁;(f) 吊车梁

Figure 4-27 Various T-shaped sections in engineering

(a) T-shaped; (b) Inverted T-shaped; (c) Beam and slab of ribbed beam floor;
(d) Hollow slab with circular hole; (e) Thin web beam; (f) Crane beam

拓展知识 4-5

4. 翼缘计算宽度 b'_f
Effective flange width b'_f

为发挥 T 形截面的作用,应充分利用翼缘受压,使混凝土受压区高度减小,减少纵向受拉钢筋用量。试验和理论分析表明,翼缘中纵向压应力分布不均匀,靠近腹板处翼缘中压应力较高,离腹板越远则翼缘中的压应力越小,如图 4-28(a)所示。实际上,与腹板共同工作,参与受压的翼缘有效宽度是有限的。

为简化计算,设计 T 形截面时应将翼缘宽度限制在一定范围内,称为翼缘计算宽度 b'_f,

如图 4-28(b)所示。认为在这个宽度范围内翼缘全部参与工作,假定压应力均匀分布。而在这个范围以外,则认为翼缘不起作用,不考虑它参与受力。

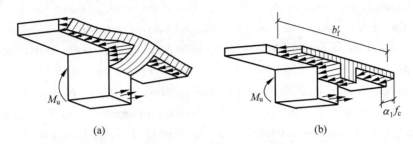

图 4-28　T 形截面应力分布

(a) 翼缘中纵向压应力分布不均匀;(b) 翼缘计算宽度 b'_f

Figure 4-28　Stress distribution of T-shaped section

(a) Uneven longitudinal compressive stress distribution in the flange; (b) Effective flange width b'_f

翼缘计算宽度 b'_f 与梁的计算跨度 l_0、梁(肋)净距 s_n、翼缘高度与截面有效高度之比 h'_f/h_0、受力条件(独立梁、现浇肋形楼盖梁)等因素有关。《混凝土结构设计标准》(GB/T 50010—2010)规定 T 形、I 形和倒 L 形截面受弯构件受压区有效翼缘计算宽度 b'_f 应取表 4-5 中各有关项的最小值。

表 4-5　T 形、I 形和倒 L 形截面受弯构件受压区有效翼缘计算宽度 b'_f

Table 4-5　Effective flange width b'_f of compression zone of T-shaped、I-shaped and inverted L-shaped section flexural members

		T 形截面、I 形截面		倒 L 形截面
	情　况	肋形梁(板)	独立梁	肋形梁(板)
1	按计算跨度 l_0 考虑	$l_0/3$	$l_0/3$	$l_0/6$
2	按梁(肋)净距 s_n 考虑	$b+s_n$	—	$b+s_n/2$
3	按翼缘高度 h'_f 考虑　$h'_f/h_0 \geqslant 0.1$	—	$b+12h'_f$	
	$0.1 > h'_f/h_0 \geqslant 0.05$	$b+12h'_f$	$b+6h'_f$	$b+5h'_f$
	$h'_f/h_0 < 0.05$	$b+12h'_f$	b	$b+5h'_f$

注:①表中 b 为梁的腹板宽度;②肋形梁在梁跨内设有间距小于纵肋间距的横肋时,可不考虑表中情况 3 的规定;③对有加腋的 T 形、I 形和倒 L 形截面,当受压区加腋的高度 $h_h \geqslant h'_f$ 且加腋长度 $b_h \leqslant 3h_h$ 时,其翼缘计算宽度可按表中情况 3 的规定分别增加 $2b_h$(T 形、I 形截面)和 b_h(倒 L 形截面);④ 独立梁受压区的翼缘板在荷载作用下经验算沿纵肋方向可能产生裂缝时,其计算宽度应取腹板宽度 b。

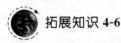拓展知识 4-6

4.6.2　T 形截面的分类和判别
Classification and discrimination of T-shaped section

T 形截面受弯构件正截面受力的分析方法与矩形截面的基本相同,T 形截面受压区混凝土的应力分布仍可按等效矩形应力图形考虑,主要区别在于 T 形截面的计算需要考虑受压翼缘的作用。根据 T 形截面受力后混凝土受压区高度 x 的大小、中和轴的位置等,可分为两类 T 形截面：第一类 T 形截面和第二类 T 形截面。

The analysis method of the normal section force of the T-shaped section flexural member is basically the same as that of the rectangular section. The stress distribution of concrete in the compression zone of the T-shaped section can still be considered as the equivalent rectangular stress block. The main difference is that the calculation of the T-shaped section needs to consider the effect of the compression flange. According to the depth of concrete compression zone x and the position of the neutral axis after the T-shaped section is subjected to force, it can be divided into two types of T-shaped sections: the first type of T-shaped section and the second type of T-shaped section.

1. 第一类 T 形截面

The first type of T-shaped section

中和轴在翼缘高度内，即混凝土受压区高度 $x \leqslant h_f'$，受压区形状为矩形，如图 4-29(a) 所示。

The neutral axis is within the flange height, that is, the depth of the concrete compression zone $x \leqslant h_f'$, and the shape of the compression zone is rectangular, as shown in Fig. 4-29(a).

2. 第二类 T 形截面

The second type of T-shaped section

中和轴在腹板内，即混凝土受压区高度 $x > h_f'$，受压区形状为 T 形，如图 4-29(b) 所示。

The neutral axis is in the web, that is, the depth of the concrete compression zone is $x > h_f'$, and the shape of the compression zone is T-shaped, as shown in Fig. 4-29(b).

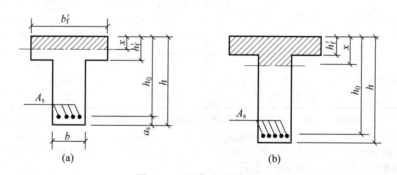

图 4-29　两类 T 形截面
(a) 第一类 T 形截面；(b) 第二类 T 形截面
Figure 4-29　Two types of T-shaped sections
(a) The first type T-shaped section; (b) The second type T-shaped section

3. 两种 T 形截面的判别

Discrimination of two T-shaped sections

要判断中和轴是否在翼缘高度范围内，首先应对中和轴在翼缘与腹板交界处的界限位置即 $x = h_f'$ 的情况进行分析，此为两类 T 形截面的分界情况，如图 4-30 所示。

To determine whether the neutral axis is within the range of flange height or not, the boundary position of the neutral axis at the junction of the flange and the web, that is $x = h'_f$, should be analyzed first, which is the boundary condition of two types of T-shaped sections, as shown in Fig. 4-30.

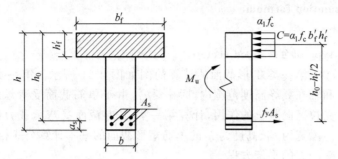

图 4-30 $x = h'_f$ 时的 T 形截面

Figure 4-30 T-shaped section when $x = h'_f$

根据截面力的平衡条件和力矩的平衡条件，$x = h'_f$ 时，得

According to the force and moment equilibrium conditions, when $x = h'_f$, one obtains

$$\alpha_1 f_c b'_f h'_f = f_y A_s \tag{4-57}$$

$$M = \alpha_1 f_c b'_f h'_f \left(h_0 - \frac{h'_f}{2}\right) \tag{4-58}$$

第一类 T 形截面 $x \leqslant h'_f$，如图 4-29(a)所示，得

The first type of T-shaped section $x \leqslant h'_f$, as shown in Fig. 4-29(a), one obtains

$$f_y A_s \leqslant \alpha_1 f_c b'_f h'_f \tag{4-59}$$

$$M \leqslant \alpha_1 f_c b'_f h'_f \left(h_0 - \frac{h'_f}{2}\right) \tag{4-60}$$

第二类 T 形截面 $x > h'_f$，如图 4-29(b)所示，得

The second type of T-shaped section $x > h'_f$, as shown in Fig. 4-29(b), one obtains

$$f_y A_s > \alpha_1 f_c b'_f h'_f \tag{4-61}$$

$$M > \alpha_1 f_c b'_f h'_f \left(h_0 - \frac{h'_f}{2}\right) \tag{4-62}$$

以上即为两类 T 形截面受弯构件的判别条件，但对不同情况需采用不同的判别条件，截面设计时，纵向受拉钢筋截面面积 A_s 未知，利用式(4-60)和式(4-62)判别；截面复核时，A_s 已知，利用式(4-59)和式(4-61)判别。

The above are the criteria for distinguishing two types of T-shaped section flexural members, but different criteria are required for different situations. When designing the section, the cross-sectional area A_s of the longitudinal tensile reinforcement is unknown, Eq. (4-60) and Eq. (4-62) are used to discriminate; when the section is reviewed, A_s is known, Eq. (4-59) and Eq. (4-61) are used to discriminate.

4.6.3 基本计算公式及适用条件
Basic calculation formulas and applicable conditions

1. 基本计算公式

Basic calculation formula

1) 第一类 T 形截面

1) The first type of T-shaped section

第一类 T 形截面的计算简图如图 4-31 所示。第一类 T 形截面的中和轴在翼缘高度范围内，即 $x \leqslant h'_f$，由于单筋截面受弯承载力计算不考虑受拉区混凝土的作用，计算第一类 T 形截面受弯承载力的基本计算公式与梁宽为 b'_f、高度为 h 的单筋矩形截面的基本计算公式相同，即：

力的平衡方程

$$\alpha_1 f_c b'_f x = f_y A_s \tag{4-63}$$

力矩平衡方程

$$M \leqslant \alpha_1 f_c b'_f x \left(h_0 - \frac{x}{2}\right) = \alpha_s \alpha_1 f_c b'_f h_0^2 \tag{4-64}$$

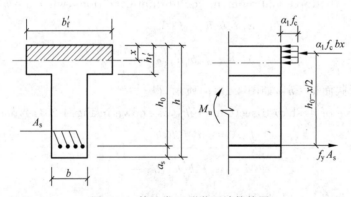

图 4-31 第一类 T 形截面计算简图

Figure 4-31　Calculation diagram of the first type of T-shaped section

2) 第二类 T 形截面

2) The second type of T-shaped section

第二类 T 形截面的计算简图如图 4-32 所示。第二类 T 形截面的中和轴在腹板内，即 $x > h'_f$，此时受压区形状为 T 形，可将 T 形截面中和轴以上的受压区分为翼缘和腹板两部分，则根据力的平衡条件和力矩平衡条件得基本计算公式：

$$\alpha_1 f_c (b'_f - b) h'_f + \alpha_1 f_c b x = f_y A_s \tag{4-65}$$

$$M \leqslant M_u = \alpha_1 f_c (b'_f - b) h'_f \left(h_0 - \frac{h'_f}{2}\right) + \alpha_1 f_c b x \left(h_0 - \frac{x}{2}\right) \tag{4-66}$$

2. 基本计算公式的适用条件

Applicable conditions of basic calculation formula

1) 防止超筋破坏：$\xi \leqslant \xi_b$

这和单筋矩形截面梁一样，是为了保证破坏始自纵向受拉钢筋的屈服，防止超筋破坏。

2）防止少筋破坏：$A_s \geqslant \rho_{\min} bh$（第一类T形截面）或$A_s \geqslant \rho_{\min}[bh+(b'_f-b)h'_f]$（第二类T形截面）

第一类T形截面x较小（$x \leqslant h'_f$），相应的纵向受拉钢筋不会太多，要验算最小配筋率，防止少筋破坏；第二类T形截面受压区已进入腹板，纵向受拉钢筋相应配置较多，一般均能满足最小配筋率的要求，不必验算。

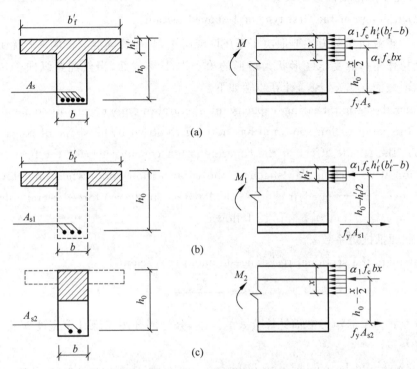

图 4-32　第二类T形截面计算简图

Figure 4-32　Calculation diagram of the second type of T-shaped section

4.6.4 截面设计
Section design

已知：T形截面的弯矩设计值M、截面尺寸、混凝土强度等级和钢筋级别，求纵向受拉钢筋截面面积A_s。

Known: design value of bending moment of T-shaped section M, cross-sectional dimensions, concrete strength grade and reinforcement grade, calculate the cross-sectional area of longitudinal tensile reinforcement A_s.

1. 判别T形截面的类型
Distinguish the type of T-shaped section

在截面设计时，首先应判别T形截面的类型。若$M \leqslant \alpha_1 f_c b'_f h'_f \left(h_0 - \dfrac{h'_f}{2}\right)$，则为第一类T形截面；若$M > \alpha_1 f_c b'_f h'_f \left(h_0 - \dfrac{h'_f}{2}\right)$，则为第二类T形截面。

In the design of section, the type of T-shaped section should be judged first. If $M \leqslant \alpha_1 f_c b_f' h_f' \left(h_0 - \dfrac{h_f'}{2}\right)$, it is the first type of T-shaped section, if $M > \alpha_1 f_c b_f' h_f' \left(h_0 - \dfrac{h_f'}{2}\right)$, it is the second type of T-shaped section.

2. 第一类 T 形截面的截面设计

Section design of the first type of T-shaped section

受弯构件正截面承载力主要取决于受压区混凝土，与受拉区混凝土的形状无关(不考虑混凝土的受拉作用)，故受压区形状为矩形的第一类 T 形截面，当仅配置纵向受拉钢筋时，可按单筋矩形截面 $b_f' \times h$ 计算，计算步骤如下。

The normal section bearing capacity of flexural members mainly depends on the concrete in the compression zone, and has nothing to do with the shape of concrete in the tension zone (the tensile effect of the concrete is not considered), for the first type of T-shaped section with rectangular shape in the compression zone, when only longitudinal tensile steel bars are arranged, it can be calculated as singly reinforced rectangular section $b_f' \times h$, and the calculation steps are as follows.

（1）求截面抵抗矩系数 α_s

(1) Calculate the sectional resistance moment coefficient α_s

$$\alpha_s = \dfrac{M}{\alpha_1 f_c b_f' h_0^2} \tag{4-67}$$

应注意此式与单筋矩形截面计算公式的区别，分母部分用翼缘计算宽度 b_f' 代替了矩形截面的梁宽 b。

Attention should be paid to the difference between this formula and the calculation formula of singly reinforced rectangular section, the denominator part uses the effective flange width b_f' instead of the beam width b of the rectangular section.

（2）求相对受压区高度 ξ

(2) Calculate the relative depth of compression zone ξ

$$\xi = 1 - \sqrt{1 - 2\alpha_s} \tag{4-68}$$

（3）求纵向受拉钢筋截面面积 A_s

(3) Calculate the cross-sectional area of longitudinal tensile reinforcement A_s

内力臂系数：

Internal arm coefficient:

$$\gamma_s = 1 - 0.5\xi \tag{4-69}$$

$$A_s = \dfrac{M}{f_y \gamma_s h_0} \tag{4-70}$$

或

$$A_s = \dfrac{\alpha_1 f_c b_f' x}{f_y} = \dfrac{\alpha_1 f_c b_f' \xi h_0}{f_y} \tag{4-71}$$

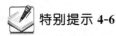

特别提示 4-6

3. 第二类 T 形截面的截面设计

Section design of the second type of T-shaped section

若为第二类 T 形截面,为避免联合方程求解,可利用计算系数法计算,计算步骤如下。

If it is the second type of T-shaped section, in order to avoid solving the joint equations, the calculation coefficient method can be used for calculation, and the calculation steps are as follows.

(1) 求截面抵抗矩系数 α_s

(1) Calculate the sectional resistance moment coefficient α_s

根据式(4-16)、式(4-18)和力矩平衡条件式(4-66),得

According to Eq. (4-16)、Eq. (4-18) and the moment equilibrium condition Eq. (4-66), one obtains

$$\alpha_s = \frac{M - \alpha_1 f_c (b'_f - b) h'_f \left(h_0 - \frac{h'_f}{2} \right)}{\alpha_1 f_c b h_0^2} \tag{4-72}$$

(2) 求相对受压区高度 ξ

(2) Calculate the relative depth of compression zone ξ

$$\xi = 1 - \sqrt{1 - 2\alpha_s} \tag{4-73}$$

(3) 求纵向受拉钢筋截面面积 A_s

(3) Calculate the cross-sectional area of longitudinal tensile reinforcement A_s

当 $\xi \leqslant \xi_b$ 时,由力的平衡条件式(4-65),得纵向受拉钢筋截面面积 A_s:

When $\xi \leqslant \xi_b$, from the force equilibrium condition Eq. (4-65), gives the cross-sectional area of longitudinal tensile reinforcement A_s:

$$A_s = \frac{\alpha_1 f_c (b'_f - b) h'_f + \alpha_1 f_c b \xi h_0}{f_y} \tag{4-74}$$

当 $\xi > \xi_b$ 时,为超筋梁,可增大梁截面高度或提高混凝土强度等级。如果有限制不能增大或提高时,可按双筋 T 形截面计算。

When $\xi > \xi_b$, for overreinforced beam, the depth of the beam section can be increased or the concrete strength grade can be improved. If there is a limit that cannot be increased or improved, it can be calculated as doubly reinforced T-shaped section.

4.6.5 截面复核

Section review

已知:T 形截面的弯矩设计值 M、混凝土强度等级和钢筋级别、截面尺寸、纵向受拉钢筋截面面积 A_s,求截面所能承担的极限弯矩 M_u,与弯矩设计值 M 比较,验算正截面受弯承载力是否足够,以判断截面是否安全。

英文翻译 4-16

1. 判别 T 形截面类型

Discrimination of the type of T-shaped section

当 $f_y A_s \leqslant \alpha_1 f_c b'_f h'_f$ 时,为第一类 T 形截面;当 $f_y A_s > \alpha_1 f_c b'_f h'_f$ 时,为第二类 T 形截面。

2. 第一类 T 形截面的截面复核

Section review of the first type of T-shaped section

若为第一类 T 形截面,可按截面尺寸为 $b'_f \times h$ 的矩形截面梁的截面复核方法计算。

3. 第二类 T 形截面的截面复核

Section review of the second type of T-shaped section

若为第二类 T 形截面,由力的平衡条件式(4-65),先计算截面受压区高度:

$$x = \frac{f_y A_s - \alpha_1 f_c (b'_f - b) h'_f}{\alpha_1 f_c b} \tag{4-75}$$

(1) 当 $x \leqslant \xi_b h_0$ 时,采用力矩平衡方程式(4-66)计算极限弯矩 M_u。

(2) 当 $x > \xi_b h_0$ 时,为超筋截面,取 $x = \xi_b h_0$,代入力矩平衡方程式(4-66),得

$$M_u = \alpha_1 f_c (b'_f - b) h'_f \left(h_0 - \frac{h'_f}{2}\right) + \alpha_1 f_c b h_0^2 \xi_b (1 - 0.5\xi_b) \tag{4-76}$$

若 $M < M_u$,则正截面受弯承载力足够,截面安全;若 $M \geqslant M_u$,则正截面受弯承载力不足,截面不安全。

【例题 4-6】 钢筋混凝土肋梁楼盖的环境类别为一类,结构安全等级为二级,弯矩设计值 $M = 410 \text{kN} \cdot \text{m}$,梁截面尺寸为 $b \times h = 200\text{mm} \times 600\text{mm}$,$b'_f = 1000\text{mm}$,$h'_f = 90\text{mm}$,混凝土强度等级为 C25,钢筋采用 HRB400。求纵向受拉钢筋截面面积,选配钢筋直径和根数,验算最小配筋率。

【Example 4-6】 The environmental category of reinforced concrete ribbed beam floor is Class Ⅰ, the structural safety level is Level 2, the bending moment design value is $M = 410 \text{kN} \cdot \text{m}$, the cross-sectional dimensions are $b \times h = 200\text{mm} \times 600\text{mm}$, $b'_f = 1000\text{mm}$, $h'_f = 90\text{mm}$, the concrete strength grade is C25, and HRB400 steel bar is used. Calculate the cross-sectional area of longitudinal tensile reinforcement, select the diameter and number of steel bars, and check the minimum reinforcement ratio.

解:根据混凝土强度等级和钢筋级别,查附表 1-3、附表 1-6、表 4-1、表 4-2 得:$f_c = 11.9 \text{N/mm}^2$,$f_y = f'_y = 360 \text{N/mm}^2$,$\alpha_1 = 1.0$,$\xi_b = 0.518$。

因弯矩设计值较大,腹板宽度较窄,预计纵向受拉钢筋需排成两层,取 $a_s = 60\text{mm}$,截面有效高度 $h_0 = h - a_s = (600 - 60)\text{mm} = 540\text{mm}$。

判别 T 形截面类型:翼缘部分所能承担的弯矩为

$$\alpha_1 f_c b'_f h'_f \left(h_0 - \frac{h'_f}{2}\right) = 1.0 \times 11.9 \times 1000 \times 90 \times \left(540 - \frac{90}{2}\right) \text{N} \cdot \text{mm}$$

$$= 530.15 \times 10^6 \text{N} \cdot \text{mm} = 530.15 \text{kN} \cdot \text{m} > M = 410 \text{kN} \cdot \text{m}$$

因此属于第一类 T 形截面。

计算纵向受拉钢筋截面面积 A_s：

$$\alpha_s = \frac{M}{\alpha_1 f_c b'_f h_0^2} = \frac{410 \times 10^6}{1 \times 11.9 \times 1000 \times 540^2} = 0.118$$

$$\xi = 1 - \sqrt{1 - 2\alpha_s} = 1 - \sqrt{1 - 2 \times 0.118} = 0.126 < \xi_b = 0.518$$

$$\gamma_s = 1 - 0.5\xi = 1 - 0.5 \times 0.126 = 0.937$$

$$A_s = \frac{M}{f_y \gamma_s h_0} = \frac{410 \times 10^6}{360 \times 0.937 \times 540} \text{mm}^2 = 2251 \text{mm}^2$$

或

$$A_s = \frac{\alpha_1 f_c b'_f x}{f_y} = \frac{\alpha_1 f_c b'_f \xi h_0}{f_y} = \frac{1.0 \times 11.9 \times 1000 \times 0.126 \times 540}{360} \text{mm}^2 = 2249 \text{mm}^2$$

选用 6 $\underline{\Phi}$ 22，$A_s = 2281 \text{mm}^2$。

验算最小配筋率：

$$\rho = \frac{A_s}{bh_0} = \frac{2281}{200 \times 540} = 2.1\% > \rho_{\min} \frac{h}{h_0} = 0.2\% \times \frac{600}{540} = 0.22\%$$

且

$$\rho = 2.1\% > 0.45 \frac{f_t}{f_y} \cdot \frac{h}{h_0} = 0.45 \times \frac{1.27}{360} \times \frac{600}{540} = 0.176\%$$

所以，最小配筋率满足要求。

本 章 小 结

（1）受弯构件的截面尺寸应满足承载力、刚度及裂缝控制的要求，还应满足使用要求、施工要求及经济要求。梁的截面高度与梁的跨度及荷载大小有关，一般情况下梁的跨度、荷载越大，梁的截面高度越高。按刚度条件，梁的截面高度可根据高跨比来估计。板的设计厚度应取构造要求的最小板厚和不需进行变形验算的跨厚比条件确定的板厚二者中较大者。

（2）混凝土保护层厚度是构件最外层钢筋（包括箍筋、构造筋、分布筋）的外表面到最近的混凝土面的垂直距离。混凝土保护层有三个作用，应满足《混凝土结构设计标准》（GB/T 50010—2010）的要求。在适筋梁的受弯过程中，梁的受力过程分为三个阶段：弹性阶段、带裂缝工作阶段、破坏阶段。

（3）根据受弯构件纵向受力钢筋配筋率的不同，梁正截面破坏时有三种破坏形态：适筋破坏、超筋破坏、少筋破坏。在实际应用中为简化计算，采用等效矩形应力图形来代换实际的曲线形应力图形，即假定混凝土的压应力在受压区是均匀分布的。

（4）界限破坏是指纵向受拉钢筋应力 σ_s 达到屈服强度 f_y 的同时，受压区边缘混凝土纤维的压应变 ε_c 也恰好达到混凝土受弯时的极限压应变 ε_{cu} 的破坏形式，界限破坏时的配筋率称为界限配筋率。适筋梁的最大配筋率就是界限破坏时纵向受拉钢筋的界限配筋率。最小配筋率是少筋梁和适筋梁的界限配筋率。

（5）单筋截面梁是只在梁的截面受拉区配置纵向受拉钢筋，在梁的截面受压区只布置固定箍筋所需要的纵向架立钢筋,架立钢筋不计其受力作用。梁中纵向钢筋的净间距应符合一定的构造要求。截面设计可采用计算系数法。截面复核是指已知受弯构件的截面设计弯矩M、混凝土强度等级、钢筋级别、截面尺寸和纵向受拉钢筋截面面积A_s,求正截面受弯承载力M_u是否足够。

（6）在梁截面受压区配置的纵向受压钢筋数量较多,不仅可以起到架立钢筋的作用,而且其受压作用在计算中不能忽略,可以协助混凝土共同承担压力,称为双筋截面。双筋矩形截面破坏时的受力特点与单筋矩形截面类似,区别在于双筋截面受压区配有纵向受压钢筋,因此,可与单筋矩形截面类似建立计算公式。

（7）为简化计算,设计T形截面时应将翼缘宽度限制在一定范围内,称为翼缘计算宽度b_f'。假定在这个宽度范围内压应力均匀分布,而在这个范围以外,认为翼缘不起作用,不考虑它参与受力。T形截面根据其受力后截面受压区高度x的大小、中和轴的位置等,可分为第一类T形截面和第二类T形截面。受压区形状为矩形的第一类T形截面,当仅配置纵向受拉钢筋时,可按截面尺寸为$b_f' \times h$的单筋矩形截面计算配筋。

第4章拓展知识和特别提示

视频：第4章小结讲解

4-1 填空题

1. 适筋梁的加载过程可分为_____、_____、_____三个阶段。

2. 适筋梁的破坏始于_____,它的破坏属于_____；超筋梁的破坏始于_____,它的破坏属于_____。

3. 钢筋混凝土受弯构件正截面破坏有_____、_____、_____三种破坏形态。

4. 在受弯构件承载力计算公式的推导过程中,受压区混凝土应力分布采用等效矩形应力图形,其等效原则是_____不变和_____不变。

5. 为便于浇筑混凝土以保证钢筋周围混凝土的密实性,梁下部纵向钢筋净距应不小于_____和_____。

6. 单筋矩形截面梁混凝土受压区高度的最大值$x_b = $_____,相应的最大配筋率$\rho_{max} = $_____,最大受弯承载力$M_u = $_____。

7. 在钢筋混凝土正截面受弯承载力计算中,防止超筋破坏的条件是_____。

8. 少筋梁与适筋梁分界处的配筋率是_____,超筋梁与适筋梁分界处的配筋率是_____。

9. 双筋矩形截面受弯构件正截面承载力计算公式的适用条件是_____、_____。

10. 若双筋矩形截面混凝土受压区高度$x < 2a_s'$,求纵向受拉钢筋截面面积A_s时,可近似假设x_____。

11. 根据 T 形截面受弯构件混凝土受压区高度 x 的不同,在 T 形截面计算时分为两种类型:当_____时为第一类 T 形截面,当_____时为第二类 T 形截面。

4-2 选择题

1. 钢筋混凝土梁受拉区边缘开始出现裂缝是因为受拉边缘(　　)。
   A. 受拉混凝土的应力达到混凝土的实际抗拉强度
   B. 受拉混凝土达到混凝土的抗拉标准强度
   C. 受拉混凝土达到混凝土的设计强度
   D. 受拉混凝土的应变超过受拉极限拉应变

2. 少筋梁正截面受弯破坏时,破坏弯矩(　　)。
   A. 小于开裂弯矩　　B. 等于开裂弯矩　　C. 大于开裂弯矩

3. 进行双筋矩形截面正截面受弯承载力计算时,纵向受压钢筋设计强度按规定不得超过 400N/mm^2,因为(　　)。
   A. 受压混凝土强度不够
   B. 结构延性
   C. 混凝土受压边缘此时已达到混凝土的极限压应变

4. 在计算钢筋混凝土梁受弯承载力时,截面受压区混凝土的抗压强度设计值是(　　)。
   A. 轴心抗压强度设计值　　　　B. 立方体抗压强度标准值
   C. 局部受压强度设计值　　　　D. 等效弯曲抗压强度设计值

5. 钢筋混凝土适筋梁在即将破坏时,纵向受拉钢筋的应力(　　)。
   A. 尚未进入屈服点　　　　　　B. 刚达到弹性极限
   C. 正处于屈服阶段　　　　　　D. 已进入强化阶段

6. 钢筋混凝土适筋梁破坏时(　　)。
   A. 纵向受拉钢筋屈服和受压区混凝土压坏必须同时发生
   B. 混凝土受压破坏先于纵向受拉钢筋屈服
   C. 纵向受拉钢筋先屈服,然后混凝土压坏
   D. 纵向受拉钢筋屈服,受压混凝土未压坏

7. 截面尺寸一定的单筋矩形截面梁,其正截面可能的最大受弯承载力与(　　)有关。
   A. 纵向受拉钢筋的配筋率　　　B. 横向钢筋的配筋率
   C. 梁的跨度　　　　　　　　　D. 梁的混凝土强度等级

8. 钢筋混凝土梁的混凝土保护层是指(　　)。
   A. 箍筋外表面至梁表面的距离　　B. 主筋外表面至梁表面的距离
   C. 主筋截面形心至梁表面的距离　D. 箍筋截面形心至梁表面的距离

9. 钢筋混凝土矩形截面梁发生超筋破坏时,其相对受压区高度 ξ 为(　　)。
   A. $\xi > \xi_b$　　B. $\xi \leqslant \xi_b$　　C. $\xi \leqslant 0.544$　　D. $\xi \geqslant 2a'_s/h_0$

10. 受弯构件中,纵向受拉钢筋的应力达到屈服强度,同时受压区边缘混凝土也达到极限压应变的破坏称为(　　)。
    A. 少筋破坏　　B. 适筋破坏　　C. 超筋破坏　　D. 界限破坏

11. 梁中配置受压纵筋后(　　)。

A. 只提高正截面受弯承载力
B. 既能提高正截面受弯承载力,又可减少构件混凝土徐变
C. 只能减少构件混凝土徐变
D. 提高正截面受弯承载力,但能加大混凝土徐变

4-3 名词解释

1. 受弯构件	2. 高跨比	3. 架立钢筋	4. 截面有效高度
5. 配筋率	6. 混凝土保护层厚度	7. 正截面破坏	8. 适筋梁
9. 超筋梁	10. 少筋梁	11. 延性破坏	12. 脆性破坏
13. 界限破坏	14. 界限配筋率	15. 平截面假定	16. 等效矩形应力图形
17. 相对受压区高度	18. 界限相对受压区高度	19. 最小配筋率	20. 单筋截面
21. 截面抵抗矩系数	22. 内力臂系数	23. 双筋截面	24. T形截面
25. 腹板	26. 翼缘计算宽度	27. 第一类T形截面	28. 第二类T形截面

4-4 思考题

1. 适筋受弯构件从加载到破坏,截面受力状况可以分成几个阶段?
2. 配筋率对受弯构件的破坏特征有什么影响?
3. 如何将受压区混凝土的应力图形代换成等效矩形应力图形?
4. 在实际工程中为什么应避免采用少筋梁和超筋梁?
5. 适筋梁、超筋梁、少筋梁的破坏特征是什么?
6. 如何防止将受弯构件设计成少筋构件和超筋构件?
7. 计算受弯构件正截面承载力时,为什么要限制最大配筋率和最小配筋率?
8. 画出单筋矩形截面受弯构件正截面受弯承载力计算简图,推导基本计算公式和配筋计算公式,并写出适用条件。
9. 矩形截面受弯构件在什么情况下采用双筋截面?
10. 画出双筋矩形截面受弯构件正截面受弯承载力计算简图,推导基本计算公式和配筋计算公式,并写出适用条件。
11. 保证双筋截面梁中纵向受压钢筋的应力达到其设计强度的条件是什么?
12. 双筋矩形截面正截面受弯承载力计算中为什么要引入 $x \geqslant 2a'_s$ 的适用条件?
13. 对于双筋矩形截面梁,当计算得出 $x < 2a'_s$ 时,应如何求纵向受拉钢筋面积?
14. 如何判别第一类T形截面和第二类T形截面?
15. T形截面梁的翼缘为什么要有翼缘计算宽度 b'_f 的规定? b'_f 应如何确定?
16. 画出第一类T形截面和第二类T形截面受弯构件正截面受弯承载力计算简图,推导基本计算公式和配筋计算公式,并写出适用条件。

4-5 计算题

1. 单筋矩形截面简支梁的截面尺寸为 $b \times h = 250 \text{mm} \times 500 \text{mm}$,混凝土强度等级为C25,纵向钢筋 HRB400,混凝土受压区高度为 $x = 170 \text{mm}$,求其最大弯矩设计值 M。

2. 矩形截面简支梁的截面尺寸为 $b \times h = 300 \text{mm} \times 800 \text{mm}$,混凝土强度等级为C25,纵向钢筋 10⏀25,判断其为超筋、少筋还是适筋梁。

3. 矩形截面梁的截面尺寸为 $b \times h = 250 \text{mm} \times 500 \text{mm}$,环境类别为一类,弯矩设计值为

$M=180$kN·m,混凝土强度等级为C30,采用HRB400钢筋。求所需的纵向受拉钢筋截面面积,选配钢筋直径和根数,并画出梁横截面配筋图。

4. 矩形截面简支梁的计算跨度 $l_0=6$m,梁上作用的均布永久荷载(不包括梁自重)标准值 $g_k=6$kN/m,均布可变荷载标准值 $q_k=12$kN/m,环境类别为一类,结构安全等级为二级,纵向受拉钢筋采用HRB400,混凝土强度等级为C30。要求:确定该梁的截面尺寸 $b\times h$,求梁所需的纵向受拉钢筋截面面积,选择钢筋直径和根数,验算最小配筋率,并绘制梁横截面配筋图。

5. 单筋矩形截面梁的截面尺寸为 $b\times h=250$mm$\times 450$mm,纵向受拉钢筋为 4⊕16,钢筋截面面积为 $A_s=804$mm^2,混凝土强度等级为C40,承受弯矩设计值 $M=120$kN·m,环境类别为一类,验算此梁截面是否安全。

6. 矩形截面简支梁的截面尺寸为 $b\times h=200$mm$\times 450$mm,混凝土强度等级为C25,弯矩设计值为 $M=150$kN·m,$a_s=35$mm,问该梁是否需要配置纵向受压钢筋?

7. 双筋矩形截面梁的截面尺寸为 $b\times h=200$mm$\times 350$mm,混凝土强度等级为C25,纵向受压钢筋为 2⊕14,纵向受拉钢筋为 3⊕22,求该梁能承受的最大弯矩设计值。

8. 矩形截面梁的截面尺寸为 $b\times h=250$mm$\times 500$mm,环境类别为一类,结构安全等级为二级,梁承受的弯矩设计值 $M=400$kN·m,纵向受力钢筋采用HRB400,混凝土强度等级为C40。求梁所需的纵向受力钢筋截面面积,选择钢筋直径和根数,并绘制梁横截面配筋图。

9. 矩形截面梁的截面尺寸为 $b\times h=300$mm$\times 500$mm,混凝土强度等级为C30,钢筋采用HRB400,截面弯矩设计值为 $M=430$kN·m,环境类别为一类,求所需纵向受压和纵向受拉钢筋截面面积。

10. T形截面简支梁的截面尺寸为 $b'_f=420$mm,$b=200$mm,$h'_f=120$mm,$h=500$mm,混凝土强度等级为C25,HRB400钢筋,纵向受拉钢筋截面面积 $A_s=1964$mm^2,判断此梁属于哪一类T形截面。

11. 肋形楼盖次梁的截面尺寸为 $b\times h=300$mm$\times 600$mm,弯矩设计值为 $M=300$kN·m,次梁跨中按T形截面计算,翼缘计算宽度和高度分别为 $b'_f=1000$mm,$h'_f=90$mm,混凝土强度等级为C25,纵向受拉钢筋采用HRB400,环境类别为一类。判断T形截面的类型,求纵向受拉钢筋截面面积,选配钢筋直径和根数,验算最小配筋率,绘制梁跨中截面配筋图。

12. T形截面梁的截面尺寸为 $b=250$mm,$h=500$mm,$b'_f=600$mm,$h'_f=100$mm,环境类别为一类,承受弯矩设计值 $M=500$kN·m,混凝土强度等级为C25,纵向受拉钢筋为 8⊕20,复核此T形截面梁是否安全。

第 5 章

受弯构件的斜截面承载力

Bearing capacity of inclined section of flexural members

教学目标：

1. 了解无腹筋梁和有腹筋梁的概念，理解斜截面受剪破坏形态及防止对策；
2. 掌握斜截面受剪承载力的计算方法及计算公式的适用条件、梁中箍筋和弯起钢筋的计算和设置；
3. 理解抵抗弯矩图的绘制，熟悉梁内纵向钢筋弯起、截断的构造要求，会确定钢筋弯起和截断的位置；
4. 熟悉钢筋锚固要求，会计算梁中钢筋锚固长度；熟悉钢筋连接的有关规定；熟悉梁内各种钢筋的构造要求。

导读：

钢筋混凝土构件的截面上除了作用弯矩（梁、板）或弯矩和轴力（柱、剪力墙）外，一般还作用有剪力。在受弯构件主要承受弯矩的区段内常产生竖向裂缝，可能导致沿竖向裂缝发生正截面受弯破坏。在剪力和弯矩共同作用的剪弯区段内常产生斜裂缝，可能导致沿斜裂缝发生无明显预兆的斜截面破坏，如图 5-1 所示。所以对受弯构件一般先进行正截面受弯承载力计算，确定截面尺寸和纵向受力钢筋后再进行斜截面承载力计算，斜截面承载力包括斜截面受剪承载力和斜截面受弯承载力两方面。为了防止梁沿斜裂缝破坏，应使梁具有一个合理的截面尺寸，按计算或按构造要求配置必要的箍筋或弯起钢筋，并符合一定的构造要求。箍筋或弯起钢筋统称为腹筋，与纵向受力钢筋、纵向构造钢筋、架立钢筋等构成梁的钢筋骨架。

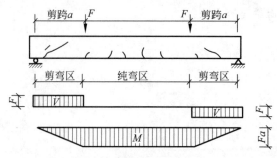

图 5-1　简支梁对称加载受力示意图

Figure 5-1　Schematic diagram of symmetrical loading force simply supported beam

引例：

第 5 章引例

核心词汇：

斜截面	inclined section	开口箍筋	open stirrup
斜裂缝	diagonal crack	封闭箍筋	closed stirrup
腹筋	web reinforcement	单肢箍筋	single-leg stirrup
有腹筋梁	beam with web reinforcement	双肢箍筋	two-leg stirrup
无腹筋梁	beam without web reinforcement	多肢箍筋	multiple-leg stirrup
弯起钢筋	bent bar	充分利用截面	fully utilized section
剪跨比	shear span ratio	不需要截面	unnecessary section
斜压破坏	diagonal compression failure	抵抗弯矩	resisting moment
剪压破坏	shear compression failure	抵抗弯矩图	moment resistance diagram
斜拉破坏	diagonal tension failure	理论切断点	theoretical cutoff point
配箍率	stirrup ratio	实际切断点	actual cutoff point

5.1 概述
Summary

为了保证受弯构件的斜截面受剪承载力，应首先根据计算截面的剪力设计值复核截面尺寸，确定适宜的混凝土强度等级，计算承受剪力所需的箍筋（箍筋强度级别、直径、间距等）。当梁中的剪力比较大时，也可增设弯起钢筋，与箍筋一同抗剪。弯起钢筋一般由梁内部分纵向受拉钢筋在某个位置弯起而成，其示意图如图 5-2 所示。根据梁中是否配置了腹筋，可将梁分为无腹筋梁和有腹筋梁。受弯构件斜截面受剪性能和破坏机理比正

英文翻译 5-1

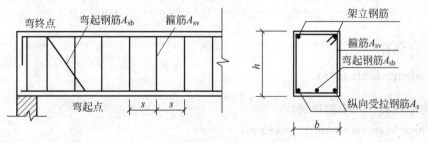

图 5-2 箍筋和弯起钢筋示意图

Figure 5-2 Schematic diagram of stirrup and bent bar

截面的情况要复杂,斜截面受剪承载力的计算方法主要是在试验研究的基础上建立的。为了保证受弯构件的斜截面受弯承载力,应使梁内纵向受力钢筋的弯起、截断、锚固等满足相应的构造要求,一般不必进行复杂的计算。

5.2 斜裂缝、剪跨比及斜截面受剪破坏形态
Diagonal crack, shear span ratio and shear destruction form of inclined section

5.2.1 腹剪斜裂缝和弯剪斜裂缝
Web-shear diagonal crack and flexural-shear diagonal crack

钢筋混凝土梁在弯矩、剪力共同作用的剪弯区段,产生的斜裂缝主要有腹剪斜裂缝和弯剪斜裂缝,如图 5-3 所示。在 I 形薄腹梁中,腹板比较薄,该处正应力小,剪应力比较大,主拉应力大致在 45°方向。当荷载增大,混凝土的拉应变达到极限拉应变时,斜裂缝首先在梁腹部中和轴附近出现,随后向梁顶和梁底部斜向发展,中间宽两头细,呈枣核形,这种斜裂缝称为腹剪斜裂缝,如图 5-3(a)所示。

在梁的纯弯段,剪应力为零,主拉应力的作用方向与梁纵轴的夹角为零,是水平的,最大主拉应力发生在截面的下边缘,当其超过混凝土的抗拉强度时,将出现较短的竖向裂缝。随着荷载的增加,竖向裂缝斜向发展,向集中荷载作用点延伸。在弯矩和剪力共同作用的剪弯区段,梁腹内主拉应力的方向是倾斜的,当主拉应力超过混凝土的抗拉强度时,将形成下宽上细的弯剪斜裂缝,如图 5-3(b)所示,这种斜裂缝是最常见的。

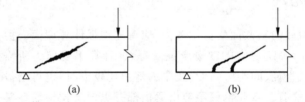

图 5-3 梁的斜裂缝

(a) 腹剪斜裂缝;(b) 弯剪斜裂缝

Figure 5-3 Diagonal cracks in beams

(a) Web-shear diagonal crack;(b) Flexural-shear diagonal crack

5.2.2 剪跨比
Shear span ratio

1. 集中荷载作用下的剪跨比

Shear span ratio under concentrated load

对集中荷载作用下的简支梁,剪跨 a 指集中荷载所在截面到临近支座截面或节点边缘的距离,如图 5-4 所示。剪跨比 λ 指剪跨 a 与截面有效高度 h_0 的比值,表达式为

For simply supported beam under concentrated load, the shear span *a* refers to the distance from the section where concentrated load is located to the edge of the adjacent support section or joint edge, as shown in Fig. 5-4. The shear span ratio λ refers to the ratio of the shear span *a* to the effective depth of section h_0, and the expression is

图 5-4 梁的剪跨

Figure 5-4 Shear span of beam

$$\lambda = \frac{a}{h_0} \tag{5-1}$$

式中, a 为剪跨, mm; h_0 为截面有效高度, mm。

Where, a is the shear span, mm; h_0 is the effective depth of section, mm.

式(5-1)只能用于计算集中荷载作用下集中荷载所在截面的剪跨比, 不能计算其他复杂荷载分布下的剪跨比。

Eq. (5-1) can only be used to calculate the shear span ratio of the section where the concentrated load is located, and the shear span ratio under other complex load distribution cannot be calculated.

2. 广义剪跨比

Generalized shear span ratio

试验研究表明, 任意荷载作用下梁的受剪性能与计算截面上的弯矩 M 与剪力 V 的相对大小有很大关系。弯矩 M 与剪力 V 分别使梁截面上产生弯曲应力 σ 和剪应力 τ, 所以, 梁的受剪性能实际上与弯曲应力 σ 与剪应力 τ 的比值 σ/τ 有关, 研究表明 σ/τ 实际上仅与 $\frac{M}{Vh_0}$ 有关。

英文翻译 5-3

任意荷载作用下任意截面的剪跨比称为广义剪跨比, 定义为计算截面上的弯矩设计值 M 与剪力设计值 V 及截面有效高度 h_0 乘积的比值, 表达式为

$$\lambda = \frac{M}{Vh_0} \tag{5-2}$$

当荷载为集中荷载时, 式(5-2)可根据式(5-3)简化为式(5-1), 此时的剪跨比与广义剪跨比相同:

$$\lambda = \frac{M}{Vh_0} = \frac{Va}{Vh_0} = \frac{a}{h_0} \tag{5-3}$$

试验研究表明, 剪跨比是一个能反映梁斜截面受剪承载力变化规律的重要参数, 对梁的斜截面受剪破坏形态和受剪承载力影响很大。剪跨比反映了截面上正应力和剪应力的相对比值, 在一定程度上反映了截面上弯矩和剪力的相对比值。在梁截面尺寸、混凝土强度等级、配箍率和纵筋配筋率基本相同的情况下, 随着剪跨比的增大, 梁的受剪承载力减小, 当剪跨比 $\lambda > 3$ 时, 对梁斜截面受剪承载力不再有明显影响, 如图 5-5 所示。所以剪跨比 $\lambda > 3$ 时, 一般取 $\lambda = 3$。

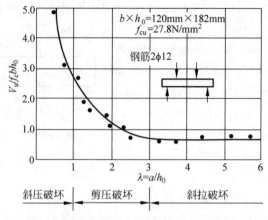

图 5-5 梁受剪承载力 V_u 与剪跨比 λ 的关系

Figure 5-5　Relationship between shear capacity V_u and shear span ratio λ of beams

5.2.3　无腹筋梁的斜截面受剪破坏形态
Shear destruction form of inclined section of beams without web reinforcement

在实际工程中，为了抗剪钢筋混凝土梁中常配有腹筋，但为了研究梁的斜裂缝发展过程和斜截面破坏形态，需要先研究相对简单的无腹筋梁的受剪性能。无腹筋梁指没有配置箍筋和弯起钢筋的梁。

1. 影响梁受剪性能的因素

Factors affecting shear performance of beams

影响梁受剪性能的因素有梁截面尺寸、混凝土和钢筋的材料强度、腹筋数量、剪跨比等。截面尺寸、斜裂缝处的骨料咬合力对无腹筋梁的斜截面受剪承载力影响较大，尺寸大的构件，破坏时的平均剪应力比尺寸小的构件低。T形截面梁的翼缘大小对受剪承载力也有影响，适当增加翼缘宽度可提高受剪承载力，但翼缘过大，增大作用趋于平缓。

英文翻译 5-4

视频 5-1

2. 无腹筋梁的斜截面受剪破坏形态

Shear destruction form of inclined section of beams without web reinforcement

无腹筋梁的斜截面受剪破坏形态及承载力与剪弯段中弯矩和剪力的组合有关。试验表明，无腹筋梁的斜截面破坏形态与剪跨比 λ 有重要关系，由于正应力和剪应力决定了主应力的大小和方向，因此就会间接影响梁斜截面破坏形态和承载力。根据剪跨比的不同，无腹筋梁的斜截面破坏有斜压破坏、剪压破坏、斜拉破坏三种主要的破坏形态，如图5-6所示。

1) 无腹筋梁的斜压破坏

1) Diagonal compression failure of beams without web reinforcement

对无腹筋梁，当剪跨比 $\lambda<1$ 时，会发生斜压破坏，如图5-6(a)所示。破坏时，集中荷载与邻近支座间的混凝土被腹剪斜裂缝分割成若干个斜向受压棱柱体而压酥破坏。斜压破坏的特点是抗剪承载力主要取决于混凝土的抗压强度 f_c，破坏性质类似于正截面中的超筋破坏，属于脆性破坏，设计时应避免。

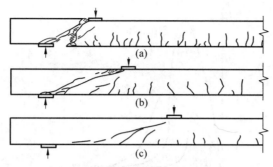

图 5-6 梁斜截面破坏形态
(a) 斜压破坏；(b) 剪压破坏；(c) 斜拉破坏

Figure 5-6　Destruction form of inclined section of beam
(a) Diagonal compression failure；(b) Shear compression failure；(c) Diagonal tension failure

2）无腹筋梁的剪压破坏

2）Shear compression failure of beams without web reinforcement

集中荷载作用下剪跨比 $1 \leqslant \lambda \leqslant 3$ 或均布荷载作用下跨高比 $3 \leqslant l_0/h_0 \leqslant 9$ 时发生剪压破坏，剪压破坏是最常见的斜截面破坏形态。剪压破坏的特点是弯剪斜裂缝随荷载增加，出现一条延伸较长、开展相对较宽的主要斜裂缝，即临界斜裂缝。临界斜裂缝向荷载作用点延伸，使剪压区混凝土达到复合应力状态下的极限强度而破坏。剪压破坏时，梁被临界斜裂缝分开的两部分有较明显的相对错动，类似于受弯破坏的压碎现象，如图 5-6(b)所示，剪压区裂缝内有混凝土碎屑。剪压破坏的抗剪承载力主要取决于混凝土在复合应力下的抗压强度。

3）无腹筋梁的斜拉破坏

3）Diagonal tension failure of beams without web reinforcement

集中荷载作用下剪跨比 $\lambda > 3$ 或均布荷载作用下跨高比 $l_0/h_0 > 9$ 时发生斜拉破坏。斜拉破坏的特点是梁下边缘的弯剪斜裂缝一出现便很快发展，伸展到梁顶的集中荷载作用点处形成临界斜裂缝，使混凝土截面沿斜向裂通，梁被分成两部分丧失承载力，如图 5-6(c)所示。

斜拉破坏的抗剪承载力取决于混凝土的抗拉强度，破坏性质类似于正截面破坏中的少筋破坏，脆性破坏性质最严重，设计中应避免。

3. 三种破坏形态的梁斜截面受剪承载力比较

Comparison of shear capacity of inclined section of beams with three destruction forms

各种斜截面破坏形态的梁剪力-跨中挠度曲线如图 5-7 所示，不同破坏形态的梁斜截面承载力各不相同：斜压破坏＞剪压破坏＞斜拉破坏。斜压破坏时，受剪承载力大而变形很小，破坏突然，曲线形状陡峭；剪压破坏时，受剪承载力较小，变形稍大，曲线形状较平缓；斜拉破坏时，受剪承载力最小，破坏很突然。在达到峰值荷载时，三种破坏形态的跨中挠度都不大，破坏后荷载都迅速下降，表明它们都属于脆性破坏。其中斜拉破坏的脆性性质最突出，斜压破坏次之，剪压破坏稍好。

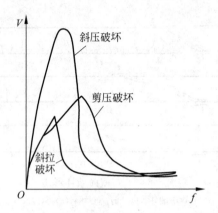

图 5-7 梁斜截面破坏的剪力-跨中挠度曲线

Figure 5-7　Shear-midspan deflection curve of inclined section failure of beam

 特别提示 5-1

5.2.4　有腹筋梁的斜截面受剪破坏形态
Shear destruction form of inclined section of beams with web reinforcement

有腹筋梁是既有纵筋又有箍筋或弯起钢筋的梁。弯起钢筋几乎与斜裂缝垂直,传力直接,但一般由部分纵筋弯起,根数少,受力不均匀,目前在建筑工程中已较少使用;箍筋不与斜裂缝正交,但分布均匀,因此使用广泛。

1. 腹筋的作用

Effect of web reinforcement

梁开裂前,有腹筋梁的受力性能与无腹筋梁相似,如图 5-8 所示,腹筋中的应力很小,当梁中拉应力 $\sigma_{tp,max}$ 大于混凝土轴心抗拉强度设计值 f_t 时,梁的剪弯段开裂,出现斜裂缝;开裂后,腹筋的应力增大,限制了斜裂缝的发展,提高了抗剪承载力。

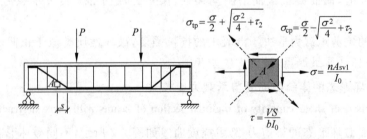

图 5-8　有腹筋梁中的应力

Figure 5-8　Stress of the beam with web reinforcement

腹筋对有腹筋梁受剪承载力的影响表现为:有腹筋梁出现斜裂缝后,腹筋不仅直接承受相当部分的剪力,还能有效抑制斜裂缝的开展和延伸,对提高剪压区混凝土的抗剪能力和纵向钢筋的销栓作用、防止梁沿斜裂缝发生脆性破坏有着积极的影响。腹筋的作用如图 5-9 所示。

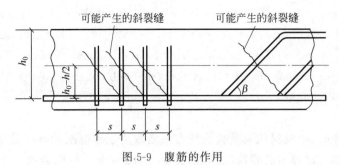

图 5-9 腹筋的作用
Figure 5-9　Effect of web reinforcement

试验表明,在配箍适当的范围内,梁的受剪承载力随配箍量的增多、箍筋强度的提高而有较大幅度的增长。所以,除跨度很小的梁,一般梁中都配有箍筋。

梁内箍筋主要承受由弯矩和剪力在梁内引起的主拉应力,同时还可以固定纵向钢筋的位置,并和其他钢筋一起形成空间骨架。箍筋的数量应根据计算及构造要求确定。

特别提示 5-2

2. 有腹筋梁的斜截面受剪破坏形态
Shear destruction form of inclined section of beams with web reinforcement

试验研究表明,有腹筋梁出现斜裂缝后,斜截面破坏形态与剪跨比、腹筋配置数量有重要关系。根据剪跨比和腹筋配置数量的不同,与无腹筋梁相似,有腹筋梁的斜截面破坏形态有斜压破坏、剪压破坏、斜拉破坏三种主要破坏形态。

The experimental research shows that the destruction form of inclined section has an important relationship with the shear span ratio and the number of web reinforcement after diagonal cracks appear in the beams with web reinforcement. According to the difference of shear span ratio and the number of web reinforcement, similar to the beam without web reinforcement, the failure mode of inclined section of the beam with web reinforcement has three main failure modes: diagonal compression failure, shear compression failure and diagonal tension failure.

1) 有腹筋梁的斜压破坏

1) Diagonal compression failure of beams with web reinforcement

当梁的剪跨比较小($\lambda<1$),或剪跨比适当($1<\lambda<3$),但梁截面尺寸过小而腹筋配置数量过多时,常发生斜压破坏。首先在梁腹部出现若干条大致互相平行的斜裂缝,随着荷载的增加,斜裂缝两端分别向支座和荷载作用点延伸,梁腹部被这些斜裂缝分割成若干个斜向受压的棱柱体,在腹筋应力未达到屈服强度前,斜裂缝间的梁腹混凝土即受压达到混凝土抗压强度,发生斜压破坏。

When the shear span ratio of the beam is relatively small ($\lambda<1$), or the shear span ratio is appropriate ($1<\lambda<3$), but cross-sectional dimensions of beams are too small and the number of web reinforcement configurations is too large, diagonal compression failure often occurs. First, several diagonal cracks roughly parallel to each other appear in the web of the beam. With the increase of the load, the two ends of the diagonal crack extend to the support and the load action point respectively, and the web of the beam is divided into several obliquely compressive prisms by these diagonal cracks. Before the web reinforcement stress reaches the yield strength, the web concrete of the beam between the diagonal cracks is compressed to the concrete compressive strength, and the diagonal compression failure occurs.

发生斜压破坏时,梁的斜截面受剪承载力主要取决于混凝土的抗压强度,破坏性质类似于正截面受弯承载力破坏中的超筋破坏,具有脆性性质,设计时应避免。在薄腹梁中,即使剪跨比较大,也会发生斜压破坏。

When diagonal compression failure occurs, the shear capacity of inclined section of the beam mainly depends on the compressive strength of concrete, and the failure property is similar to the overreinforced failure in flexural capacity failure of normal section. Failure is brittle and should be avoided in design. In a thin web beam, diagonal compression failure will occur even if the shear span ratio is relatively large.

2) 有腹筋梁的剪压破坏

2) Shear compression failure of beams with web reinforcement

当梁的剪跨比中等($1 \leqslant \lambda \leqslant 3$),且腹筋配置数量适当,或梁的剪跨比较大($\lambda > 3$),但腹筋配置数量适当时,梁的剪弯区段下边缘首先出现垂直裂缝。随着荷载的增加,这些垂直裂缝大致沿着主拉应力轨迹向集中荷载作用点延伸,剪压区的高度减小,其中一条裂缝发展成临界斜裂缝,与临界斜裂缝相交的腹筋应力增大,限制了斜裂缝的延伸开展;随荷载增加,腹筋应力先达到屈服强度,其限制斜裂缝开展的作用消失,斜裂缝宽度增大,斜裂缝上端剩余截面的高度减小,最后剪压区混凝土在剪应力、正应力的剪压复合应力作用下达到混凝土复合受力强度而发生剪压破坏。

When the shear span ratio of the beam is medium($1 \leqslant \lambda \leqslant 3$), and the number of web reinforcement is appropriate or when the shear span ratio of the beam is large($\lambda > 3$), but the number of web reinforcement is appropriate, the lower edge of the shear-bending section of the beam first appears vertical cracks, as the load increases, these vertical cracks extend roughly along the principle tensile stress trajectory towards the action point of concentrated load, the height of shear compression zone decreases, and one of the cracks develops into critical diagonal crack, the stress in the web reinforcement intersecting with the critical diagonal crack increases, restricting the extension and development of the diagonal crack; with the increase of the load, the stress in the web reinforcement first reaches the yield strength, the effect of restricting the development of diagonal cracks disappears, the width of diagonal cracks increases, the depth of remaining section at the upper end of the diagonal crack decreases, and finally, the concrete in shear compression zone reaches the composite strength of concrete under the action of shear compression composite stress of the shear stress and the normal stress, and the shear compression failure occurs.

剪压破坏是斜截面受剪破坏中最常见的一种破坏形式,梁被临界斜裂缝分开的两部分有较明显的相对错动和类似于受弯破坏的压碎现象,剪压区裂缝内有混凝土碎屑,破坏性质类似于正截面受弯承载力破坏中的适筋破坏。

Shear compression failure is the most common failure mode in shear failure of inclined section. The two parts of beam separated by critical diagonal crack have relatively obvious relative dislocation and crushing phenomenon similar to flexural failure. There are concrete fragments in the cracks in the shear compression zone, and the failure property is similar to the underreinforced failure in the failure of normal section flexural capacity.

3) 有腹筋梁的斜拉破坏

3) Diagonal tension failure of beams with web reinforcement

当梁的剪跨比过大($\lambda > 3$),且腹筋配置数量过少时,梁下边缘的弯剪斜裂缝一出现,迅速延伸到梁顶的集中荷载作用点,很快形成临界斜裂缝,因为腹筋配置数量过少,腹筋应力很快达到屈服强度,变形很大,腹筋对斜裂缝开展的限制作用不再存在,梁斜向被拉裂成两部分而突然破坏。因为是混凝土在正应力和剪应力共同作用下发生的主拉应力破坏,所以称为斜拉破坏。

When the shear span ratio of beam is too large ($\lambda > 3$) and the number of web reinforcement is too small, as soon as the flexural-shear diagonal crack at the lower edge of the beam appears, it quickly extends to the action point of concentrated load at the top of the beam, forming critical diagonal crack quickly, because if the number of web reinforcement is too small, the stress of web reinforcement will quickly reach the yield strength, and the deformation will be very large. The limiting effect of web reinforcement on diagonal crack development no longer exists, and the beam broke suddenly when it was pulled into two parts. The principle tensile stress failure occurs under the combined action of normal stress and shear stress, so it is called diagonal tension failure.

斜拉破坏的斜截面受剪承载力主要取决于混凝土的抗拉强度,破坏性质类似于正截面受弯承载力破坏中的少筋破坏,破坏过程迅速,破坏前梁变形很小,具有明显的脆性性质,设计中应避免。

The shear capacity of inclined section of diagonal tension failure mainly depends on the tensile strength of concrete. The failure property is similar to the lightly reinforced failure of normal section flexural capacity. The failure process is rapid, and the deformation of beam before failure is very small. It has obvious brittle properties and should be avoided in design.

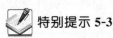 特别提示 5-3

5.3 斜截面受剪承载力的计算
Calculation of shear capacity of inclined section

5.3.1 斜截面受剪承载力计算原则
Calculation principle of shear capacity of inclined section

对于梁的三种斜截面破坏形态,在工程设计时都应设法避免。斜压破坏和斜拉破坏可采取一定的构造措施避免。斜压破坏是因为梁截面尺寸过小或配箍率太高而发生的,可以通过控制梁截面尺寸不能过小、控制最大配箍率来避免斜压破坏;斜拉破坏是由于腹筋数量过少而发生的,可以通过配置足够数量的箍筋、保证适当的箍筋间距、控制最小配箍率来避免斜拉破坏;剪压破坏因其承载力变化幅度较大,必须给出受剪承载力计算公式,通过计算,使构件满足一定的斜截面受剪承载力,以确定所需配置的箍筋及弯起钢筋。

For three types of destruction forms of inclined section of beams, all should be avoided in the engineering design. The diagonal compression failure and the diagonal tension failure can be avoided by taking certain structural measures. The diagonal compression failure occurs because the cross-sectional dimensions of the beam is too small or the stirrup ratio is too high. The diagonal compression failure can be avoided by controlling the cross-sectional dimensions of the beam not too small and the maximum stirrup ratio. The diagonal tension failure occurs because the number of web reinforcement is too small, and can be avoided by configuring a sufficient number of stirrups, ensuring proper stirrup spacing and controlling the minimum stirrup ratio. Due to the large amplitude of variation in bearing capacity of shear compression failure, the formula for calculating the shear capacity must be given. By calculation the member can satisfy the certain shear capacity to determine the required configuration of stirrups and bent bars.

《混凝土结构设计标准》(GB/T 50010—2010)的斜截面受剪承载力计算公式是根据剪压破坏的受力特征建立的,采用的是理论与试验相结合的方法。假定配有箍筋和弯起钢筋的简支梁发生斜截面剪压破坏时,取出被破坏斜截面所分割的一段梁作为隔离体,如图5-10所示,隔离体上作用的力有剪压区混凝土的剪力和压力、箍筋和弯起钢筋的拉力、纵筋的拉力等。假定梁发生剪压破坏时,与斜裂缝相交的箍筋和弯起钢筋的拉应力都达到其屈服强度,但要考虑拉应力可能不均匀,特别是靠近剪压区的箍筋有可能达不到屈服强度。

The formula for calculating the bearing capacity of inclined section in the "Standard for design of concrete structures" (GB/T 50010—2010) is established according to the stress characteristics of shear compression failure, and adopts the method of combining theory and experiment. It is assumed that when the simply supported beam with stirrup and bent bar undergoes shear compression failure of inclined section, a segment of the beam divided by the damaged inclined section is taken out as an isolation body, as shown in Fig. 5-10, the forces acting on the isolation body include the shear force and compressive force of concrete in the shear compression zone, the tensile force of stirrup and bent bar, and the tensile force of longitudinal bar. It is assumed that when the beam undergoes shear compression failure, the tensile stress of the stirrups and the bent bars intersecting with the diagonal cracks reach their yield strength, but it should be considered that the tensile stress may not be uniform, especially the stirrups near the shear compression zone may not reach their yield strength.

为了简化计算并便于应用,《混凝土结构设计标准》(GB/T 50010—2010)考虑了一些主要因素,忽略了次要因素或将其归于其他因素中,采用半理论半经验的方法建立受剪承载力计算公式。梁发生剪压破坏时,斜截面受剪承载力设计值 V_u 由三部分组成,根据图5-10所

示隔离体竖向力的平衡条件,可列出斜截面受剪承载力计算公式如下:

In order to simplify calculation and facilitate application, "Standard for design of concrete structures"(GB/T 50010—2010) considers some main factors, ignores the secondary factors or belongs them to other factors, and adopts the semi-theoretical and semi-empirical method to establish the calculation formula of shear capacity. When the beam undergoes shear compression failure, the design value V_u of shear capacity of inclined section is composed of three parts. According to the equilibrium condition of the vertical force of the isolation body shown in Fig. 5-10, the calculation formula of the shear capacity of the inclined section can be listed as follows:

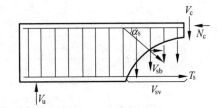

图 5-10 梁斜截面受剪承载力的组成

Figure 5-10 Composition of shear capacity of inclined section of beam

$$V_u = V_c + V_{sv} + V_{sb} = V_{cs} + V_{sb} \tag{5-4}$$

式中,V_u 为梁斜截面受剪承载力设计值,kN;V_c 为剪压区混凝土承担的剪力,kN;V_{sv} 为与斜裂缝相交的箍筋承担的剪力,kN;V_{sb} 为与斜裂缝相交的弯起钢筋承担的拉力的竖向分力,kN;V_{cs} 为梁斜截面上混凝土和箍筋的受剪承载力设计值,kN。

Where, V_u is the design value of the shear capacity of the inclined section of the beam, kN; V_c is the shear force borne by the concrete in the shear compression zone, kN; V_{sv} is the shear force borne by the stirrups intersecting with the diagonal cracks, kN; V_{sb} is the vertical component of the tensile force borne by the bent bar intersecting with the diagonal cracks, kN; V_{cs} is the design value of the shear capacity of concrete and stirrups on the inclined section of the beam, kN.

5.3.2 无腹筋梁混凝土剪压区受剪承载力的试验结果及计算公式
Experimental results and calculation formula of shear capacity of concrete shear compression zone of beams without web reinforcement

试验研究表明,梁中配置箍筋后,其对受剪承载力影响规律的研究比较困难,不像无腹筋梁的试验结果那样明确,所以,《混凝土结构设计标准》(GB/T 50010—2010)提出的受弯构件斜截面受剪承载力计算公式主要是以无腹筋梁的试验结果为基础,分为两种情况:情况 1 是大量无腹筋简支浅梁、短梁、深梁和连续浅梁的试验结果,以支座处的剪力值 V_u 作为混凝土剪压区的承载力进行分析,如图 5-11(a)所示,图中横坐标 l_0/h 中的 l_0 表示梁的计算跨度,h 表示梁截面高度;情况 2 是大量集中荷载作用下的独立浅梁、短梁、深梁的试验结果,如图 5-11(b)所示,图中横坐标 λ 为剪跨比。

图 5-11 中试验结果的点很分散,从安全角度考虑,无腹筋梁受剪承载力取图中黑线所示的下包线,即取试验结果的偏小值。

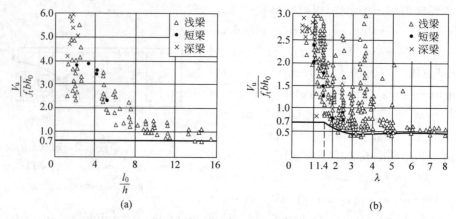

图 5-11 无腹筋梁混凝土剪压区受剪承载力的试验结果与计算结果对比
(a) 均布荷载作用下；(b) 集中荷载作用下

Figure 5-11 Comparison between the experimental results and the calculated results of shear capacity of concrete shear compression zone of beams without web reinforcement
(a) Under uniform load；(b) Under concentrated load

均布荷载作用下的梁：

$$V_c = 0.7 f_t b h_0 \tag{5-5}$$

集中荷载作用下的独立梁：

$$V_c = \frac{1.75}{\lambda + 1} f_t b h_0 \tag{5-6}$$

5.3.3 板类受弯构件的斜截面受剪承载力计算公式
Calculation formula for shear capacity of inclined section of flexural members of plate type

普通厚度的板跨高比通常比较大，且大多承受均布荷载，相对于正截面承载力，板的斜截面承载力一般是足够的。但是高层建筑、水利工程、港口工程中的某些基础底板、转换层板厚达几米，由于板类构件难以配置箍筋，因此截面尺寸是影响受剪承载力的重要因素。随着板厚的增加，斜裂缝宽度也会增加，因此，计算厚板的受剪承载力时，应考虑尺寸效应的影响。

《混凝土结构设计标准》(GB/T 50010—2010)6.3.3 条规定，不配置箍筋和弯起钢筋的一般板类受弯构件的斜截面受剪承载力为

$$V \leqslant V_u = 0.7 \beta_h f_t b h_0 \tag{5-7a}$$

$$\beta_h = \left(\frac{800}{h_0}\right)^{1/4} \tag{5-7b}$$

式中，V 为一般板类受弯构件的剪力设计值，kN；V_u 为一般板类受弯构件的受剪承载力设计值，kN；β_h 为截面高度影响系数，当 $h_0 < 800$ mm 时，取 800 mm，当 $h_0 > 2000$ mm 时，取 2000 mm；f_t 为混凝土轴心抗拉强度设计值，N/mm²。

5.3.4 有腹筋梁斜截面受剪承载力计算公式
Calculation formula for shear capacity of inclined section of beams with web reinforcement

梁内腹筋的配置方法有仅配置箍筋不设弯起钢筋和既配置箍筋也配置弯起钢筋两种情况。

There are two methods to configure web reinforcement in the beam: one is to configure only stirrups without bent bars and the other is to configure both stirrups and bent bars.

1. 仅配置箍筋,不设弯起钢筋梁的斜截面受剪承载力
Shear capacity of inclined section of beams with stirrups only and without bent bars

当梁仅配置箍筋,无弯起钢筋而发生剪压破坏时,由式(5-4)得混凝土和箍筋的斜截面受剪承载力为

When the beam is only equipped with stirrups without bent bars and shear compression failure occurs, from Eq. (5-4) the shear capacity of inclined section of concrete and stirrups is:

$$V_u = V_c + V_{sv} = V_{cs} \tag{5-8}$$

式中,V_u 为斜截面受剪承载力设计值,kN;V_c 为剪压区混凝土承担的剪力,kN;V_{sv} 为与斜裂缝相交的箍筋承担的剪力,kN;V_{cs} 为斜截面上混凝土和箍筋的受剪承载力设计值,kN。

Where, V_u is the design value of shear capacity of inclined section, kN; V_c is the shear force borne by concrete in shear compression zone, kN; V_{sv} is the shear force borne by the stirrups intersecting with the diagonal cracks, kN; V_{cs} is the design value of shear capacity of concrete and stirrups on inclined section, kN.

由于影响梁斜截面受剪承载力的因素很多,因此受剪承载力计算是一个极其复杂的问题。国内外学者进行了大量试验和理论研究,但至今仍未能提出一个能被普遍认可、适用于各种情况的计算理论和破坏模式。《混凝土结构设计标准》(GB/T 50010—2010)采用的是大量试验数据统计分析得出的半经验、半理论实用受剪承载力计算公式。

Because there are many factors affecting the shear capacity of inclined section of the beam, the calculation of shear capacity is an extremely complex problem. Domestic and foreign scholars have carried out a lot of experimental and theoretical research, but so far they have not been able to propose a computational theory and failure mode that can be generally recognized and applicable to various situations. "Standard for design of concrete structures" (GB/T 50010—2010) adopts semi-empirical and semi-theoretical practical shear capacity calculation formula obtained by statistical analysis of a large number of experimental data.

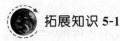
拓展知识 5-1

试验结果表明,$V_{cs}/(bh_0)$ 与混凝土轴心抗拉强度 f_t 和配箍强度 $\rho_{sv} f_{yv}$ 之间为线性

关系：

The experimental results show that $V_{cs}/(bh_0)$ has a linear relationship with the axial tensile strength of concrete f_t and the stirrup strength $\rho_{sv}f_{yv}$:

$$\frac{V_{cs}}{bh_0} = \alpha_{cv} f_t + \alpha_{sv}\rho_{sv} f_{yv} \tag{5-9}$$

式中，V_{cs} 为构件斜截面上混凝土和箍筋的受剪承载力设计值，kN；α_{cv} 为斜截面上受剪承载力系数；α_{sv} 为与荷载形式、截面形状等因素有关的系数；ρ_{sv} 为箍筋的配筋率；f_{yv} 为箍筋的抗拉强度设计值，N/mm²。

Where, V_{cs} is the design value of the shear capacity of concrete and stirrups on the inclined section of the member, kN; α_{cv} is the shear capacity coefficient on the inclined section; α_{sv} is the coefficient related to the load form, cross-sectional shape and other factors; ρ_{sv} is the stirrup ratio; f_{yv} is the design value of tensile strength of stirrups, N/mm².

式(5-9)两边同除以混凝土轴心抗拉强度设计值 f_t 得

Dividing both sides of Eq. (5-9) by the design value of axial tensile strength of concrete f_t, one obtains

$$\frac{V_{cs}}{f_t bh_0} = \alpha_{cv} + \alpha_{sv}\frac{\rho_{sv} f_{yv}}{f_t} \tag{5-10}$$

将配箍率 $\rho_{sv} = A_{sv}/bs$ 代入式(5-10)得

Substituting $\rho_{sv} = A_{sv}/bs$ into Eq. (5-10), one obtains

$$V_{cs} = \alpha_{cv} f_t bh_0 + \alpha_{sv} f_{yv}\frac{A_{sv}}{s}h_0 \tag{5-11}$$

式(5-11)中，V_{cs} 由两项组成，第一项是混凝土剪压区承担的剪力，第二项中的大部分是箍筋承担的剪力，小部分是混凝土承担的剪力。配置箍筋后能抑制斜裂缝的开展，使混凝土剪压区高度增加，提高混凝土剪压区的受剪承载力，但很难确定提高了多少，也没有必要把它从第二项中分离出。

In Eq. (5-11), V_{cs} consists of two terms, the first term is the shear force borne by the concrete shear compression zone, and most of the second term is the shear force borne by the stirrups, but a small part is the shear force borne by the concrete. The configuration of stirrups can inhibit the development of diagonal cracks, increase the depth of concrete shear compression zone, and improve the shear capacity of the concrete shear compression zone, but by how much it has been increased makes it difficult and unnecessary to separate that from the second term.

《混凝土结构设计标准》(GB/T 50010—2010)分两种情况给出了斜截面受剪承载力计算公式。

The calculation formula of shear capacity of inclined section is given in two cases in "Standard for design of concrete structures"(GB/T 50010—2010).

1) 一般受弯构件的斜截面受剪承载力设计值

1) Design value of shear capacity of inclined section of general flexural members

当仅配置箍筋时，对矩形、T形和I形截面的一般受弯构件，将 $\alpha_{cv} = 0.7, \alpha_{sv} = 1$ 代入

式(5-11)中,得斜截面受剪承载力的计算公式为

When only stirrups are configured, for general flexural member with rectangular, T-shaped and I-shaped section, substituting $\alpha_{cv}=0.7$, $\alpha_{sv}=1$ into Eq. (5-11), gives the formula for calculating the shear capacity of inclined section

$$V \leqslant V_u = V_{cs} = V_c + V_{sv} = 0.7 f_t b h_0 + f_{yv} \frac{A_{sv}}{s} h_0 \tag{5-12}$$

式中,V 为构件斜截面上的最大剪力设计值,kN;V_{cs} 为构件斜截面上混凝土和箍筋的受剪承载力设计值,kN;f_t 为混凝土轴心抗拉强度设计值,N/mm²;b 为矩形截面的宽度、T 形或 I 形截面的腹板宽度,mm;h_0 为截面有效高度,mm;f_{yv} 为箍筋的抗拉强度设计值,N/mm²,一般可取 $f_{yv}=f_y$,但当 $f_y>360\text{N/mm}^2$,应取 $f_{yv}=360\text{N/mm}^2$;A_{sv} 为配置在同一截面内箍筋各肢的全部截面面积,$A_{sv}=nA_{sv1}$,mm²;n 为同一截面内箍筋的肢数,如图 5-12 所示;A_{sv1} 为一个箍筋单肢的截面面积,如图 5-13 所示,mm²;s 为沿构件长度方向的箍筋间距,mm。

Where, V is the maximum shear design value on the inclined section of the member, kN; V_{cs} is the design value of the shear capacity of concrete and stirrups on the inclined section of the member, kN; f_t is the axial tensile strength design value of concrete, N/mm²; b is the width of the rectangular section, web width of T-shaped or I-shaped section, mm; h_0 is the effective depth of the section, mm; f_{yv} is the design value of tensile strength of stirrups, N/mm², generally $f_{yv}=f_y$, but when $f_y>360\text{N/mm}^2$, it should be $f_{yv}=360\text{N/mm}^2$; A_{sv} is the total cross-sectional area of each limb of the stirrup configured in the same section, $A_{sv}=nA_{sv1}$, mm²; n is the number of legs in the same section, as shown in Fig. 5-12; A_{sv1} is the cross-sectional area of one leg of a stirrup, as shown in Fig. 5-13, mm²; s is the stirrup spacing along the length direction of the member, mm.

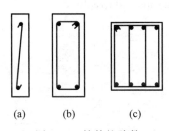

图 5-12 箍筋的肢数
(a) 单肢箍筋;(b) 双肢箍筋;
(c) 多肢(四肢)箍筋

Figure 5-12 Number of legs of stirrup
(a) Single-leg stirrup; (b) Two-leg stirrup;
(c) Multiple-leg(four legs) stirrup

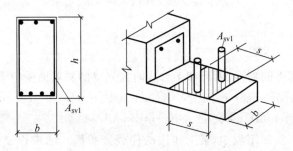

图 5-13 箍筋有关计算参数示意图

Figure 5-13 Schematic diagram of releated calculation parameters of stirrup

满足式(5-12)即满足抗剪承载力要求,正常使用极限状态下斜裂缝宽度不超过允许值。

Satisfying Eq. (5-12) means meeting shear capacity requirements, and the width of the diagonal crack under the serviceability limit state does not exceed the allowable value.

图 5-14(a)所示为均布荷载作用下有腹筋梁受剪承载力的试验值(图中的三角形点)与式(5-12)计算值(图中实线)的比较,可见计算值不是试验结果的统计平均值,而是试验值的偏下限值,是由满足设计可靠指标要求的破坏强度下包络线求得的。计算结果偏安全。

Fig. 5-14(a) shows the comparison between the experimental value (triangle points in the figure) and the calculated value (solid line in the figure) of Eq. (5-12) of the shear capacity of the beam with web reinforcement under the uniform load, and it can be seen that the calculated value is not the statistical average of the experimental results, but is the lower limit value of the experimental value, it is obtained from the lower envelope of failure strength that meets the requirements of design reliability index. The calculation results are relatively safe.

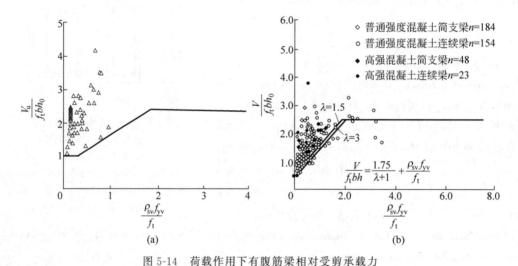

图 5-14 荷载作用下有腹筋梁相对受剪承载力
(a) 均布荷载;(b) 集中荷载

Figure 5-14 Relative shear capacity of beams with web reinforcement under load
(a) Uniform load;(b) Concentrated load

特别提示 5-4

2) 集中荷载作用下的矩形、T形和I形截面独立梁的斜截面受剪承载力设计值

2) Design value of shear capacity of inclined section of rectangular, T-shaped and I-shaped section isolated beams under concentrated load

实际工程中,梁上作用的荷载分布形式通常很复杂,当梁上作用有多种荷载,其中集中荷载对支座截面或节点边缘所产生的剪力值占总剪力值的 75% 以上时,梁的受剪性能与仅承受集中荷载的梁类似,可以按承受集中荷载的情况计算。现浇混凝土楼盖和装配整体式混凝土楼盖中的主梁

虽然主要承受集中荷载,但不是独立梁。建筑工程中除吊车梁、试验梁外,独立梁是很少见的。

集中荷载作用下的独立梁仅配置箍筋时,式(5-11)中 $\alpha_{cv}=1.75/(\lambda+1)$,$\alpha_{sv}=1$,则斜截面受剪承载力的计算公式为

$$V \leqslant V_u = V_{cs} = \frac{1.75}{\lambda+1} f_t b h_0 + f_{yv} \frac{A_{sv}}{s} h_0 \tag{5-13}$$

式中,λ 为计算截面的剪跨比,可取 $\lambda=a/h_0$,a 为剪跨,是集中荷载作用点至邻近支座截面或节点边缘的距离;当 $\lambda<1.5$ 时,取 $\lambda=1.5$;当 $\lambda>3$ 时,取 $\lambda=3$。集中荷载作用点至邻近支座之间的箍筋应均匀布置。

式(5-12)和式(5-13)都适用于矩形、T 形和 I 形截面,并不说明截面形状对受剪承载力没有影响,只是影响不大。

图 5-14(b)表示集中荷载作用下有腹筋梁受剪承载力的试验值(图中的点)与式(5-13)计算值(图中实线)的比较,可见计算值不是试验结果的统计平均值,而是试验值的偏下限值,是由满足设计可靠指标要求的破坏强度下包络线求得的。计算结果偏安全。

讨论 5-1:相同截面的梁承受集中荷载作用时的斜截面受剪承载力和承受均布荷载时的孰高孰低?

讨论 5-2:纵向受力钢筋配筋率 ρ 对梁斜截面受剪承载力是否有影响?为什么?

Discussion 5-1:For beams of the same section, which is the higher and which is the lower the shear capacity of the inclined section under the concentrated load and the uniform load?

Discussion 5-2:Does the longitudinal stressed reinforcement ratio ρ have an effect on the shear capacity of the inclined section of the beam? Why?

2. 同时配置箍筋和弯起钢筋时,矩形、T 形和 I 形截面受弯构件的斜截面承载力设计值
Design value of inclined section bearing capacity of rectangular, T-shaped and I-shaped section flexural members with stirrups and bent bars configured at the same time

当梁计算截面承受的剪力设计值较大时,若仅用箍筋和混凝土抗剪,会使箍筋直径 d 很大,间距 s 很小,箍筋数量多但可能仍不满足截面抗剪要求,还需配置弯起钢筋与箍筋一起抗剪。弯起钢筋承担的剪力设计值等于弯起钢筋的拉力在垂直于梁纵轴方向的分力:

英文翻译 5-9

$$V_{sb} = 0.8 f_y A_{sb} \sin\alpha_s \tag{5-14}$$

式中,V_{sb} 为弯起钢筋承担的剪力设计值,kN;0.8 为弯起钢筋受剪承载力折减系数,取这一系数是考虑到弯起钢筋与斜裂缝顶端相交时,有可能已接近剪压区,斜截面受剪破坏时应力达不到屈服强度;f_y 为弯起钢筋的抗拉强度设计值,N/mm²;A_{sb} 为同一弯起平面内弯起钢筋的截面面积,mm²;α_s 为弯起钢筋与构件纵轴线的夹角,一般取 45°,梁截面高度 $h>800$mm 时取 60°。

同时配有箍筋和弯起钢筋时梁的斜截面受剪承载力表达式为

$$V \leqslant V_u = V_{cs} + V_{sb} = V_{cs} + 0.8 f_y A_{sb} \sin\alpha_s \tag{5-15}$$

式中,V 为配置弯起钢筋处的剪力设计值。

特别提示 5-5

5.3.5 计算公式的适用范围
Scope of application of calculation formula

梁的斜截面受剪承载力计算公式是根据剪压破坏的受力特征和试验结果建立的,仅适用于剪压破坏情况,它有一定的适用范围,即公式有其上、下限值。为了防止发生斜压破坏,必须控制梁的截面尺寸不能过小;为了防止发生斜拉破坏,必须控制梁的配箍率不能过小,《混凝土结构设计标准》(GB/T 50010—2010)给出了相应的控制条件。

The formula for calculating the inclined section shear capacity of the beam is established according to the mechanical characteristics and experimental results of shear compression failure. It is only suitable for shear compression failure, and has a certain range of application, that is, the formula has its upper and lower limiting value. In order to prevent the diagonal compression failure, the cross-sectional dimensions of the beam must not be too small; in order to prevent the diagonal tension failure, the stirrup ratio of the beam must not be too small, and the "Standard for design of concrete structures" (GB/T 50010—2010) gives the corresponding control conditions.

1. 截面尺寸限制条件(上限值)
   Restriction conditions of section size (upper limit value)

当梁的截面尺寸确定后,斜截面受剪承载力并不能随配箍量的增大而无限提高。当梁承受的剪力较大,而截面尺寸较小时,若配箍量超过一定数值,破坏时箍筋的拉应力达不到屈服强度,箍筋不能充分发挥作用,梁的斜截面承载力几乎不再增大,还可能发生斜压破坏,即使多配置箍筋也不起作用。所以,为避免斜压破坏,梁的截面尺寸不宜过小,配箍量不能太大,这也是为了防止梁在使用阶段裂缝过宽(主要是薄腹梁)。

When the cross-sectional dimensions of the beam are determined, the shear capacity of inclined section cannot increase infinitely with the increase of the number of stirrups. When the shear force of the beam is large and the section size is small, if the amount of stirrups exceeds a certain value, the tensile stress of the stirrup can not reach the yield strength during failure, the stirrup cannot fully exert its function, and the inclined section bearing capacity of the beam almost no longer increases, and diagonal compression failure may occur, even multi-configuration stirrups do not work. Therefore, in order to avoid diagonal compression failure, the section size of the beam should not be too small, and the amount of stirrups should not be too large, and it is also to prevent cracks from being too wide during the service stage of the beam (mainly thin web beams).

矩形、T 形和 I 形截面梁的受剪截面限制条件(上限值)如下:
The shear section restriction conditions of rectangular, T-shaped, and I-shaped beams (upper limit value) are:

当 $\dfrac{h_w}{b} \leqslant 4$ 时(厚腹梁,即一般梁),截面尺寸应满足

When $\dfrac{h_w}{b} \leqslant 4$ (thick web beam, i.e. general beam), section size must satisfy

$$V \leqslant 0.25\beta_c f_c b h_0 \qquad (5\text{-}16\text{a})$$

当 $\dfrac{h_w}{b} \geqslant 6$ 时（薄腹梁），截面尺寸应满足

When $\dfrac{h_w}{b} \geqslant 6$ (thin web beam), section size must satisfy

$$V \leqslant 0.2\beta_c f_c b h_0 \qquad (5\text{-}16\text{b})$$

当 $4 < \dfrac{h_w}{b} < 6$ 时，按线性内插法确定。

When $4 < \dfrac{h_w}{b} < 6$, it is determined by linear interpolation.

式中，h_w 为截面的腹板高度，对矩形截面，取截面有效高度 h_0；对 T 形截面，取截面有效高度减翼缘高度（即 $h_0 - h_f'$）；对 I 形截面，取腹板净高（即 $h - h_f' - h_f$），如图 5-15 所示，mm。V 为构件斜截面上的最大剪力设计值，kN。β_c 为混凝土强度影响系数，当混凝土强度等级不超过 C50 时取 1.0，混凝土强度等级为 C80 时取 0.8，其间按线性内插法确定。b 为矩形截面的宽度、T 形截面或 I 形截面的腹板宽度，mm。

Where, h_w is the web depth of the section; for rectangular section, take the effective depth of section h_0; for T-shaped section, take the effective depth of section minus the flange thickness (i.e. $h_0 - h_f'$); for I-shaped section, take the clear depth of the web (i.e. $h - h_f' - h_f$), as shown in Fig. 5-15. mm. V is the maximum shear force design value on inclined section of member, kN. β_c is the influence coefficient of concrete strength, take 1.0 when concrete strength grade does not exceed C50, and take 0.8 when concrete strength grade is C80, during which is determined by linear interpolation method. b is the width of rectangular section, web width of T-shaped section or I-shaped section, mm.

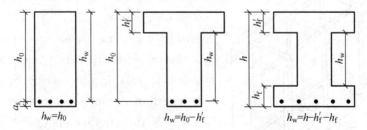

图 5-15　各类截面腹板高度 h_w 的取值

Figure 5-15　Value of web height h_w of various cross-sections

以上公式表示受弯构件在相应情况下斜截面受剪承载力的上限值，相当于限制了最小截面尺寸。在工程设计中，若不满足上述要求，说明根据正截面受弯承载力计算确定的梁截面尺寸不合适，应加大梁截面尺寸或提高混凝土强度等级。

Above formula represents the upper limit value of the shear capacity of the inclined section of the flexural member in corresponding conditions, which is equivalent to limiting the minimum cross-sectional dimension. In engineering design, if the above requirements are not met, it means that the section size of the beam determined by the calculation of the flexural capacity of normal section is not suitable, and the section size of the beam should be increased or the concrete strength grade should be improved.

拓展知识 5-2

2. 最小配箍率（下限值）

Minimum stirrup ratio (lower limit value)

1) 配箍率

1) Stirrup ratio

梁中箍筋配置数量的多少用配箍率表示，梁内箍筋的配筋率（也称配箍率）是指沿梁长，在箍筋的一个间距 s 范围内，箍筋各肢的全部截面面积与混凝土水平截面面积的比值。配箍率的计算公式为

The number of stirrup configuration in the beam is represented by stirrup ratio. The reinforcement ratio of stirrup in the beam (also called stirrup ratio) refers to along the length of the beam, within a spacing range s of stirrups, total cross-sectional area of each leg of the stirrup to the horizontal cross-sectional area of the concrete. The formula for calculating the stirrup ratio is

$$\rho_{sv} = \frac{A_{sv}}{bs} = \frac{nA_{sv1}}{bs} \tag{5-17}$$

式中，A_{sv} 为配置在同一截面内箍筋各肢的全部截面面积，$A_{sv}=nA_{sv1}$，mm^2；b 为矩形截面梁的宽度，T 形截面或 I 形截面梁的腹板宽度，mm；s 为沿构件长度方向箍筋的间距，如图 5-16 所示，mm；n 为同一截面内箍筋的肢数；A_{sv1} 为一个箍筋单肢的截面面积，mm^2。

Where, A_{sv} is the total cross-sectional area of each leg of the stirrup arranged in the same section, $A_{sv}=nA_{sv1}$, mm^2; b is the width of the beam with rectangular section, the web width of the beam with T-shaped section or I-shaped section, mm; s is the spacing of stirrups along the length of the member, as shown in Fig. 5-16, mm; n is the number of legs of stirrups in the same section; A_{sv1} is the cross-sectional area of one leg of a single stirrup, mm^2.

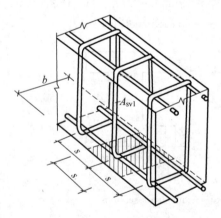

图 5-16 配箍率有关参数示意图

Figure 5-16 Schematic diagram of related parameters of stirrup ratio

试验表明，如果钢筋混凝土梁的配箍率在适当范围内，则梁的受剪承载力随配箍量的增多、箍筋强度的提高而有较大幅度的增长，配箍率的大小会影响有腹筋梁的破坏形态。在相同配箍率的条件下，间距较密的箍筋可以减小裂缝的宽度；但是，较细的箍筋又可能使梁的钢筋骨架没有足够的刚度，因此箍筋的直径和间距应综合考虑。

Experiments show that in reinforced concrete beams, if the stirrup ratio is in the appropriate range, then the shear capacity of the beam increases greatly with the increase of the amount of stirrups and the strength of the stirrups. The stirrup ratio will affect the destruction form of beams with web reinforcement. Under the condition of the same stirrup ratio, more closely spaced stirrups can reduce the width of the cracks; however, thinner stirrups may also make the steel skeleton of the beam less rigid, so the diameter and spacing of stirrups should be considered comprehensively.

2) 最小配箍率

2) Minimum stirrup ratio

梁出现斜裂缝后，斜裂缝处原来由混凝土承担的拉力全部转给箍筋承担，箍筋中的拉应力突然增大；若箍筋配置过少，则斜裂缝一旦出现，箍筋中的拉应力可能很快达到屈服强度f_{yv}，不能有效抑制斜裂缝的开展，造成裂缝的加速发展，甚至箍筋被拉断，导致斜拉破坏。因此当梁内配有一定数量的箍筋，且间距又不过大时，能保证与斜裂缝相交，即可防止斜拉破坏。为了避免斜拉破坏，当$V>0.7f_t bh_0$时，《混凝土结构设计标准》(GB/T 50010—2010)通过规定配箍率的下限值即最小配箍率来防止斜拉破坏。最小配箍率的表达式为

After diagonal cracks appear in the beam, all the tensile force originally borne by the concrete at the diagonal cracks is transferred to the stirrup, and the tensile stress in the stirrup increases suddenly; if the stirrup configuration is too little, the tensile stress in the stirrup may reach the yield strength f_{yv} quickly once the diagonal crack appears, and the development of diagonal cracks cannot be effectively inhibited, causing accelerated development of cracks, and even breaking of stirrups, resulting in diagonal tension failure. Therefore, when there are a certain number of stirrups inside the beam, and the spacing is not too large, ensuring intersection with diagonal cracks can prevent the diagonal tension failure. In order to avoid the diagonal tension failure, when $V>0.7f_t bh_0$, "Standard for design of concrete structures" (GB/T 50010—2010) prevents the diagonal tension failure by specifying the lower limit value of the stirrup ratio, that is, the minimum stirrup ratio. The expression of minimum stirrup ratio is

$$\rho_{sv,min} = 0.24\frac{f_t}{f_{yv}} \tag{5-18}$$

3) 配箍率的验算

3) Checking calculation of stirrup ratio

当需要按计算配置箍筋时，配箍率应满足以下要求：

When it is necessary to configure stirrups according to calculation, the stirrup ratio should meet the following requirement:

$$\rho_{sv} = \frac{A_{sv}}{bs} = \frac{nA_{sv1}}{bs} \geqslant \rho_{sv,min} \tag{5-19}$$

若满足此条件，则表明梁中箍筋的数量满足最小配箍率的要求。

If this condition is met, it indicates that the number of stirrups in the beam meets the requirement of minimum stirrup ratio.

为了充分发挥箍筋的作用，除满足最小配箍率条件外，尚需对箍筋最小直径（见表 5-1）

和最大间距(见表 5-2)加以限制。因为如果箍筋间距过大,斜裂缝有可能在箍筋之间出现,此时箍筋将不能有效地限制斜裂缝的开展。

In order to give full play to the role of stirrups, in addition to meeting the condition of minimum stirrup ratio, it is necessary to limit the minimum diameter (see Table 5-1) and maximum spacing of stirrups(see Table 5-2). Because if the stirrup spacing is too large, the diagonal cracks may appear between the stirrups, and the stirrups will not be able to effectively limit the development of diagonal cracks.

表 5-1 梁中箍筋最小直径

Table 5-1 Minimum diameter of stirrup in beam

梁高 h/mm	箍筋直径 d/mm
$h \leqslant 800$	不宜小于 6
$h > 800$	不宜小于 8

表 5-2 梁中箍筋最大间距 s_{max}

Table 5-2 Maximum stirrup spacing in beam s_{max}

单位:mm

梁高 h	$V > 0.7 f_t b h_0$	$V \leqslant 0.7 f_t b h_0$
$150 < h \leqslant 300$	150	200
$300 < h \leqslant 500$	200	300
$500 < h \leqslant 800$	250	350
$h > 800$	300	400

5.3.6 斜截面受剪承载力的计算截面
Calculation section of shear capacity of inclined section

根据《混凝土结构设计标准》(GB/T 50010—2010)6.3.2 条规定,按以下要求确定斜截面受剪承载力计算截面的位置。

(1)支座边缘处的斜截面,如图 5-17(a)中 1—1 截面,取该处的剪力设计值。

(2)受拉区弯起钢筋弯起点处的斜截面,如图 5-17(a)中 2—2、3—3 截面。计算第一排(从支座算起)弯起钢筋时,取支座边缘处的剪力设计值;计算以后每一排弯起钢筋时,取前一排(对支座而言)弯起钢筋弯起点处的剪力设计值。

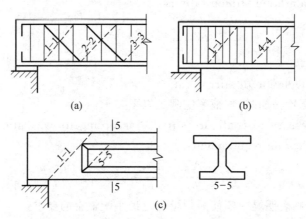

图 5-17 斜截面受剪承载力的计算截面位置

(a) 1—1,2—2,3—3 截面位置;(b) 1—1,4—4 截面位置;(c) 1—1,5—5 截面位置

Figure 5-17 Calculated section position of inclined section shear capacity

(a) 1—1,2—2,3—3 section position;(b) 1—1,4—4 section position;(c) 1—1,5—5 section position

(3) 箍筋截面面积或间距改变处的斜截面,如图 5-17(b)中 4—4 截面,取箍筋间距开始改变处的剪力设计值。

(4) 构件截面改变处的截面,如图 5-17(c)中 5—5 截面,梁截面由矩形截面变为 I 形截面,由于腹板宽度变小,必然使梁的斜截面受剪承载力受到影响。

5.3.7 截面设计
Section design

通过截面设计,确定有腹筋梁中箍筋、弯起钢筋的步骤如下。

Through the section design, the steps of determining the stirrups and bent bars in the beam with web reinforcement are as follows

1. 根据力学方法求梁的剪力设计值,确定计算截面,绘制梁的剪力图,确定计算斜截面的剪力设计值

Calculate the shear design value of the beam according to the mechanical method, determine the calculated section, draw the shear diagram of the beam, and determine the shear design value of the calculated inclined section

2. 验算截面尺寸限制条件,防止斜压破坏

Check the constraint conditions of section size to prevent diagonal compression failure

按式(5-16a)和式(5-16b)验算截面尺寸是否合适,当不满足要求时,应加大截面尺寸或提高混凝土强度等级。对于板类受弯构件,按式(5-7a)和式(5-7b)验算板厚,一般不需计算腹筋。

Check and calculate whether the section size is appropriate or not according to Eq. (5-16a) and Eq. (5-16b). When the requirements are not met, the section size should be increased or concrete strength grade should be improved. For plate type flexural members, the slab thickness should be checked according to Eq. (5-7a) and Eq. (5-7b), and generally it is not necessary to calculate the web reinforcement.

3. 判断按计算还是按构造要求配置箍筋

Determine whether to configure stirrups according to calculation or structural requirements

当梁承受的剪力较小而截面尺寸较大,满足式(5-20)和式(5-21)的条件时,梁内可不按计算配置箍筋,应按构造要求配置箍筋,即箍筋的最大间距 s_{max} 和最小直径 d_{min} 应满足相应的构造要求,否则,应按计算配置箍筋。

When the shear force borne by the beam is small and the cross-sectional dimensions are large, and the conditions of Eq. (5-20) and Eq. (5-21) are satisfied, the stirrups in the beam may not be configured according to the calculation, and the stirrups should be configured according to the structural requirements, that is, the maximum spacing s_{max} and minimum diameter d_{min} should meet the corresponding structural requirements. Otherwise, stirrups should be configured according to the calculation.

1) 对于矩形、T 形和 I 形截面的一般受弯构件

1) For general flexural members with rectangular, T-shaped and I-shaped sections

$$V \leqslant 0.7 f_t b h_0 \tag{5-20}$$

2) 对于集中荷载作用下的矩形、T形和I形截面独立梁

2) For rectangular, T-shaped and I-shaped section isolated beams under concentrated load

$$V \leqslant \frac{1.75}{\lambda+1} f_t b h_0 \tag{5-21}$$

4. 当需要按计算配置腹筋时,按两种情况计算

When it is necessary to configure the web reinforcement according to the calculation, the calculation can be carried out in two situations

1) 仅配置箍筋,不配置弯起钢筋的矩形、T形和I形截面梁

1) Rectangular, T-shaped and I-shaped section beams with only stirrups and no bent bars

对均布荷载作用下的一般受弯构件,由式(5-12)得:

For general flexural members under uniform load, it can be obtained from Eq. (5-12)

$$\frac{A_{sv}}{s} = \frac{nA_{sv1}}{s} \geqslant \frac{V - 0.7 f_t b h_0}{f_{yv} h_0} \tag{5-22}$$

对集中荷载作用下的独立梁,由式(5-13)得

For isolated beams under concentrated load, it can be obtained from Eq. (5-13)

$$\frac{A_{sv}}{s} = \frac{nA_{sv1}}{s} \geqslant \frac{V - \dfrac{1.75}{\lambda+1} f_t b h_0}{f_{yv} h_0} \tag{5-23}$$

nA_{sv1}/s 中的未知量有3个:箍筋肢数n、一个箍筋单肢的截面面积A_{sv1}和箍筋间距s。可以先按构造要求选用箍筋肢数n和箍筋直径d,再根据箍筋直径d查附表2-1或计算得出一个箍筋单肢的截面面积A_{sv1},根据式(5-22)或式(5-23)求出nA_{sv1}/s,再根据nA_{sv1}/s求出箍筋间距s,并符合构造要求$s \leqslant s_{\max}$($s_{\max}$为构造要求规定的最大箍筋间距,查表5-2确定),最后根据式(5-19)验算最小配箍率$\rho_{sv,\min}$。

There are 3 unknown quantities in nA_{sv1}/s: the number of stirrup legs n, the cross-sectional area of one leg of a stirrup A_{sv1} and the stirrup spacing s. The number of stirrup legs n and the stirrup diameter d can be selected according to the structural requirements. Check the Attached table 2-1 based on the diameter of the stirrup or calculate the cross-sectional area of one leg of a stirrup A_{sv1}, calculate nA_{sv1}/s according to Eq. (5-22) or Eq. (5-23), and then calculate the stirrup spacing s according to nA_{sv1}/s, and meet the structural requirements $s \leqslant s_{\max}$ ($s_{\max}$ is the maximum stirrup spacing specified by the structural requirements, see Table 5-2 to determine), and finally check and calculate the minimum stirrup ratio $\rho_{sv,\min}$ according to Eq. (5-19).

2) 同时配置箍筋和弯起钢筋的矩形、T形和I形截面梁

2) Rectangular, T-shaped and I-shaped section beams with both stirrups and bent bars

可以先根据经验和构造要求配置箍筋(选定箍筋肢数n、直径d、间距s),按式(5-12)或式(5-13)求出V_{cs},再根据式(5-24)计算弯起钢筋的截面面积A_{sb},如图5-18所示。也可先选定弯起钢筋的截面面积(可由正截面受弯承载力计算所得纵向受拉钢筋中的弯起钢筋的

截面面积确定),再由式(5-15)等计算箍筋的数量。

Stirrups can be configured according to experience and structural requirements (select the number of stirrup legs n, diameter d, spacing s), obtain V_{cs} according to Eq.(5-12) or Eq.(5-13), and then according to Eq.(5-24) calculate the cross-sectional area A_{sb} of the bent bar, as shown in Fig. 5-18. It is also possible to first select the section area of the bent bar (it can be determined

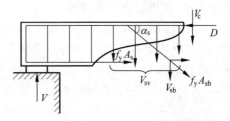

图 5-18 梁中箍筋、弯起钢筋计算简图

Figure 5-18 Calculation diagram of stirrups and bent bars in beams

by the section area of the bent bar in the longitudinal tensile steel bar calculated from the flexural capacity of normal section), then the number of stirrups is calculated by Eq.(5-15),etc.

$$A_{sb} = \frac{V - V_{cs}}{0.8 f_y \sin\alpha_s} \tag{5-24}$$

5. 验算配箍率

Check the stirrup ratio

按式(5-17)计算配箍率,按式(5-18)计算最小配箍率,再根据式(5-19)验算配箍率。

Calculate the stirrup ratio according to Eq.(5-17), calculate the minimum stirrup ratio according to Eq.(5-18), and then check the stirrup ratio according to Eq.(5-19).

6. 标注箍筋

Label stirrups

应标注出箍筋的钢筋级别、直径以及相邻箍筋的中心距,例如,ϕ8@150 表示直径为 8mm 的 HPB300 级箍筋沿梁的纵轴线平行排列,箍筋间距为 150mm。

The labeling of stirrups should mark the reinforcement grade, diameter and center distance of adjacent stirrups. For example:ϕ8@150 means that HPB300 stirrups with diameter of 8mm are arranged in parallel along the longitudinal axis of the beam, and the stirrup spacing is 150mm.

5.3.8 截面复核

Section review

已知构件截面的剪力设计值、混凝土强度等级、钢筋级别、构件截面尺寸、箍筋配置情况、弯起钢筋截面面积及弯起角度,求斜截面受剪承载力设计值 V_u,并验算是否 $V \leqslant V_u$。

(1) 根据式(5-16a)或式(5-16b)验算截面限制条件。若不满足,应修改构件截面尺寸或提高混凝土强度等级;若满足则进行下一步计算。

(2) 验算箍筋直径和间距。当 $V > 0.7 f_t b h_0$ 时,检查是否满足式(5-12);当 $V > \frac{1.75}{\lambda+1} f_t b h_0$ 时,检查是否满足式(5-13)。若不满足,说明箍筋配置不符合规范要求,应修改箍筋配置;若满足要求,则进行下一步计算。

(3) 若同时配置有箍筋和弯起钢筋,将已知数据代入式(5-15),若仅配有箍筋,则代入式(5-12)或式(5-13),检验不等式条件是否满足。若满足则斜截面抗剪承载力足够,若不满足则需修改设计,重新复核。

【例题 5-1】 受均布荷载作用的矩形截面简支梁,净跨 $l_n = 3.6\text{m}$,截面尺寸、支承情况如图 5-19 所示,结构安全等级为二级,梁上均布荷载设计值 $q = 70\text{kN/m}$(含梁自重),C25 混凝土,纵筋为 HRB400,箍筋为 HPB300,确定此梁需配置的箍筋。

【Example 5-1】 A simply supported rectangular section beam subjected to uniform load, with a clear span of $l_n = 3.6\text{m}$, the cross-sectional dimensions and support conditions are shown in Fig. 5-19, the structural safety level is Level 2, and the design value of uniform load on the beam is $q=70\text{kN/m}$ (including self-weight of the beam), C25 concrete, HRB400 longitudinal bars, HPB300 stirrups. Determine the stirrups that need to be configured for this beam.

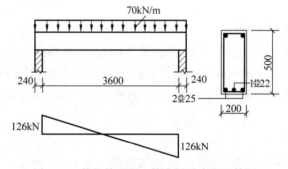

图 5-19 某钢筋混凝土简支梁受力和配筋图

Figure 5-19　Force and reinforcement diagram of a reinforced concrete simply supported beam

解:(1) 计算支座边缘截面的剪力设计值

梁的剪力如图 5-19 所示,支座边缘截面的剪力设计值为

$$V = \frac{1}{2}ql_n = \frac{1}{2} \times 70 \times 3.6\text{kN} = 126\text{kN}$$

(2) 验算梁截面尺寸

对矩形截面,$h_w = h_0 = h - a_s = (500 - 35)\text{mm} = 465\text{mm}$,$h_w/b = 465/200 = 2.325 < 4$,属厚腹梁。

梁混凝土强度等级为 C25,低于 C50,取混凝土强度影响系数 $\beta_c = 1$,则

$$0.25\beta_c f_c b h_0 = 0.25 \times 1 \times 11.9 \times 200 \times 465\text{N} = 226675\text{N}$$
$$\approx 226.7\text{kN} > V = 126\text{kN}$$

所以,截面尺寸符合要求。

(3) 验算是否需要按计算配置箍筋

$$0.7 f_t b h_0 = 0.7 \times 1.27 \times 200 \times 465\text{N} = 82677\text{N} \approx 82.7\text{kN} < V = 126\text{kN}$$

所以,需要按计算进行配箍。

(4) 计算配箍量

$$V_{cs} = V_c + V_{sv} = 0.7 f_t b h_0 + f_{yv} \frac{A_{sv}}{s} h_0$$

$$126000 = 0.7 \times 1.27 \times 200 \times 465 + 270 \times \frac{nA_{sv1}}{s} \times 465$$

则

$$\frac{nA_{sv1}}{s} = \frac{126000 - 82677}{270 \times 465} \text{mm}^2/\text{mm} = 0.345 \text{mm}^2/\text{mm}$$

根据梁的截面宽度和受压钢筋的情况，选用双肢箍筋，HPB300 级，直径 $d=6\text{mm}$，则

$$s = \frac{nA_{sv1}}{0.345} = \frac{2 \times 28.3}{0.345} \text{mm} = 164 \text{mm}$$

取箍筋间距 $s=160\text{mm}$（一般取末位为 0，小于且最接近箍筋间距 s 计算值的整数），且 $s \leqslant s_{max} = 200\text{mm}$（$s_{max}$ 查表 5-2）。

(5) 验算最小配箍率

配箍率

$$\rho_{sv} = \frac{nA_{sv1}}{bs} = \frac{2 \times 28.3}{200 \times 160} = 0.177\%$$

最小配箍率

$$\rho_{sv,min} = 0.24 \frac{f_t}{f_{yv}} = 0.24 \times \frac{1.27}{270} = 0.11\% < \rho_{sv}$$

所以，配筋率满足要求。

【例题 5-2】 条件同例题 5-1，设箍筋配置为双肢 $\phi 8@200$，确定此梁需配置的弯起钢筋。

【Example 5-2】 The conditions are the same as Example 5-1, set stirrup configuration as double legs $\phi 8@200$, and determine the bent bar to be configured for this beam.

解：(1) 求混凝土与箍筋的受剪承载力设计值 V_{cs}

$$\begin{aligned} V_{cs} &= V_c + V_{sv} = 0.7 f_t b h_0 + f_{yv} \frac{nA_{sv1}}{s} h_0 \\ &= \left(0.7 \times 1.27 \times 200 \times 465 + 270 \times \frac{2 \times 28.3}{200} \times 465\right) \text{N} \\ &= (82677 + 35530.65) \text{N} = 118207.65 \text{N} \end{aligned}$$

(2) 计算弯起钢筋的截面面积 A_{sb}

根据图 5-19 中梁底部已配的 3 根纵向钢筋，可将中间一根直径为 22mm 的纵筋弯起，弯起钢筋截面面积 $A_{sb} = 380 \text{mm}^2$，梁高 $h = 500 \text{mm} < 800 \text{mm}$，设弯起角 $\alpha_s = 45°$，则

$$A_{sb} = \frac{V - V_{cs}}{0.8 f_y \sin\alpha_s} = \frac{126000 - 118207.65}{0.8 \times 360 \times 0.707} \text{mm}^2 = 38.3 \text{mm}^2 < A_{sb} = 380 \text{mm}^2$$

满足要求。

【例题 5-3】 钢筋混凝土简支梁如图 5-20 所示，截面尺寸为 $b \times h = 200\text{mm} \times 400\text{mm}$，混凝土强度等级为 C30，纵向受力钢筋 $2\underline{\Phi}22 + 1\underline{\Phi}20$，箍筋 HPB300。跨中两个集中荷载设计值 $F = 100\text{kN}$（不计自重），环境类别为一类。(1) 求受剪箍筋（无弯起钢筋）；(2) 将纵向受拉钢筋弯起作为弯起钢筋时，求所需箍筋。

【Example 5-3】 The reinforced concrete simply supported beam is shown in Fig. 5-20, cross-sectional dimensions are $b \times h = 200\text{mm} \times 400\text{mm}$, concrete strength grade is C30, longitudinally stressed reinforcement is $2 \underline{\Phi} 22 + 1 \underline{\Phi} 20$, and stirrup is HPB300. The design value of two concentrated loads at midspan is $F = 100\text{kN}$ (excluding self-weight), and the environment category is Class I. (1) Calculate the shear stirrup (no bent bar); (2) When the longitudinal tensile steel bar is used as the bent bar, calculate the required stirrups.

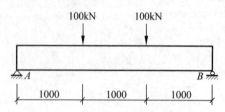

图 5-20 某承受集中荷载的简支梁
Figure 5-20 A simply supported beam subjected to concentrated load

解：环境类别一类，查附表 3-2 得 C30 混凝土最小保护层厚度 $c = 20\text{mm}$，设下部纵向受拉钢筋为单层放置，则 $a_s = 35\text{mm}$，$h_w = h_0 = h - a_s = (400 - 35)\text{mm} = 365\text{mm}$。

（1）求不设弯起钢筋时的受剪箍筋

① 求剪力设计值。由剪力图得 $V_{max} = 100\text{kN}$。

② 验算截面尺寸。$h_w/b = h_0/b = 365/200 = 1.825 < 4$，属厚腹梁。混凝土强度等级 C30，小于 C50，所以 $\beta_c = 1.0$，则

$$0.25\beta_c bh_0 = 0.25 \times 1.0 \times 14.3 \times 200 \times 365\text{N} = 260975\text{N}$$
$$\approx 260.975\text{kN} > V_{max} = 100\text{kN}$$

所以，截面尺寸符合要求。

③ 计算箍筋用量。对只有集中荷载作用的独立梁，先验算是否需要按计算配置箍筋。

剪跨比 $\lambda = a/h_0 = 1000/365 = 2.74$，$\dfrac{1.75}{\lambda+1} f_t bh_0 = \dfrac{1.75}{2.74+1} \times 1.43 \times 200 \times 365\text{N} = 48845.6\text{N} < V_{max} = 100\text{kN}$。

所以，需要按计算配置箍筋。根据以下公式计算箍筋用量：

$$V \leqslant \dfrac{1.75}{\lambda+1} f_t bh_0 + f_{yv} \dfrac{A_{sv}}{s} h_0$$

$$100000 = \dfrac{1.75}{2.74+1} \times 1.43 \times 200 \times 365 + 270 \times \dfrac{nA_{sv1}}{s} \times 365$$

$$\dfrac{nA_{sv1}}{s} = \dfrac{100000 - 48845.6}{270 \times 365}\text{mm}^2/\text{mm} = 0.519\text{mm}^2/\text{mm}$$

选用直径 8mm 的双肢箍筋，则箍筋间距为

$$s = \dfrac{nA_{sv1}}{0.519} = \dfrac{2 \times 50.3}{0.519}\text{mm} = 193.8\text{mm}$$

取 $s = 190\text{mm} < s_{max} = 200\text{mm}$（$s_{max}$ 查表 5-2）满足最大箍筋间距 s_{max} 的要求。所以，选用ϕ8@190 的双肢箍筋。

④ 验算最小配箍率。配箍率：

$$\rho_{sv} = \dfrac{nA_{sv1}}{bs} = \dfrac{2 \times 50.3}{200 \times 190} = 0.265\%$$

最小配箍率：
$$\rho_{sv,min} = 0.24 \frac{f_t}{f_y} = 0.24 \times \frac{1.43}{270} = 0.127\% < \rho_{sv}$$

所以，配箍率满足要求。

(2) 利用现有纵筋为弯起钢筋，求所需箍筋

① 若弯起 1Φ20 钢筋（取弯起角为 45°），则弯起钢筋承担的剪力
$$V_{sb} = 0.8 f_y A_{sb} \sin\alpha_s = 0.8 \times 360 \times 314.2 \times \sin45°\text{N} = 76997.9\text{N}$$

② 混凝土和箍筋承担的剪力
$$V_{cs} = V - V_{sb} = (100000 - 76997.9)\text{N} = 23002.1\text{N}$$

③ 计算箍筋用量。选用ϕ6@200 双肢箍筋，根据公式：
$$V_{cs} = \frac{1.75}{\lambda+1} f_t b h_0 + f_{yv} \frac{A_{sv}}{s} h_0$$

可得混凝土和箍筋所能承担的剪力
$$V_{cs} = \left(\frac{1.75}{2.74+1} \times 1.43 \times 200 \times 365 + 270 \times \frac{2 \times 28.3}{200} \times 365\right)\text{N}$$
$$= (48845.6 + 27889.65)\text{N} = 76735.25\text{N} > 23002.1\text{N}$$

满足要求。

④ 验算最小配箍率。配箍率
$$\rho_{sv} = \frac{nA_{sv1}}{bs} = \frac{2 \times 28.3}{200 \times 200} = 0.142\%$$

最小配箍率
$$\rho_{sv,min} = 0.24 \frac{f_t}{f_y} = 0.24 \times \frac{1.43}{270} = 0.127\% < \rho_{sv}$$

所以，配箍率满足要求。

(3) 弯起钢筋弯起点处的验算

由于在集中荷载与支座之间剪力设计值均为 100kN，而三根纵筋只能弯起一根（不能覆盖 1000mm 的范围），弯起钢筋的弯起点处用ϕ6@200 不能满足抗剪要求，因此，弯起钢筋弯起点处及以外的范围仍然需要配置ϕ8@190 的双肢箍筋才能符合要求。

5.4 保证斜截面受弯承载力的构造要求

Structural requirements to ensure the flexural capacity of inclined section

在受弯构件正截面受弯承载力和斜截面受剪承载力的计算中，钢筋强度的充分发挥应建立在可靠的配筋构造基础上。在钢筋混凝土结构设计中，构造要求与设计计算同样重要，不能忽视钢筋的构造要求。

In the calculation of flexural capacity of normal section and the shear capacity of inclined section of flexural members, the full exertion of the reinforcement strength should be established on the basis of reliable reinforcement structure. In the design of reinforced concrete structures, the structural requirements are as important as design and calculation, and the structural requirements of reinforcement cannot be ignored.

5.4.1 箍筋的构造要求
Structural requirements for stirrups

在梁内配置箍筋除了承受剪力,提供斜截面承载力,抑制斜裂缝的开展外,还能固定纵筋位置,与纵筋形成钢筋骨架,防止纵筋压屈;联系梁的受拉区和受压区,增加受压区混凝土的延性。

The configuration of stirrups in the beam not only bears shear force, provides the bearing capacity of inclined section, restrains the development of diagonal cracks, but also fixes the position of longitudinal bars and forms steel skeleton with the longitudinal bars to prevent the longitudinal bars from buckling; connect the tension zone and the compression zone of the beam, and increase the ductility of the concrete in the compression zone.

1. 箍筋的形式
Form of stirrups

箍筋的形式有开口式和封闭式,如图 5-21 所示,一般情况下采用封闭式箍筋。采用封闭式箍筋可以方便固定纵筋,提高梁的抗扭能力。《混凝土结构设计标准》(GB/T 50010—2010)9.2.9 条规定,当梁中配有按计算需要的纵向受压钢筋时,必须采用封闭式箍筋。现浇 T 形截面梁不受扭矩和动荷载时,在跨中截面上部受压区的区段内可采用开口式箍筋。

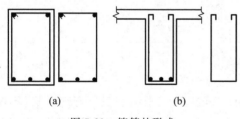

图 5-21 箍筋的形式
(a) 封闭式箍筋;(b) 开口式箍筋
Figure 5-21 Form of stirrups
(a) Closed stirrup;(b) Open stirrup

The forms of stirrups are open type and closed type, as shown in Fig. 5-21. Generally, closed stirrup is used. The use of closed stirrup can easily fix the longitudinal bars and improve the torsion resistance of the beam. Article 9.2.9 of "Standard for design of concrete structures"(GB/T 50010—2010) stipulates that closed stirrups must be used when the beam is equipped with longitudinal compressive bars according to the calculation requirements. When the cast-in-place T-shaped section beam is not subject to torque and dynamic loads, open stirrups can be used in the upper compression zone of the mid-span section.

2. 箍筋的肢数
Number of legs of stirrups

箍筋按肢数分为单肢箍筋、双肢箍筋和多肢箍筋,一般按以下情况选用。单肢箍筋:梁宽 $b \leqslant 100$ mm 时采用或用于拉结筋;双肢箍筋:梁宽 $b \leqslant 400$ mm,且一层内的纵向受压钢筋不多于 4 根时采用;多肢箍筋:梁宽 $b > 400$ mm,且一层内的纵向受压钢筋多于 3 根时,或梁宽 $b \leqslant 400$ mm,但一层内的纵向受压钢筋多于 4 根时采用。

The stirrups are divided into single-leg stirrups, two-leg stirrups and multiple-leg stirrups according to the number of legs, which are generally selected according to the following conditions. Single-leg stirrup: the beam width b is less than or equal to 100mm, it is used or used for tie bar; Two-leg stirrup: used when the beam width b is less than or equal to 400mm, and there are no more than four longitudinal compression bars in one layer; multiple-leg stirrup: the beam width $b > 400$mm, and there are more than 3 longitudinal compression bars in one layer, or when the beam width $b \leqslant 400$mm, but there are more than 4 longitudinal compression bars in one layer.

3. 箍筋的强度
Strength of stirrups

剪切破坏属于脆性破坏，为增加截面延性，不宜采用高强度钢筋做箍筋。

Shear failure belongs to brittle failure. In order to increase the ductility of the section, it is not suitable to use high strength steel bars for stirrups.

4. 箍筋的直径
Diameter of stirrups

箍筋直径过小会使钢筋骨架没有足够的刚度，为了使钢筋骨架具有一定的刚度，便于制作安装，箍筋直径不应太小。《混凝土结构设计标准》(GB/T 50010—2010)9.2.9条规定的箍筋最小直径见表5-1。当梁中配有计算需要的纵向受压钢筋时，箍筋直径尚不应小于$d/4$（d为纵向受压钢筋的最大直径）。

If the diameter of the stirrup is too small, the steel skeleton will not have adequate rigidity. In order to make the steel skeleton have a certain rigidity and facilitate production and installation, the diameter of the stirrup should not be too small. The minimum diameter of stirrup specified in Article 9.2.9 of "Standard for design of concrete structures"(GB/T 50010—2010)is shown in Table 5-1. When the beam is equipped with the longitudinal compression reinforcement required by the calculation, the diameter of the stirrup should not be less than $d/4$ (d is the maximum diameter of the longitudinal compression reinforcement).

5. 箍筋的间距
Spacing of stirrups

当梁中配有按计算需要的纵向受压钢筋时，为了使箍筋的设置与受压钢筋协调，防止受压钢筋压屈，箍筋间距s应满足下述要求，如图5-22所示。

When the beam is equipped with longitudinal compression reinforcement according to the calculation requirements, in order to coordinate the setting of the stirrup with the compression reinforcement and prevent the compression reinforcement from buckling, the stirrup spacing s should meet the following requirements, as shown in Fig. 5-22.

(1) $s \leqslant 15d$，且不应大于400mm（d为纵向受压钢筋的最小直径）；

(2) 当一层内的纵向受压钢筋多于5根且直径$d > 18$mm时，箍筋间距尚应符合$s \leqslant 10d$。

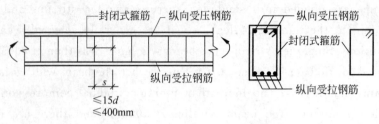

图 5-22 封闭式箍筋的构造要求
Figure 5-22 Structural requirements of closed stirrups

(1) $s \leqslant 15d$, and should not be greater than 400mm (d is the minimum diameter of longitudinal compression reinforcement);

(2) When there are more than 5 longitudinal compression bars in one layer and the diameter $d > 18$mm, the stirrup spacing should also conform to $s \leqslant 10d$.

若箍筋间距很大,则破坏时斜裂缝不与箍筋相交,或相交在不能充分发挥作用的位置(如非常靠近剪压区),箍筋不能抑制斜裂缝的开展,出现斜拉破坏。箍筋间距除了应满足计算要求外,《混凝土结构设计标准》(GB/T 50010—2010)9.2.9条规定梁中箍筋最大间距应符合表 5-2 的规定。

If the stirrup spacing is large, the diagonal cracks at the time of failure do not intersect with the stirrup, or the intersection is in a position that cannot fully function (such as very close to the shear compression zone), and the stirrup cannot inhibit the development of diagonal cracks, resulting in diagonal tension failure. In addition to meeting the calculation requirements, Article 9.2.9 of "Standard for design of concrete structures" (GB/T 50010—2010) stipulates that the maximum stirrup spacing in the beam should meet the requirements of Table 5-2.

当梁中绑扎骨架内纵向钢筋为非焊接搭接时,在搭接长度内,应配置直径不小于搭接钢筋直径 0.25 倍的箍筋。当纵筋受拉时,箍筋间距不应大于 5d,且不应大于 100mm;当纵筋受压时,箍筋间距不应大于 10d(d 为搭接钢筋中的最小直径),且不应大于 200mm。当受压钢筋直径大于 25mm 时,应在搭接接头两个端面外 100mm 范围内各设置两道箍筋。

When the longitudinal bars in the binding skeleton in the beam are non-welded lapped splices, stirrups with diameter not less than 0.25 times the diameter of the lapped bars should be arranged within the lap length. When the longitudinal bars are in tension, the stirrup spacing should not be greater than $5d$, and should not be larger than 100mm; when the longitudinal bars are in compression, the stirrup spacing should not be larger than $10d$ (d is the minimum diameter of the lapped bars), and it should not be greater than 200mm. When the diameter of compression bar is greater than 25mm, two stirrups should be set within 100mm outside the two faces of the head of the lapped splice.

当采用机械锚固措施时,锚固长度范围内的箍筋不应少于 3 个,直径不应小于纵向钢筋直径的 0.25 倍,间距不应大于纵向钢筋直径的 5 倍。当纵向钢筋的混凝土保护层厚度不小于钢筋直径或等效直径的 5 倍时,可不配置上述箍筋。

When mechanical anchoring measures are adopted, the number of stirrups within the anchorage length should not be less than 3, the diameter should not be less than 0.25 times the diameter of the longitudinal bars, and the spacing should not be greater than 5 times the diameter of the longitudinal bars. When the thickness of concrete cover of the longitudinal bars is not less than 5 times the diameter of the reinforcement or the equivalent diameter, the above stirrups may not be arranged.

6. 箍筋的设置范围

Setting range of stirrups

《混凝土结构设计标准》(GB/T 50010—2010)9.2.9 条规定，对按承载力计算不需要箍筋的梁：

Article 9.2.9 of "Standard for design of concrete structures" (GB/T 50010—2010) stipulates that for beams that do not need stirrups according to bearing capacity calculation:

(1) 当截面高度 $h>300$mm 时，应沿梁全长设置构造箍筋。

(2) 当截面高度 $h=150\sim300$mm 时，可仅在构件端部 $l_0/4$ (l_0 为梁的跨度)范围内设置构造箍筋；但当在构件中部 $l_0/2$ 范围内有集中荷载作用时，则应沿梁全长设置箍筋。

(3) 当截面高度 $h<150$mm 时，可以不设置箍筋。

(1) When the section depth $h>300$mm, structural stirrups should be set along the full length of the beam.

(2) When the section depth $h=150\sim300$mm, structural stirrups can be set only in the range of $l_0/4$ (l_0 is the span of the beam) at the end of the member; but when there is concentrated load in the range of $l_0/2$ in the middle of the member, stirrups should be set along the full length of the beam.

(3) When the section depth $h<150$mm, stirrups may not be provided.

5.4.2 弯起钢筋的构造要求

Structural requirements for bent bars

弯起钢筋一般由纵向受力钢筋弯起而成，它除了在跨中承受由弯矩产生的拉力外，还在靠近支座的弯起段承受弯矩和剪力共同产生的主拉应力。试验研究表明，箍筋对抑制斜裂缝开展的效果比弯起钢筋要好，所以梁中宜优先采用箍筋作为承受剪力的钢筋，然后再考虑弯起钢筋。当采用弯起钢筋时，弯起钢筋的数量、位置由计算确定。

1. 弯起角、弯起顺序和位置

Bending angle, bending order and position

梁中弯起钢筋的弯起角一般宜取 45°，当梁的截面高度 $h>700$mm 时，宜采用 60°。钢筋弯起的顺序一般是先内层后外层、先内侧后外侧。由于弯起钢筋承受的拉力比较大且集中，可能引起弯起点处的混凝土劈裂，因此放置在梁侧边缘的钢筋不宜弯起，梁底部位于箍筋转角处的钢筋不应弯起,顶层钢筋中的角部钢筋不应弯下,以便和箍筋绑扎成钢筋骨架。

2. 弯起钢筋的间距
Spacing of bent bars

弯起钢筋的间距指前一排弯起钢筋的弯起点到后一排弯起钢筋的弯终点之间的距离。按抗剪设计需设置弯起钢筋时,弯起钢筋的最大间距 s_{max} 同表 5-2 中 $V>0.7f_t bh_0$ 的情况一样。靠近梁端的第一根弯起钢筋的弯起点到支座边缘的距离不小于 50mm。当梁宽较大时(如 $b>350mm$),为使弯起钢筋在整个宽度范围内受力均匀,宜在一个截面内同时弯起两根钢筋,如图 5-23 所示。

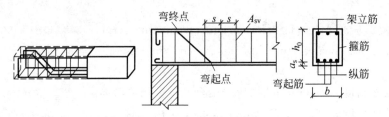

图 5-23 弯起钢筋示意图

Figure 5-23　Schematic diagram of bent bars

3. 吊筋
Hanging bar

当采用弯起钢筋抗剪后不能满足抵抗弯矩图的要求时,可单独设置抗剪斜筋抗剪,将斜筋布置成吊筋,如图 5-24(a)所示。

4. 鸭筋
Duck shape bar

当不能弯起纵向受力钢筋抗剪时,可设单独的抗剪弯筋,应将弯筋布置成两端均锚固在受压区的鸭筋形式,如图 5-24(a)所示。因为弯起钢筋的作用是将斜裂缝之间的混凝土斜向压力传递给受压区混凝土,以加强不同部位混凝土之间的共同工作,形成一拱形桁架,因而不允许采用浮筋,如图 5-24(b)所示。浮筋在受拉区只有一小段水平长度,锚固性能不如鸭筋可靠,一旦发生滑移,将使斜裂缝开展过大,因此不应采用。

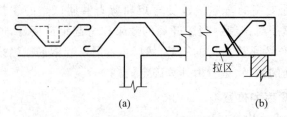

图 5-24　吊筋、鸭筋和浮筋

(a) 吊筋和鸭筋；(b) 浮筋

Figure 5-24　Hanging bar, duck shape bar and floating bar

(a) Hanging bar and duck shape bar; (b) Floating bar

5.4.3 抵抗弯矩图
Moment resistance diagram

构件斜截面承载力包括斜截面受剪承载力和斜截面受弯承载力两个方面。梁的斜截面受弯承载力是当斜截面破坏时,斜截面上的纵向受拉钢筋、箍筋、弯起钢筋等各自所提供的拉力(如图 5-25 所示)对剪压区 A 的内力矩之和,表达式为:

The bearing capacity of inclined section of the member includes two aspects: the shear capacity of inclined section and the flexural capacity of inclined section. The flexural capacity of inclined section of the beam is that when the inclined section is damaged, the sum of the internal moments of the tensile force provided by the longitudinal tensile reinforcement, stirrup and bent bar on the inclined section as shown in Fig. 5-25, about the shear compression zone A, the expression is:

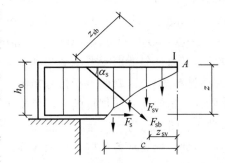

图 5-25 梁的斜截面受弯承载力
Figure 5-25 Flexural capacity of inclined section

$$M_u = F_s z + F_{sv} z_{sv} + F_{sb} z_{sb} \quad (5\text{-}25)$$

式中,M_u 为斜截面受弯承载力,kN·m; $F_s z$、$F_{sv} z_{sv}$、$F_{sb} z_{sb}$ 分别为纵筋、箍筋、弯起钢筋的斜截面受弯承载力,kN·m。

Where, M_u is the flexural capacity of inclined section, kN·m; $F_s z$、$F_{sv} z_{sv}$、$F_{sb} z_{sb}$ are the flexural capacity of inclined section of longitudinal reinforcement, stirrup and bent bar respectively, kN·m.

受弯构件出现斜裂缝后,斜截面上同时存在弯矩和剪力,斜截面和正截面所承受的弯矩一样,如果按跨中最大弯矩 M_{max} 计算配置的纵筋沿梁全长既不弯起也不截断,一定能满足斜截面的抗弯要求。但在实际工程设计中,纵筋有时要弯起或截断,此时斜截面的受弯承载力有可能无法满足,因此,必须考虑斜截面的受弯承载力问题。通常斜截面受弯承载力是不进行计算的,而是用梁内纵向钢筋的弯起、截断、锚固及箍筋的间距等构造措施来保证。

After diagonal cracks appear in flexural members, the inclined section has both bending moment and shear force, and the bending moment of inclined section and normal section is the same. If the longitudinal bars calculated and configured according to the maximum moment M_{max} at midspan is neither bent nor truncated along the full length of the beam, it must meet the flexural resistance requirements of inclined section. However, in actual engineering design, the longitudinal bars are sometimes bent or cut off. At this time, the flexural capacity of inclined section may not be satisfied. Therefore, the flexural capacity of inclined section must be considered. Usually, the flexural capacity of inclined section is not calculated, but is guaranteed by structural measures such as bending, truncation, anchoring of longitudinal bars and the stirrup spacing in the beam.

1. 抵抗弯矩图的确定
Determination of moment resistance diagram

弯矩图是根据梁的支承条件和荷载作用形式，用力学方法求出荷载对梁的各个正截面产生的弯矩设计值所绘出的弯矩分布图形，如图 5-26(a)所示的 M 图。

The moment diagram is the bending moment distribution diagram drawn by the mechanical method to obtain the design value of bending moment generated by the load on each normal section of the beam according to the supporting conditions of the beam and the form of load action, as the M diagram shown in Fig. 5-26(a).

为了保证斜截面受弯承载力和钢筋的可靠锚固，一般要通过绘制正截面的抵抗弯矩图（或称材料图）予以判断。抵抗弯矩图是按梁实际配置的纵向受力钢筋所确定的沿梁纵轴线方向各个正截面所能抵抗的弯矩所绘出的图形（也称正截面受弯承载力图），如图 5-26(a)所示的 M_u 图。因为 M_u 是由钢筋、混凝土两种材料提供的，所以 M_u 图也称材料图。抵抗弯矩图中纵坐标表示各相应正截面实际配置的纵向受力钢筋所能抵抗的弯矩值，横坐标表示相应的正截面位置。

In order to ensure the flexural capacity of inclined section and the reliable anchorage of reinforcement, it is generally necessary to draw the moment resistance diagram(or material diagram) of normal section to judge. The moment resistance diagram is the bending moment diagram that can be resisted by each normal section along the longitudinal axis of the beam determined according to the actual configuration of the longitudinally stressed reinforcement of the beam (also known as the flexural capacity diagram of normal section), as the M_u diagram shown in Fig. 5-26 (a). Because M_u is provided by reinforcement and concrete, the M_u diagram is also called the material diagram. The ordinate in the moment resistance diagram represents the bending moment value that can be resisted by the longitudinally stressed reinforcement actually configured in each corresponding normal section, and the abscissa represents the position of the corresponding normal section.

图 5-26(a)所示为承受均布荷载的简支梁，按跨中最大弯矩计算，需配置纵筋 2 Φ 22 + 1 Φ 20。如果实际配置的纵筋截面面积等于根据跨中最大弯矩计算的纵筋截面面积，则 M_u 图的外围水平线 cd 正好与 M 图上最大弯矩点相切。如果实际配置的纵筋截面面积大于根据跨中最大弯矩计算的纵筋截面面积，则可根据实际配筋量，求得 M_u 图外围水平线 cd 的位置。

Fig. 5-26(a) shows a simply supported beam uniformly loaded, calculated according to the maximum moment in the midspan, and longitudinal bars 2 Φ 22 + 1 Φ 20 are needed. If the cross-sectional area of the longitudinal bars actually configured is equal to the cross-sectional area of the longitudinal bars calculated according to the maximum moment in the mid-span, the peripheral horizontal line cd of the M_u diagram is just tangent to the point of maximum moment on the M diagram. If the cross-sectional area of the longitudinal bars actually configured is larger than the cross-sectional area of the longitudinal bars calculated according to the maximum moment in the midspan, the position of the horizontal line cd

on the periphery of the M_u diagram can be obtained according to the actual reinforcement amount.

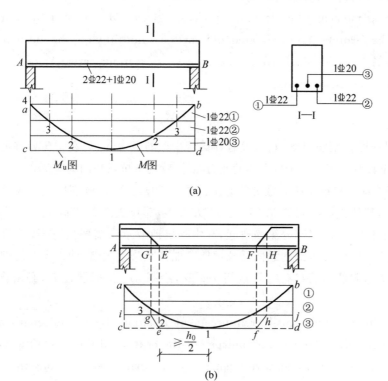

图 5-26　简支梁的抵抗弯矩图
(a) 纵筋不弯起时；(b) 纵筋弯起时

Figure 5-26　Moment resistance diagram of simply supported beam
(a) When the longitudinal reinforcement does not bend up; (b) When the longitudinal reinforcement bends up

由单筋矩形截面力的平衡方程

From the force equilibrium equation of singly reinforced rectangular section

$$\alpha_1 f_c b x = f_y A_s \tag{5-26}$$

得混凝土受压区高度 x：

The depth of concrete compression zone x can be written as

$$x = \frac{f_y A_s}{\alpha_1 f_c b} \tag{5-27}$$

将式(5-27)代入单筋矩形截面的力矩平衡方程，得所有纵向受拉钢筋所提供的受弯承载力 M_u，即 M_u 图外围水平线 cd 的位置：

Substituting Eq. (5-27) into the moment equilibrium equation of singly reinforced rectangular section, gives the flexural capacity M_u provided by all longitudinal tensile bars, that is, the position of the peripheral horizontal line cd of the M_u diagram：

$$M_u = f_y A_s \left(h_0 - \frac{x}{2}\right) = f_y A_s \left(h_0 - \frac{f_y A_s}{2\alpha_1 f_c b}\right) \tag{5-28}$$

每根纵向受拉钢筋所抵抗的弯矩（提供的受弯承载力）M_{ui}可近似按该根钢筋的截面面积A_{si}与所有纵向受拉钢筋总截面面积A_s的比值乘以M_u得到，即

The bending moment (the flexural capacity provided) M_{ui} resisted by each longitudinal tensile steel bar can be approximated obtained by multiplying the ratio of the cross-sectional area A_{si} of the steel bar to the total cross-sectional area A_s of all longitudinal tensile steel bars by M_u, that is:

$$M_{ui} = \frac{A_{si}}{A_s} M_u \tag{5-29}$$

当全部纵向受拉钢筋沿梁通长，没有在梁跨中截断或弯起，并在支座处有足够的锚固长度时，则沿梁全长各个正截面抵抗弯矩的能力相等，梁的抵抗弯矩图为矩形$abdc$，如图5-26(a)所示。每一根纵向受拉钢筋所能抵抗的弯矩按式(5-29)计算，也分别用水平线示意于图5-26中。除跨中外，其他正截面处的M_u都比M大得多，邻近支座处正截面受弯承载力大大富余。在实际设计中，往往将部分纵筋弯起，利用其抗剪或承受支座负弯矩，达到充分利用钢筋、经济的效果。因为梁底部的纵向受拉钢筋不能在梁跨中截断，伸入支座的钢筋也不能少于2根，所以能弯起的钢筋只有③号钢筋1$\underline{\Phi}$20，绘制M_u图时要注意，必须将它画在M_u图的外侧。

If all the longitudinal tensile reinforcement is along the whole length of the beam without being cut off or bent in the midspan of the beam, and there is sufficient anchorage length at the support, the flexural resistance of each normal section along the whole length of the beam is equal, and the resistance of the beam is equal. The moment resistance diagram is rectangle $abdc$, as shown in Fig. 5-26(a). The bending moment that each longitudinal tensile steel bar can resist is calculated according to Eq. (5-29), and it is also indicated by the horizontal line in Fig. 5-26. Except for the midspan, M_u at other normal section is much larger than M, and the flexural capacity of normal section near the support is greatly spared. In the actual design, some longitudinal steel bars are often bent up, and its shear resistance or bearing negative bending moment is used to achieve the effect of making full use of reinforcement and economical effect. Because the longitudinal tensile steel bars at the bottom of the beam cannot be cut off in the midspan, and the number of steel bars protruding into the support should not be less than 2, so the only steel bar that can be bent is the steel bar No. ③ 1 $\underline{\Phi}$ 20. When drawing the M_u diagram, it should be noted that it must be drawn on the outside of the M_u diagram.

任务 5-1：计算图 5-26 中①、②、③号钢筋所提供的受弯承载力。

Task 5-1: Calculate the flexural capacity provided by the steel bars ①, ② and ③ in Fig. 5-26.

2. 抵抗弯矩图的作用

   Function of moment resistance diagram

（1）反映构件中材料的利用程度。为了满足$M_u \geq M$，即抵抗弯矩M_u不小于设计弯矩

M,M_u 图必须将 M 图包纳在内,才能保证梁的各个正截面受弯承载力。M_u 图越接近 M 图,表明纵向钢筋的利用越充分,构件设计越经济。

(1) Reflect the utilization degree of materials in the member. In order to satisfy $M_u \geqslant M$, that is, the resisting moment M_u is not less than the design moment M, the M_u diagram must include the M diagram in order to ensure the flexural capacity of each normal section of the beam. The closer the M_u diagram is to the M diagram, it shows that the more fully the longitudinal bars are used, and the more economical the member design is.

(2) 确定弯起钢筋的弯起位置。为节约钢筋,可将一部分纵筋在不需要该纵筋的受弯承载力处予以弯起,用于斜截面抗剪和抵抗支座负弯矩,如图 5-26(b)所示。

(2) Determine the bending position of the bent bar. In order to save steel bars, a part of the longitudinal bars can be bent up at the place where the flexural capacity of the longitudinal bars is not needed, which is used for the shear resistance of the inclined section and the resistance to the negative bending moment of the support, as shown in Fig. 5-26(b).

(3) 确定纵筋的截断位置。可在不需要该纵筋的受弯承载力处考虑将纵筋截断,从而确定纵筋的实际截断位置。

(3) Determine the truncation position of longitudinal bars. The longitudinal bars can be considered to be truncated at the place where the longitudinal bars are not needed for flexural capacity, so as to determine the actual cutting position of the longitudinal bars.

3. 充分利用截面

Fully utilized section

视频 5-3

充分利用截面是从受拉边算起,某根钢筋的抵抗弯矩图 M_{ui} 第一次与弯矩图 M 相交的截面,过了此截面,钢筋强度就不需要充分发挥。如图 5-26(a)所示,抵抗弯矩图中跨中截面 1 处①、②、③号钢筋的强度都被充分利用;跨中截面 2 处,①、②号钢筋的强度被充分利用,而③号钢筋不再需要;跨中截面 3 处,①号钢筋的强度被充分利用,而②、③号钢筋不再需要。通常把截面 1 称为③号钢筋的充分利用截面,截面 2 称为②号钢筋的充分利用截面,截面 3 称为①号钢筋的充分利用截面。

Fully utilized section is the section where the moment resistance diagram M_{ui} of a steel bar intersects with the moment diagram M for the first time, starting from the tension side. After this section, the strength of steel bars does not need to be fully exerted. As shown in Fig. 5-26(a), the strength of steel bars ①, ②, and ③ in the mid-span section 1 in the moment resistance diagram are fully utilized; in the mid-span section 2, the strength of steel bars ① and ② are fully utilized, while No. ③ steel bar is no longer needed; in the mid-span section 3, the strength of No. ① steel bar is fully utilized, while No. ② and ③ steel bars are no longer needed. Usually, section 1 is called the fully utilized section of No. ③ steel bar, section 2 is called the fully utilized section of No. ② steel bar, and section 3 is called the fully utilized section of No. ① steel bar.

4. 不需要截面（理论截断截面）
Unnecessary section (theoretical cutoff section)

不需要截面是从梁受拉边算起，某根钢筋的抵抗弯矩图 M_{ui} 第二次与弯矩图 M 相交的截面。对正截面受弯承载力要求来说，这根钢筋在此处既然是多余的，理论上便可以切断，实际切断截面还将延伸过此点一定长度。

The unnecessary section is the section where the moment resistance diagram M_{ui} of the certain steel bar intersects with the moment diagram M for the second time, starting from the tension side of the beam. For the flexural capacity requirement of normal section, since the steel bar is redundant here, it can theoretically be cut off, and the actual cutoff section will extend over this point for a certain length.

如图 5-26(a)所示，抵抗弯矩图中跨中截面 2 处，①、②号钢筋的强度被充分利用，而③号钢筋不再需要，通常把截面 2 称为③号钢筋的不需要截面或理论截断截面；截面 3 处，①号钢筋的强度被充分利用，而②、③号钢筋不再需要，通常把截面 3 称为②号钢筋的不需要截面或理论截断截面；截面 a 处①号钢筋不再需要，通常把截面 a 称为①号钢筋的不需要截面或理论截断截面。

As shown in Fig. 5-26(a), the strength of No. ①、② steel bars are fully utilized in the mid-span section 2 in the moment resistance diagram, while No. ③ steel bar is no longer needed. Usually, section 2 is called the unnecessary section of No. ③ steel bar or theoretical section of cutoff; at section 3, the strength of No. ① steel bar is fully utilized, while No. ② and ③ steel bars are no longer needed, and section 3 is usually referred to as the unnecessary section or theoretical section of cutoff of No. ② steel bar; No. ① steel bar is no longer needed at section a, usually section a is called the unnecessary section or theoretical cutoff section of No. ① steel bar.

特别提示 5-6

5.4.4 纵向钢筋的弯起
Bending of longitudinal bars

为保证梁正截面、斜截面承载力，纵筋的弯起必须考虑以下三方面的要求。

1. 保证正截面受弯承载力
Ensure the flexural capacity of the normal section

一部分纵筋弯起后，剩下的纵筋数量减少，正截面受弯承载力降低。为了保证正截面受弯承载力，纵筋的弯起点必须位于按正截面受弯承载力计算该钢筋的充分利用截面之外，使抵抗弯矩图包在设计弯矩图的外面，不能切入设计弯矩图以内。

2. 保证斜截面受弯承载力
Ensure the flexural capacity of the inclined section

为了保证斜截面受弯承载力，至少要求斜截面受弯承载力与正截面受弯承载力等强，

《混凝土结构设计标准》(GB/T 50010—2010)规定纵向受力钢筋应伸过其充分利用截面至少 $0.5h_0$ 才能弯起。弯起点可设在按正截面受弯承载力计算不需要该钢筋的截面之前。并且,梁中弯起钢筋与梁纵轴线的交点 a 应位于该钢筋正截面抗弯的理论截断截面(不需要截面)之外,如图 5-27 所示。连续梁中,把跨中承受正弯矩的纵向钢筋弯起,且作为承担支座负弯矩的钢筋时也必须遵循这些规定。

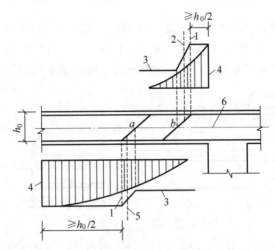

图 5-27 弯起钢筋弯起点与弯矩图的关系

1—受拉区的弯起点;2—按计算不需要"b"的截面;3—正截面受弯承载力图;
4—按计算充分利用钢筋"a"或"b"强度的截面;5—按计算不需要钢筋"a"的截面;6—梁中心线

Figure 5-27　Relationship between bending starting point and moment diagram of bent bar
1—Bending starting point of tension zone;2—The section that does not need "b" according to the calculation;3—Flexural capacity diagram of normal section;4—The section that fully utilizes the strength of the steel bar "a" or "b" according to the calculation;5—The section that does not need the steel bar "a" according to the calculation;6—Beam centerline

3. 弯起钢筋弯终点的位置
Position of the end point of bent bars

弯起纵向钢筋的目的是用于斜截面抗剪和抵抗支座处弯矩,弯起钢筋的数量是由斜截面受剪承载力计算确定的。为了使每根弯起钢筋都能与斜裂缝相交,以保证斜截面的受剪和受弯承载力,弯起钢筋的弯终点到支座边或到前一排弯起钢筋弯起点的距离均应小于箍筋的最大间距 s_{max},其值见表 5-2 中 $V>0.7f_tbh_0$ 一栏的规定。

5.4.5　钢筋的锚固
Anchorage of reinforcement

1. 锚固的作用
Role of anchoring

当简支梁在支座边缘截面出现斜裂缝时,混凝土上的力转移给钢筋承担,纵向钢筋的应力突然增加,此时梁的抗弯能力还取决于纵向钢筋在支座处的锚固。若无足够的锚固长度，纵筋与混凝土将发生相对滑移，导致斜裂缝宽度显著增大，甚至使纵筋从支座中拔出而导致支座处的黏结锚固破坏，降低梁的承载力，这种情况容易发生在靠近支座处有较大集中荷载时。

钢筋的锚固是通过钢筋埋置段或机械措施将钢筋所受的力传给混凝土，钢筋能否可靠地锚固在混凝土中直接影响到这两种材料的共同工作，关系到结构和构件的安全和材料强度的充分利用。

为了保证钢筋混凝土构件正常可靠地工作，防止纵向受力钢筋在支座处被拔出导致构件破坏，梁纵向受力钢筋应伸入支座一定的锚固长度l_a，如图 5-28 所示。考虑到支座处同时又存在横向压应力的有利作用，支座处的锚固长度可比基本锚固长度略小。

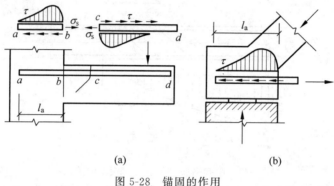

图 5-28 锚固的作用
（a）悬挑梁；（b）屋架

Figure 5-28 The role of anchoring
(a) Cantilever beam; (b) Roof truss

2. 钢筋的锚固长度
Anchorage length of reinforcement

1) 基本锚固长度

1) Basic anchorage length

基本锚固长度是指钢筋的拉应力σ_s达到屈服强度f_y时，尚未产生黏结破坏所需的锚固长度。《混凝土结构设计标准》(GB/T 50010—2010)8.3.1 条规定，当计算中充分利用钢筋的抗拉强度时(如悬挑梁的上部受力钢筋)，根据钢筋拔出试验(如图 5-29 所示)列出锚固钢筋的力的平衡方程：

The basic anchorage length refers to the anchorage length required for bond failure when the tensile stress σ_s of the reinforcement reaches the yield strength f_y. Article 8.3.1 of "Standard for design of concrete structures"(GB/T 50010—2010)stipulates that when the tensile strength of reinforcement is fully utilized in the calculation (such as the upper stressed steel bars of cantilever beams), according to the pull-out test of the steel bar(as shown in Fig. 5-29), the force equilibrium equation of the anchoring reinforcement is listed：

$$\tau_u \pi d l_{ab} = f_y A_s \tag{5-30}$$

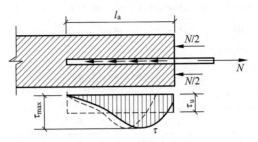

图 5-29 钢筋拔出试验

Figure 5-29 Pull-out test of steel bar

式中,τ_u 为锚固钢筋与混凝土之间的平均剪应力,N/mm²;d 为锚固钢筋的直径,mm;l_{ab} 为锚固钢筋的基本锚固长度,见表 5-3,mm;f_y 为锚固钢筋的抗拉强度设计值,N/mm²;A_s 为锚固钢筋的截面面积,mm²。

Where, τ_u is the average shear stress between anchoring steel bar and concrete, N/mm²; d is the diameter of anchoring reinforcement, mm; l_{ab} is the basic anchorage length of anchoring reinforcement, see Table 5-3, mm; f_y is the design value of tensile strength of anchoring reinforcement, N/mm²; A_s is the cross-sectional area of anchoring reinforcement, mm².

表 5-3 受拉钢筋基本锚固长度 l_{ab}、l_{abE}

Table 5-3 Basic anchorage length l_{ab}、l_{abE} of tensile reinforcement

钢筋种类	抗震等级	混凝土强度等级							
		C25	C30	C35	C40	C45	C50	C55	>C60
HPB300	一、二级	$39d$	$35d$	$32d$	$29d$	$28d$	$26d$	$25d$	$24d$
	三级	$36d$	$32d$	$29d$	$26d$	$25d$	$24d$	$23d$	$22d$
	四级、非抗震	$34d$	$30d$	$28d$	$25d$	$24d$	$23d$	$22d$	$21d$
HRB400 HRBF400 RRB400	一、二级	$46d$	$40d$	$37d$	$33d$	$32d$	$31d$	$30d$	$29d$
	三级	$42d$	$37d$	$34d$	$30d$	$29d$	$28d$	$27d$	$26d$
	四级、非抗震	$40d$	$35d$	$32d$	$29d$	$28d$	$27d$	$26d$	$25d$
HRB500 HRBF500	一、二级	$55d$	$49d$	$45d$	$41d$	$39d$	$37d$	$36d$	$35d$
	三级	$50d$	$45d$	$41d$	$38d$	$36d$	$34d$	$33d$	$32d$
	四级、非抗震	$48d$	$43d$	$39d$	$36d$	$34d$	$32d$	$31d$	$30d$

注:①混凝土强度等级应取锚固区的混凝土强度等级;②l_{abE} 为受拉钢筋抗震基本锚固长度。

由式(5-30)得纵向受拉钢筋的基本锚固长度 l_{ab} 为

From Eq. (5-30), the basic anchorage length l_{ab} of the longitudinal tensile reinforcement is:

$$l_{ab} = \frac{f_y A_s}{\tau_u \pi d} = \frac{1}{4} \cdot \frac{f_y}{\tau_u} d = \alpha \frac{f_y}{f_t} d \quad (5\text{-}31)$$

式中,α 为锚固钢筋的外形系数,见表 5-4;f_t 为混凝土轴心抗拉强度设计值,N/mm²,当混凝土强度等级高于 C60 时,按 C60 取值。

Where, α is the shape coefficient of anchorage steel bar, see Table 5-4; f_t is the axial tensile strength design value of concrete, N/mm², when the concrete strength grade is higher than C60, the value is C60.

表 5-4　锚固钢筋的外形系数 α

Table 5-4　Shape coefficient of anchorage steel bar α

钢筋类型	光圆钢筋	带肋钢筋	螺旋肋钢丝	三股钢绞线	七股钢绞线
α	0.16	0.14	0.13	0.16	0.17

注：光圆钢筋末端应做 180°弯钩，弯后平直段长度不应小于 $3d$，但作为受压钢筋时可不做弯钩。

2）锚固长度修正系数

2) Anchorage length correction coefficient

按式 (5-31) 计算出的基本锚固长度 l_{ab} 应乘以下修正系数 ζ_a：

（1）当带肋钢筋的公称直径大于 25mm 时，修正系数取 1.1。

（2）具有环氧树脂涂层的带肋钢筋，修正系数取 1.25。

（3）施工过程中易受扰动的纵向受拉钢筋，修正系数取 1.1。

（4）当纵向受力钢筋的实际配筋面积大于其设计计算面积时，修正系数取设计计算面积与实际配筋面积的比值，但对有抗震设防要求及直接承受动力荷载的结构构件，不应考虑此项修正。

（5）锚固长度范围内纵向受力钢筋周边保护层厚度为 $3d$ 时，修正系数取 0.8；保护层厚度不小于 $5d$ 时，修正系数取 0.7；中间时按内插取值。此处 d 为锚固钢筋直径。

The basic anchorage length l_{ab} calculated according to Eq. (5-31) should be multiplied by the following correction coefficient ζ_a:

(1) When the nominal diameter of the ribbed bar is greater than 25mm, the correction coefficient is taken as 1.1.

(2) For epoxy-coated ribbed reinforcement, the correction coefficient is 1.25.

(3) For longitudinal tensile reinforcement which is easily disturbed during construction, the correction coefficient should be 1.1.

(4) When the actual reinforcement area of the longitudinally stressed reinforcement is larger than the design calculation area, the correction coefficient should be the ratio of the design calculation area to the actual reinforcement area, but for structural members with seismic fortification requirements and directly subjected to dynamic loads, it should not be considered.

(5) When the thickness of concrete cover around the longitudinally stressed reinforcement within the range of the anchorage length is $3d$, the correction coefficient is taken as 0.8; when the thickness of concrete cover is not less than $5d$, the correction coefficient is 0.7; in the middle the interpolation value is taken, where d is the diameter of the anchoring reinforcement.

对普通钢筋，当锚固长度修正系数多于一项时，可按连乘计算。但不宜小于 0.6；对预应力筋，可取 1.0。经修正后的锚固长度不应小于 200mm。

For ordinary steel bars, when the anchorage length correction coefficient is more than one item, it can be calculated by continuous multiplication, but not less than 0.6; for prestressed bars, 1.0 can be taken. The modified anchorage length should not be less than 200mm.

3) 受拉钢筋的锚固长度 l_a 和抗震锚固长度 l_{aE}

3) Tensile reinforcement anchorage length l_a and seismic anchorage length l_{aE}

受拉钢筋的锚固长度 l_a 和抗震锚固长度 l_{aE} 应根据锚固条件按下列公式计算,且不应小于 200mm:

The anchorage length l_a and the seismic anchorage length l_{aE} of the tensile reinforcement should be calculated by the following formula according to the anchorage condition, and should not be less than 200mm:

非抗震设计时: $$l_a = \zeta_a l_{ab} \tag{5-32a}$$

抗震设计时: $$l_{aE} = \zeta_{aE} l_a \tag{5-32b}$$

Non-seismic design: $$l_a = \zeta_a l_{ab} \tag{5-32a}$$

Seismic design: $$l_{aE} = \zeta_{aE} l_a \tag{5-32b}$$

式中,l_a 为受拉钢筋的锚固长度,mm；l_{ab} 为受拉钢筋的基本锚固长度,mm；l_{aE} 为抗震锚固长度,mm；ζ_a 为锚固长度修正系数；ζ_{aE} 为抗震锚固长度修正系数,对一、二级抗震等级取 1.15,对三级抗震等级取 1.05,对四级抗震等级取 1.0。

Where, l_a is the anchorage length of tensile reinforcement, mm; l_{ab} is the basic anchorage length of tensile reinforcement, mm; l_{aE} is the seismic anchorage length, mm; ζ_a is the anchorage length correction coefficient; ζ_{aE} is the correction coefficient of seismic anchorage length, which is 1.15 for the first and second seismic grades, 1.05 for the third seismic grade, and 1.0 for the fourth seismic grade.

当锚固钢筋的保护层厚度不大于 $5d$ 时,锚固钢筋长度范围内应设置横向构造钢筋,其直径不应小于 $d/4$(d 为锚固钢筋的最大直径)；对梁、柱等构件间距不应大于 $5d$,对板、墙等构件间距不应大于 $10d$,且均不应大于 100mm。此处 d 为锚固钢筋的最小直径。

When the thickness of concrete cover of anchoring reinforcement is not more than $5d$, the transverse structural reinforcement should be arranged within the anchorage length, and its diameter should not be less than $d/4$ (d is the maximum diameter of anchoring reinforcement); For beams, columns, etc., the distance should not be greater than $5d$; for slabs and walls, the distance should not be greater than $10d$, and should not be greater than 100mm. Here d is the minimum diameter of anchoring reinforcement.

若纵向受力钢筋伸入支座范围的锚固长度不符合上述要求,应采取弯钩或机械锚固措施,如在梁端将钢筋向上弯,加焊横向短钢筋(贴焊锚筋)、镦头、焊锚固钢板、将钢筋端部焊接在梁端预埋件上等,如图 5-30 所示。

If the anchorage length of the longitudinally stressed reinforcement extending into the support does not meet the above requirements, hooks or mechanical anchoring measures should be taken, such as bending the steel bar upward at the beam end, welding the transverse short steel bars (welding anchor bars), heading, welding the anchoring steel plate, welding the end of the steel bar on the embedded parts at the end of the beam, etc., as shown in Fig. 5-30.

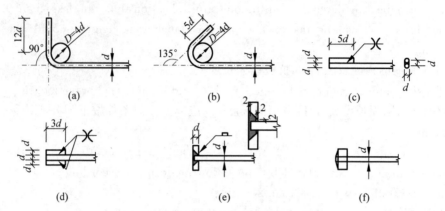

图 5-30 纵向钢筋弯钩和机械锚固措施

(a) 末端带 90°弯钩；(b) 末端带 135°弯钩；(c) 末端一侧贴焊锚筋；(d) 末端两侧贴焊锚筋；
(e) 末端与钢板穿孔塞焊；(f) 末端带螺栓锚头

Figure 5-30　Hook and mechanical anchoring measures of longitudinal bars

(a) End with 90° hook；(b) End with 135° hook；(c) Welding anchoring bars on one side of the end；
(d) Welding anchoring bars on both sides of the end；(e) Perforation plug welding with steel plate at the end；
(f) Anchor head with bolt at the end

当纵向受拉普通钢筋末端采用弯钩或机械锚固措施时，包括弯钩或锚固端头在内的锚固长度(投影长度)可取基本锚固长度 l_{ab} 的 60%。弯钩和机械锚固的形式和技术要求应符合有关规定。

When hook or mechanical anchoring measures are used at the end of the longitudinal tensile ordinary reinforcement, the anchorage length (projection length) including the hook or anchorage end can be taken as 60% of the basic anchorage length l_{ab}. The form and technical requirements of hook and mechanical anchorage should meet the relevant regulations.

500MPa 级带肋钢筋末端采用弯钩锚固措施时，当直径 $d \leqslant 25\text{mm}$ 时，钢筋弯折的弯钩内直径不应小于钢筋直径的 6 倍；当直径 $d > 25\text{mm}$ 时，钢筋弯折的弯钩内直径不应小于钢筋直径的 7 倍。

When the end of 500MPa ribbed steel bar adopts hook anchoring measure, when the diameter $d \leqslant 25\text{mm}$, the inner diameter of the bending hook of the steel bar should not be less than 6 times the diameter of the steel bar; when the diameter $d > 25\text{mm}$, the inner diameter of the bending hook of the steel bar should not be less than 7 times the diameter of the steel bar.

4) 受压钢筋的锚固长度

4) Anchorage length of compression reinforcement

混凝土结构中的纵向受压钢筋，当充分利用其抗压强度并需锚固时，其锚固长度不应小于受拉钢筋锚固长度的 70%。受压钢筋不应采用末端弯钩的锚固形式。

When the longitudinal compression reinforcement in the concrete structure is fully utilized its compressive strength and needs to be anchored, its anchorage length should not be less than 70% of the anchorage length of the tensile reinforcement. The anchorage form of the end hook should not be used for the compression reinforcement.

3. 节点或支座处纵筋的锚固

Anchorage of longitudinal reinforcement at nodes or supports

1) 简支端支座处纵筋的锚固

1) Anchorage of longitudinal reinforcement at simply supported end supports

如图 5-31 所示,简支梁和连续梁的简支端下部纵向受力钢筋伸入支座范围内不应少于两根。伸入支座范围内的锚固长度 l_{as} 应符合下列条件:

当 $V \leqslant 0.7 f_t b h_0$ 时:$l_{as} \geqslant 5d$(d 为锚固钢筋直径)。

当 $V > 0.7 f_t b h_0$ 时:$l_{as} \geqslant 12d$(带肋钢筋);$l_{as} \geqslant 15d$(光圆钢筋)。

梁简支端支座截面上部应配负弯矩钢筋,其数量不小于下部纵向受力钢筋的 1/4,且不少于两根。

2) 支承在砌体结构上的钢筋混凝土独立梁的锚固

2) Anchorage of reinforced concrete isolated beams supported on masonry structures

由于梁端约束较小,在纵筋的锚固长度范围内应加强配箍,应配置不少于两个箍筋,直径不宜小于纵向受力钢筋最大直径的 0.25 倍,间距不宜大于纵向受力钢筋最小直径的 10 倍。当采取机械锚固措施时,机械锚固的锚固力较多集中在锚头附近,有较大的挤压力,此处混凝土容易破碎,应加以约束,因此箍筋间距不宜大于纵向受力钢筋最小直径的 5 倍。

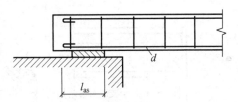

图 5-31 简支梁和连续梁简支端纵筋的锚固

Figure 5-31 Anchorage of longitudinal reinforcement at simply supported end of simply supported beam and continuous beam

3) 非框架梁纵筋的锚固

3) Anchorage of longitudinal reinforcement of non-frame beams

(1) 非框架梁上部纵向钢筋在端支座的锚固要求。当设计按铰接时,钢筋平直段伸至端支座对边后弯折,且平直段长度 $\geqslant 0.35 l_{ab}$,弯折段长度为 $15d$(d 为纵向钢筋直径);当充分利用钢筋的抗拉强度时,平直段伸至端支座对边后弯折,且平直段长度 $\geqslant 0.6 l_{ab}$,弯折段长度为 $15d$,如图 5-32 所示。设计者应在施工图中注明采用何种构造,当多数采用同种构造时可在图注中统一写明,并将少数不同之处在图中注明。

(2) 非框架梁下部纵向钢筋在中间支座和端支座的锚固要求。非框架连续梁下部纵向钢筋在中间支座和端支座的锚固长度,对于带肋钢筋为 $12d$(d 为纵向钢筋直径),如图 5-32 所示。当计算中需要充分利用下部纵向钢筋的抗压强度或抗拉强度,或具体工程有特殊要求时,其锚固长度应由设计者按照《混凝土结构设计标准》(GB/T 50010—2010)的相关规定另行变更。

当非框架连续梁下部纵向钢筋在端支座的锚固长度不满足直锚 $12d$ 要求时,可伸至支座对边弯折,弯折前的平直段长度,对于带肋钢筋至少为 $7.5d$(d 为纵向钢筋直径),如图 5-33 所示。

伸至支座对边弯折且设计按铰接时：$\geqslant 0.35l_{ab}$；充分利用钢筋的抗拉强度时：$\geqslant 0.6l_{ab}$；伸入端支座直段长度满足l_a时，可直锚。

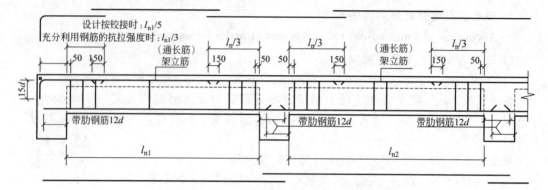

图 5-32　非框架梁纵向钢筋在支座处的锚固

Figure 5-32　Anchorage of longitudinal reinforcement of non-frame beam at support

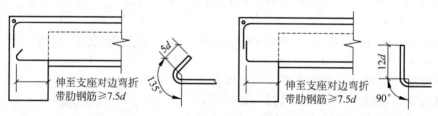

图 5-33　非框架梁下部纵筋在端支座处弯锚构造

Figure 5-33　Bending anchor structure of lower longitudinal reinforcement of non-frame beam at end support

（3）当非框架梁配有受扭纵向钢筋时，受扭纵筋锚入支座的长度为l_a，如图 5-34（a）所示，当纵筋伸入端支座的直线段锚固长度满足l_a时可直锚。在端支座直锚长度不足时可伸至端支座对边后弯折，且平直段长度$\geqslant 0.6l_{ab}$，弯折段长度为$15d$（d为纵向钢筋直径），如图 5-34（b）所示，设计者应在图中注明。

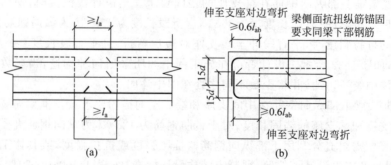

图 5-34　非框架梁受扭纵筋的锚固
(a) 直线锚固；(b) 弯折锚固

Figure 5-34　Anchorage of torsional longitudinal reinforcement of non-frame beam
(a) Linear anchorage; (b) Bending anchorage

(4) 当梁纵筋兼做温度应力钢筋时,其锚入支座的长度由设计确定。

4) 非抗震设计时框架梁下部纵向钢筋在节点或中间支座的锚固

4) Anchorage of lower longitudinal bars of frame beam at nodes or intermediate supports in non-seismic design

非抗震设计时框架梁在中间支座处承受负弯矩作用,上部纵筋受拉,应贯穿中间支座,下部纵筋受压,下部纵向钢筋在中间支座处应满足下列锚固要求。

(1) 当设计中不利用支座下部纵向钢筋强度时,其伸入节点或支座的锚固长度对于带肋钢筋不小于 $12d$(d 为纵向钢筋的最大直径)。

(2) 当设计中充分利用支座下部纵向钢筋的抗拉强度时,钢筋可采用直线方式锚固在节点或支座内,锚固长度不应小于钢筋的受拉锚固长度 l_a。

(3) 当设计中充分利用支座下部纵向钢筋的抗压强度时,应受压钢筋锚固在节点或中间支座内,直线锚固长度不应小于 $0.7l_a$。这是考虑到在实际结构中,压力主要靠混凝土传递,受压钢筋作用较小,对锚固长度要求不高。

5) 抗震设计时框架梁纵筋在支座处的锚固

5) Anchorage of longitudinal reinforcement of frame beam at support in seismic design

抗震设计时框架梁纵向钢筋在支座处的锚固要求如图 5-35 所示。图 5-35 中跨度值 l_n 为左跨 l_{ni} 和右跨 l_{ni+1} 的较大值,其中 $i=1,2,\cdots$。h_c 为柱截面沿框架方向的高度。梁上部不通长钢筋与非贯通钢筋直径相同时,连接位置宜位于跨中 $l_n/3$ 范围内;梁下部钢筋连接位置宜位于支座 $l_n/3$ 范围内,且在同一连接区段内钢筋接头面积百分率不宜大于 50%。当上柱截面尺寸小于下柱截面尺寸时,梁上部纵筋锚固长度的起算位置应为上柱内边缘,梁下部纵筋锚固长度的起算位置为下柱内边缘。

当梁下部纵筋不能在柱内锚固时,也可在节点或支座外梁中弯矩较小处设置搭接接头,搭接长度的起始点至节点或支座边缘的距离不应小于 $1.5h_0$,如图 5-36 所示。相邻跨钢筋直径不同时,搭接位置位于钢筋较小直径一跨。

当柱截面尺寸不足,梁下部纵筋在柱中的直线锚固无法保证锚固长度时,可采用 90°弯折锚固的形式,或采取在钢筋端部加锚头(锚板)等机械锚固措施,如图 5-37 所示。

6) 悬挑梁纵筋的锚固

6) Anchorage of longitudinal reinforcement of cantilever beam

悬挑梁是承受负弯矩的受弯构件,梁上部的纵向受拉钢筋是承受负弯矩的钢筋。悬挑梁上部第一排应有不少于两根纵向受力钢筋,应从钢筋强度被充分利用的截面(即支座边缘截面)起延伸至悬挑梁外端,且竖直向下弯折不小于 $12d$,其余钢筋不应在梁的上部截断。当具体工程需将部分上部纵筋从规定的弯起点位置斜向下弯时,在弯折钢筋的终点外应留有平行于梁轴线方向的锚固长度,在受压区不应小于 $10d$,在受拉区不应小于 $20d$,应由设计者另外注明,如图 5-38 所示。

悬挑梁上部第二排钢筋延伸至 $0.75l$ 位置,l 为自柱(梁)边算起的悬挑净长。若梁的下部纵向受力钢筋在计算上作为受压钢筋时,伸入支座中的长度≥$15d$,如图 5-39 所示。

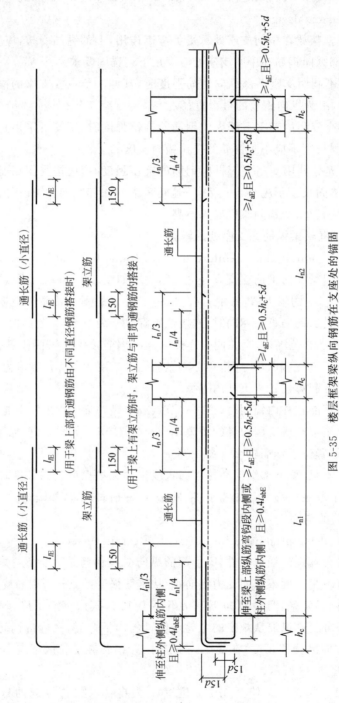

图 5-35 楼层框架梁纵向钢筋在支座处的锚固

Figure 5-35 Anchorage of longitudinal reinforcement of frame beam at the support on the floor

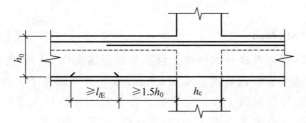

图 5-36 中间层中间节点框架梁下部纵筋在节点外搭接

Figure 5-36　Lower longitudinal reinforcement of frame beam of intermediate node of middle layer lapped outside the node

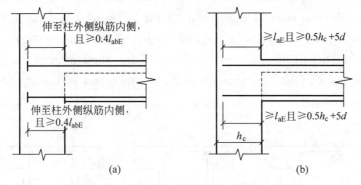

图 5-37　框架梁下部纵筋在端支座中的锚固
（a）端支座加锚头（锚板）锚固；(b) 端支座直锚

Figure 5-37　Anchorage of lower longitudinal reinforcement of frame beam at end support
(a) End support with anchor head (anchor plate) anchorage；(b) End support with straight anchor

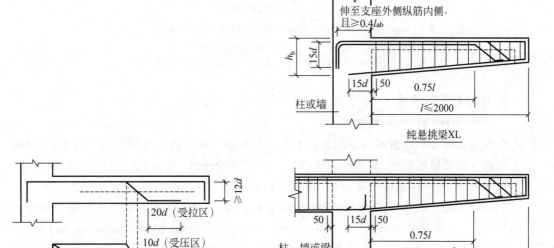

图 5-38　悬挑梁纵筋的锚固要求

Figure 5-38　Anchorage requirements of longitudinal reinforcement of cantilever beam

图 5-39　悬挑梁的锚固要求

Figure 5-39　Anchorage requirements of cantilever beam

4. 箍筋的锚固
Anchorage of stirrups

箍筋是受拉钢筋，必须有良好的锚固。箍筋端部应锚固在受压区内，封闭式箍筋末端常用 135°弯钩，且弯钩端头直线段长度不小于 50mm 或 5 倍箍筋直径。若采用 90°弯钩，则箍筋受拉时弯钩会翘起，导致混凝土保护层崩裂。若梁两侧有楼板与梁整浇时，也可采用 90°弯钩，但弯钩端部直线段长度不小于 10 倍箍筋直径，如图 5-40 所示。

Stirrups are tensile reinforcement which must be well anchored. The end of the stirrup should be anchored in the compression zone, 135° hook is commonly used at the end of the closed stirrup, and the length of the straight section at the end of the hook is not less than 50mm or 5 times the diameter of the stirrup. If the 90° hook is used, the hook will warp when the stirrup is tensioned, causing the collapse of the concrete cover. If there are floors on both sides of the beam, 90° hook can also be used, but the length of the straight section at the end of the hook is not less than 10 times the diameter of the stirrup, as shown in Fig. 5-40.

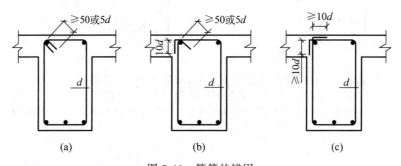

图 5-40 箍筋的锚固
（a）135°弯钩；（b）135°和 90°弯钩；（c）90°弯钩

Figure 5-40 Anchorage of stirrups
(a) 135° hook；(b) 135° and 90° hook；(c) 90° hook

5. 弯起钢筋的锚固
Anchorage of bent bars

弯起钢筋的端部应留有一定的锚固长度。弯起钢筋的弯终点外应留有平行于梁纵向轴线方向的直线段锚固长度，其长度在受拉区$\geqslant 20d$，在受压区$\geqslant 10d$（d 为弯起钢筋的直径）。对于光圆钢筋，在末端还应设置弯钩，如图 5-41 所示。

There should be a certain anchorage length at the end of the bent bar. The anchorage length of the straight section parallel to the longitudinal axis of the beam should be left outside the bending end point of the bent bar. Its length in the tension zone is $\geqslant 20d$, in the compression zone is $\geqslant 10d$ (d is the diameter of the bent bar). For the round bar, a hook should also be set at the end, as shown in Fig. 5-41.

位于梁底或梁顶的角筋以及梁截面高度两侧的钢筋不宜弯起。

The corner bars at the bottom or top of the beam and the bars on both sides of the beam section height should not be bent up.

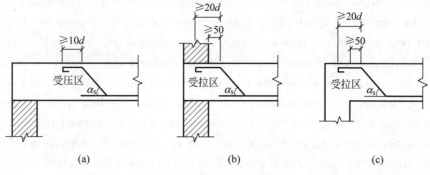

图 5-41 弯起钢筋的锚固
(a) 受压区锚固；(b),(c) 受拉区锚固
Figure 5-41 Anchorage of bent bars
(a) Anchorage in compression zone；(b),(c) Anchorage in tension zone

 拓展知识 5-3

5.4.6 纵向钢筋的截断
Cutoff of longitudinal reinforcement

1. 纵筋的截断原则

Cutoff principle of longitudinal reinforcement

承受正弯矩的梁下部纵筋不能在梁跨中受拉区截断，因为梁的正弯矩图形的范围比较大，受拉区几乎覆盖梁的整个跨度，且截断处钢筋截面面积骤减，应力突增，将过早出现裂缝。可以将梁下部承受正弯矩的纵向受力钢筋在不需要的截面处弯起，作为抗剪钢筋或作为承受支座负弯矩的钢筋，而不在跨中截断。

 视频 5-4

The lower longitudinal bars of the beam bearing the positive bending moment cannot be truncated in the tension zone at midspan of the beam, because the range of the positive bending moment diagram of the beam is relatively large, the tension zone almost covers the entire span of the beam, and the cross-sectional area of the reinforcement at the truncation point decreases sharply. A sudden increase in stress will cause premature cracks. The longitudinally stressed reinforcement bearing the positive bending moment at the lower part of the beam is bent at the unnecessary section as the shear reinforcement or the reinforcement bearing the negative bending moment of the support, instead of being truncated in the midspan.

对于在支座附近的负弯矩区段内梁顶的纵向受拉钢筋，因为负弯矩区段的范围不大，往往采用截断的方式减少纵筋的数量，但不宜在受拉区截断。连续梁和框架梁承受支座负弯矩的纵向受拉钢筋向跨内延伸，当需要截断时，可根据弯矩图的变化在适当部位通过计算将不需要的纵向受拉钢筋分批截断。每批钢筋应延伸至正截面受弯承载力计算不需要该钢筋的截面（理论截断截面）以外，实际截断点还应再延伸一段距离。

For the longitudinal tensile reinforcement on the top of the beam in the negative bending moment region near the support, because the negative bending moment region is not large, the number of longitudinal bars is often reduced by truncation, but it is not suitable for truncation in the tension zone. The longitudinal tensile reinforcement bearing negative bending moment of continuous beam and frame beam support extends into the span. When truncation is required, the unnecessary longitudinal tensile reinforcement can be cut off in batches by computation at proper position according to the change of the moment diagram. Each batch of steel bars should be extended beyond the section where the normal section flexural capacity calculation does not require the reinforcement (theoretical cutoff section), and the actual cutoff point should be extended for a further distance.

纵筋截断后,在截断位置其抵抗弯矩等于零。钢筋截断前后抵抗弯矩发生突变。纵筋截断点应满足以下两个控制条件。

After the longitudinal reinforcement is cut off, at the truncation position its resistance bending moment is equal to zero. The resistance bending moment of the reinforcement before and after truncation changes abruptly. The cutoff point of longitudinal bars should satisfy the following two control conditions.

(1) 从该钢筋充分利用截面到截断点的长度称为伸出长度,为了钢筋的可靠锚固,负弯矩钢筋截断时必须满足伸出长度的要求。

(1) The length from the fully utilized section to the cutoff point of the reinforcement is called the extension length. For the reliable anchorage of the steal bar, the negative bending moment bar must meet the requirement of extension length when it is cut off.

(2) 从不需要该钢筋的截面到截断点的长度称为延伸长度,为了保证斜截面受弯承载力,负弯矩钢筋截断时还必须满足延伸长度的要求。

(2) The length from the section that does not require this reinforcement to the cutoff point is called the development length. In order to ensure the flexural capacity of inclined section, the reinforcement bearing negative bending moment must also meet the requirement of development length when it is cut off.

2. 梁支座截面负弯矩纵向受拉钢筋分批截断的规定

Provisions on batch truncation of negative bending moment longitudinal tensile reinforcement at the support section of the beam

为了使负弯矩钢筋的截断不影响它在各个截面中发挥抗弯能力,《混凝土结构设计标准》(GB/T 50010—2010)9.2.3 条对钢筋混凝土梁支座截面负弯矩纵向受拉钢筋分批截断的规定如下:

1) 当 $V \leqslant 0.7 f_t b h_0$ 时

1) When $V \leqslant 0.7 f_t b h_0$

梁剪弯区在使用阶段一般不出现斜裂缝,延伸长度和伸出长度都只与正截面受弯承载力有关,而与斜截面受弯承载力无关。钢筋应延伸至按正截面受弯承载力计算不需要该钢筋的截面以外不小于 $20d$ 处截断,且从该钢筋强度充分利用截面伸出的长度不应小于 $1.2l_a$,即实际延伸长度$\geqslant 20d$(从不需要截面起算),且$\geqslant 1.2l_a$(从充分利用截面起算),如

图 5-42(a)所示。

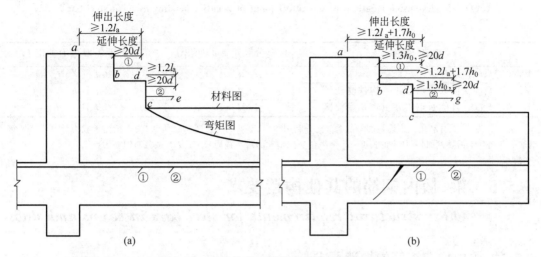

图 5-42 支座负弯矩钢筋截断时的延伸长度
(a) $V \leqslant 0.7 f_t bh_0$ 时；(b) $V > 0.7 f_t bh_0$ 时

Figure 5-42 Extension length of negative bending moment steel bar at support when truncated
(a) When $V \leqslant 0.7 f_t bh_0$; (b) When $V > 0.7 f_t bh_0$

2) 当 $V > 0.7 f_t bh_0$，且截断点已进入构件受压区时

2) When $V > 0.7 f_t bh_0$, and the cutoff point has entered the compression zone of the member

梁剪弯区在使用阶段可能会出现斜裂缝，斜裂缝顶端的弯矩增大，使未截断纵筋的拉应力超过钢筋屈服强度发生破坏，因此延伸长度不仅要满足 $20d$ 的要求，还应考虑斜裂缝水平投影长度这段距离。试验表明，斜裂缝的水平投影长度大致为 $(0.75 \sim 1.0) h_0$，为安全起见，取其上限 h_0。伸出长度也应在原来 $1.2 l_a$ 的基础上增加斜裂缝的水平投影长度 h_0。

所以，纵筋应延伸至按正截面受弯承载力计算不需要该钢筋的截面以外不小于 h_0 且不小于 $20d$ 处截断，且从该钢筋的充分利用截面伸出的长度不应小于 $1.2 l_a + h_0$，即实际延伸长度 $\geqslant h_0$，且 $\geqslant 20d$（从不需要截面起算），且 $\geqslant 1.2 l_a + h_0$（从充分利用截面起算）。

3) 当负弯矩区段相对长度较大，按前面 1)、2) 条确定的支座负弯矩钢筋截断点仍位于支座最大负弯矩对应的负弯矩受拉区时

3) When the relative length of the negative moment section is large, the cutoff point of the negative bending moment bar at the support determined according to the previous items 1) and 2) is still located in the negative bending moment tension zone corresponding to the maximum negative bending moment at the support

这时，对延伸长度和伸出长度的要求就更高了。纵筋应延伸至按正截面受弯承载力计算不需要该钢筋的截面以外不小于 $1.3 h_0$ 且不小于 $20d$ 处截断，且从该钢筋强度充分利用截面伸出的长度不应小于 $1.2 l_a + 1.7 h_0$，如图 5-42(b) 所示。

钢筋混凝土梁支座截面的负弯矩纵向受拉钢筋截断时的延伸长度 l_d 可按表 5-5 中 l_{d1} 和 l_{d2} 中取外伸长度较长者确定。

表 5-5 负弯矩钢筋实际截断点的延伸长度 l_d

Table 5-5 Extension length of actual cutoff point of negative bending moment steel bar l_d

单位：mm

截 面 条 件	充分利用截面伸出 l_{d1}	计算不需要截面伸出 l_{d2}
$V \leqslant 0.7 f_t b h_0$	$1.2 l_a$	$20d$
$V > 0.7 f_t b h_0$ 且截断点已进入构件受压区	$1.2 l_a + h_0$	$20d$ 且 h_0
$V > 0.7 f_t b h_0$ 且截断点仍在负弯矩受拉区	$1.2 l_a + 1.7 h_0$	$20d$ 且 $1.3 h_0$

注：l_{d1} 为从"充分利用该钢筋强度的截面"（充分利用截面）延伸出的长度；l_{d2} 为从"按正截面承载力计算不需要该钢筋的截面"（不需要截面）延伸出的长度；l_a 为纵向受拉钢筋的最小锚固长度；d 为被截断钢筋的直径。

5.5 梁、板内钢筋的其他构造要求

Other structural requirements for steel bars in beams and slabs

5.5.1 梁中构造钢筋的构造要求

Structural requirements of structural steel bars in beams

1. 梁上部纵向构造钢筋

Longitudinal structural steel bars on the upper part of the beam

《混凝土结构设计标准》(GB/T 50010—2010) 9.2.6 条规定梁的上部纵向构造钢筋应符合下列要求：

Article 9.2.6 of "Standard for design of concrete structures" (GB/T 50010—2010) stipulates that the upper longitudinal structural steel bars of the beam should meet the following requirements：

当梁端按简支计算但实际受到部分约束时，应在支座区上部设置纵向构造钢筋，其截面面积不应小于梁跨中下部纵向受力钢筋计算所需截面面积的 1/4，且不应少于两根。该纵向构造钢筋自支座边缘向跨内伸出的长度不应小于 $l_0/5$，l_0 为梁的计算跨度。

When the beam end is calculated as simple support but is actually partially constrained, longitudinal structural reinforcement should be set in the upper part of the support area, and the cross-sectional area should not be less than 1/4 of the cross-sectional area required for the calculation of longitudinally stressed reinforcement in the lower part of the beam at midspan and should not be less than 2. The length of the longitudinal structural reinforcement extending from the support edge into the span should not be less than $l_0/5$, where l_0 is the effective span of the beam.

2. 梁侧面纵向构造钢筋

Longitudinal structural reinforcement on the side of the beam

为了抑制由荷载作用、混凝土收缩等原因在梁腹板高度范围内产生的垂直裂缝，如图 5-43 所示，同时增加钢筋骨架的刚度，《混凝土结构设计标准》(GB/T 50010—2010) 9.2.13 条规定，当梁的腹板高度 $h_w \geqslant 450$mm 时，在梁的两个侧面应沿高度配置纵向构造钢筋，又称腰筋，如图 5-44 所

示。每侧纵向构造钢筋（不包括梁上、下部纵向受力钢筋及架立钢筋）的间距不宜大于200mm，截面面积不应小于腹板截面面积bh_w的0.1%，但当梁宽较大时可以适当放松，如图5-45所示。此处，腹板高度h_w按式(5-16a)和式(5-16b)参数说明中的规定取用。

In order to suppress vertical cracks within the height range of the beam web due to load action, concrete shrinkage, etc., as shown in Fig. 5-43, and to increase the stiffness of the steel skeleton, Article 9.2.13 of "Standard for design of concrete structures" (GB/T 50010—2010) stipulates that when the web height of the beam h_w is greater than or equal to 450mm, longitudinal structural reinforcement, also known as waist bars, should be arranged on both sides of the beam along the height, as shown in Fig. 5-44. The spacing of longitudinal structural reinforcement on each side (excluding longitudinally stressed reinforcement on the upper and lower part of the beam and erection bar) should not be greater than 200mm, and the cross-sectional area should not be less than 0.1% of the web cross-sectional area bh_w, but when the beam width is large, it can be appropriately relaxed, as shown in Fig. 5-45. Here, The height h_w of the web is taken according to the provisions in the formula parameter description of Eq. (5-16a) and Eq. (5-16b).

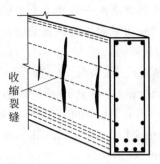

图5-43　梁侧纵向构造钢筋不足产生的竖向裂缝

Figure 5-43　Vertical cracks caused by insufficient longitudinal structural reinforcement on the side of the beam

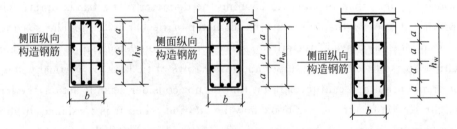

图5-44　梁侧面纵向构造钢筋和拉结筋

Figure 5-44　Longitudinal structural steel bars and tie bars on the side of the beam

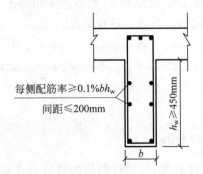

图5-45　梁侧面纵向构造钢筋的构造要求

Figure 5-45　Structural requirements of longitudinal structural steel bars on the side of the beam

当梁侧面配有直径不小于纵向构造钢筋的受扭纵筋时,受扭纵筋可以代替纵向构造钢筋。梁侧面纵向构造钢筋的搭接及锚固长度可取 $15d$。梁侧面受扭纵筋的搭接长度为 l_{lE} 或 l_l,其锚固长度为 l_{aE} 或 l_a,锚固方式同框架梁下部钢筋。

When the side of the beam is provided with torsional longitudinal bars with the diameter not less than the longitudinal structural bars, the torsional longitudinal bars can replace the longitudinal structural bars. The lap and anchorage length of longitudinal structural bars of beam side can be taken as $15d$. The lap length of torsional longitudinal bars of beam side is l_{lE} or l_l, and its anchorage length is l_{aE} or l_a, and the anchoring method is the same as that of the lower reinforcement of the frame beam.

拓展知识 5-4

3. 拉结筋
Tie bar

梁两侧纵向构造钢筋之间宜用拉结筋联系起来,拉结筋也称联系筋,如图 5-44 所示。当梁宽 $b \leqslant 350\text{mm}$ 时,拉结筋直径为 6mm;梁宽 $b > 350\text{mm}$ 时,拉结筋直径为 8mm。拉结筋的间距常取非加密区箍筋间距的 2 倍。当设有多排拉结筋时,上下两排拉结筋竖向错开设置。非框架梁及不考虑地震作用的悬挑梁,拉结筋弯钩平直段长度可为 $5d$,当其受扭时,应为 $10d$。拉结筋弯钩构造做法采用何种形式设计时根据图集指定。

The longitudinal structural reinforcement on both sides of the beam should be connected by tie bars, which is also called connecting bars, as shown in Fig. 5-44. When the beam width b is less than or equal to 350mm, the diameter of tie bar is 6mm; when the beam width b is greater than 350mm, the diameter of tie bar is 8mm. The spacing of tie bars is often taken as twice of the stirrup spacing in the non densification area. When there are multiple rows of tie bars, the upper and lower rows of tie bars are vertically staggered. For non-frame beam and cantilever beam that do not consider seismic action, the length of the straight section of the tie bar hook may be $5d$, and when it is torsioned, it should be $10d$. The form of the tie bar hook construction method is specified according to the atlas during design.

任务 5-2:现浇钢筋混凝土梁板结构的梁高为 600mm,要求分析现浇板厚分别为 120mm、110mm 时梁侧面纵向构造钢筋的设置情况。

Task 5-2: The beam height of cast-in-place reinforced concrete beam-slab structure is 600mm. It is required to analyze the setting of longitudinal structural reinforcement on the side of the beam when the cast-in-situ slab thickness is 120mm and 110mm respectively.

5.5.2 钢筋的连接
Connection of steel bars

实际工程中,当构件的长度较大,且超过钢筋的供货长度时,要通过连接接头将钢筋接长,将一根钢筋所受的力传给另一根钢筋。钢筋连接可采用绑扎搭接、焊接或机械连接。机械接头能产生较牢固的连接力,因此应优先采用机械连接。机械连接接头及焊接接头的类型及质量应符合国家有关标准的规定。

In actual engineering, when the length of the member is large and exceeds the supply length of the steel bar, the steel bar should be lengthened through splices, and transfer the force on one steel bar to the other. Steel may be spliced by binding lap, welding or mechanical connection. Mechanical joint can produce stronger connection force, and mechanical connection should be preferred. The type and quality of mechanical connection joints and welded joints should comply with the relevant national standards.

混凝土结构中受力钢筋的连接接头宜设置在受力较小处,同一根钢筋上宜少设接头。在结构的重要构件和关键传力部位,纵向受力钢筋不宜设置连接接头。纵向受力钢筋连接位置宜避开梁端、柱端箍筋加密区,如必须在此连接时,应采用机械连接或焊接。需进行疲劳验算的构件,其纵向受拉钢筋不得采用绑扎搭接接头,也不宜采用焊接接头,除端部锚固外不得在钢筋上焊有附件。疲劳验算时,焊接接头应符合疲劳应力幅限值的规定。

The splices of the stressed reinforcement in concrete structure should be set at the places with less stress, and fewer joints should be set on the same steel bar. In important structural members and key force transmission parts of the structure, splices should not be set for longitudinally stressed reinforcement. The connection position of longitudinally stressed reinforcement should avoid the densification area of stirrups at the beam end and column end. If this connection is necessary here, mechanical connection or welding should be used. For members that need fatigue calculation, the longitudinal tensile reinforcement should not adopt binding lap splice or welded splice, and no accessories should be welded to the steel bars except for end anchoring. For fatigue calculation, welded splice should conform to the provisions of the limit value of fatigue stress amplitude.

1. 绑扎搭接
Binding lap

轴心受拉构件及小偏心受拉杆件(如桁架或拱的拉杆)中纵向受力钢筋不应采用绑扎搭接。其他构件中的钢筋采用绑扎搭接时,受拉钢筋直径不宜大于25mm,受压钢筋直径不宜大于28mm。HPB300级钢筋末端应做180°弯钩。

The longitudinally stressed reinforcement of axial tension members and small eccentric tension members (such as truss or tensile rods of arch) should not adopt binding lap. When the steel bars in other members adopt binding lap, the diameter of the tensile reinforcement should not be greater than 25mm, and the diameter of the compression reinforcement should not be greater than 28mm. The end of HPB300 steel bar should be made of 180° hook.

同一构件中相邻纵向受力钢筋的绑扎搭接接头宜相互错开。

The binding lap splices of adjacent longitudinally stressed reinforcement in the same member should be staggered from each other.

1) 绑扎搭接接头连接区段

1) Binding lap splice connection section

绑扎搭接接头连接区段的长度为1.3倍搭接长度l_l,凡接头中点位于连接区段长度内的连接接头均属于同一连接区段,如图5-46所示。

The length of the connection section of binding lap splice is 1.3 times the lap length l_l. All connection joints whose midpoints are within the length of the connection section belong to the same connection section, as shown in Fig. 5-46.

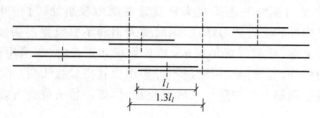

图5-46 同一连接区段内纵向受拉钢筋绑扎搭接长度和连接区段长度

Figure 5-46 Binding lap length of longitudinal tensile reinforcement in the same connection section and length of connection section

2) 受拉钢筋搭接接头面积百分率

2) Percentage of lap splice area of tensile reinforcement

同一连接区段内纵向钢筋搭接接头面积百分率为该区段内有连接接头的纵向受力钢筋截面面积与全部纵向钢筋截面面积的比值。对梁类、板类及墙类构件,不宜大于25%;对柱类构件,不宜大于50%。当工程中确有必要增大受拉钢筋搭接接头面积百分率时,对梁类构件,不宜大于50%;对板类、墙类及柱类构件,可根据实际情况放宽。

Percentage of lap splice area of longitudinal bars in the same connection section is the ratio of the cross-sectional area of longitudinally stressed reinforcement with connection joints in this section to the cross-sectional area of all longitudinal bars. For beams, slabs and wall members, it should not exceed 25%; for column members, it should not exceed 50%. When it is really necessary to increase percentage of lap splice area of tensile reinforcement in the project, it should not be greater than 50% for beam members; for slab, wall and column members, it can be relaxed according to the actual situation.

3) 绑扎搭接长度l_l

3) Binding lap length l_l

纵向受拉钢筋绑扎搭接接头的搭接长度l_l应根据位于同一连接区段内的钢筋搭接接头的面积百分率按式(5-33)计算,且在任何情况下搭接长度都不应小于300mm。

The lap length l_l of the binding lap splice of longitudinal tensile reinforcement should be calculated according to the percentage of lap splice area of longitudinal bars in the same connection section by Eq. (5-33), and should not be less than 300mm in any case.

$$l_l = \zeta_l l_a \tag{5-33}$$

式中,l_l为纵向受拉钢筋的搭接长度,mm,如图5-46所示;l_a为纵向受拉钢筋的锚固长度,mm;ζ_l为纵向受拉钢筋搭接长度修正系数,根据位于同一连接区段内的钢筋搭接接头的面积百分率按表 5-6 取用。

Where, l_l is the lap length of longitudinal tensile reinforcement, mm, as shown in Fig. 5-46; l_a is the anchorage length of longitudinal tensile reinforcement, mm; ζ_l is the correction coefficient of the lap length of longitudinal tensile reinforcement, based on the percentage of lap splice area of reinforcement in the same connection section, according to Table 5-6.

表 5-6　纵向受拉钢筋搭接长度修正系数 ζ_l

Table 5-6　Correction coefficient of lap length of longitudinal tensile reinforcement ζ_l

纵向钢筋搭接接头面积百分率/%	≤25	50	100
搭接长度修正系数 ζ_l	1.2	1.4	1.6

绑扎搭接长度 l_l 还和带肋钢筋是否有环氧树脂涂层、纵向受拉钢筋施工过程中是否易受扰动、搭接长度范围内纵向受力钢筋周边的保护层厚度有关，分别乘以相应的修正系数。

The binding lap length l_l is also related to whether the ribbed bar has an epoxy resin coating, whether the longitudinal tensile reinforcement is easily disturbed during the construction process, and the thickness of surrounding concrete cover of longitudinally stressed reinforcement within the lap length range, is multiplied by the corresponding correction coefficient respectively.

4）纵向受压钢筋的搭接长度

4) Lap length of longitudinal compression reinforcement

构件中的纵向受压钢筋采用搭接连接时，其受压搭接长度不应小于纵向受拉钢筋搭接长度 l_l 的 70%，且在任何情况下都不应小于 200mm。

When the longitudinal compression reinforcement in the member adopts lap splice, the compression lap length should not be less than 70% of the lap length l_l of longitudinal tensile reinforcement, and in any case should not be less than 200mm.

2. 机械连接

Mechanical connection

机械连接的形式有多种，目前我国使用较多的是冷轧直螺纹套筒连接。纵向受力钢筋的机械连接接头宜相互错开。

There are many forms of mechanical connection. At present, cold-rolled straight thread sleeve connection is widely used in China. The mechanical connection joints of longitudinally stressed reinforcement should be staggered.

钢筋机械连接区段的长度为 $35d$，d 为连接钢筋的较大直径，凡接头中点位于该连接区段长度内的机械连接接头均属于同一连接区段。在受力较大处设置机械连接接头时，位于同一连接区段内的纵向受拉钢筋接头面积百分率不宜大于 50%；但对板、墙、柱及预制构件的连接处，可根据实际情况放宽。纵向受压钢筋的接头面积百分率可不受限制。

The length of mechanical connection section of reinforcement is $35d$, d is the larger diameter of connecting reinforcement. All mechanical connection joints whose joint midpoint is within the length of the connection section belong to the same connection section. While mechanical connection joints are set at places with high stress, the area percentage of longitudinal tensile reinforcement joints located in the same connection section should not be greater than 50%; however, the connection of the slab, wall, column and prefabricated members can be relaxed according to the actual situation. The joint area percentage of longitudinal compression reinforcement is not limited.

机械连接套筒的保护层厚度宜满足有关钢筋最小保护层厚度的规定。机械连接套筒的横向净间距不宜小于 25mm。套筒处箍筋的间距仍应满足相应的构造要求。

The thickness of concrete cover of mechanical connection sleeve should meet the provisions on the minimum concrete cover thickness of reinforcement. The lateral clear spacing of the mechanical connection sleeve should not be less than 25mm. The spacing of stirrups at the sleeve should still meet the corresponding structural requirements.

直接承受动力荷载结构构件中的机械连接接头,除应满足设计要求的抗疲劳性能外,位于同一连接区段内的纵向受力钢筋接头面积百分率不应大于 50%。

The mechanical connection joints in structural members directly subjected to dynamic loads, in addition to the fatigue resistance that should meet the the design requirements, the percentage of longitudinally stressed reinforcement joint area located in the same connection section should not be greater than 50%.

3. 焊接

Welding

细晶粒热轧带肋钢筋以及直径大于 28mm 的带肋钢筋,其焊接应经试验确定;余热处理钢筋不宜焊接。当纵向受力钢筋采用焊接接头时,接头位置应相互错开,钢筋焊接接头连接区段的长度为 $35d$(d 为连接钢筋的较大直径),且不小于 500mm,凡接头中点位于该连接区段长度内的焊接接头均属于同一连接区段。对位于同一连接区段纵向受拉钢筋的接头,面积百分率不应大于 50%,纵向受压钢筋的接头面积百分率可不受限制。

The welding of fine grain hot-rolled ribbed steel bar and ribbed steel bar with a diameter greater than 28mm should be determined by tests; the residual heat treatment reinforcement is not suitable for welding. When welded joints are adopted for longitudinally stressed reinforcement, the position of joints should be staggered with each other. The length of the connection section of the welded joint of the steel bar is $35d$ (d is the larger diameter of the connecting reinforcement) and not less than 500mm. All welded joints whose joint midpoint is within the length of the connection section belong to the same connection section. For the joint of longitudinal tensile reinforcement located in the same connection section, the area percentage should not be greater than 50%, and the percentage of joint area of longitudinally compressed steel bars is not limited.

(1) 斜截面受剪承载力、斜截面受弯承载力和正截面承载力同样重要,要引起高度重视。斜裂缝主要有两种:腹剪斜裂缝和弯剪斜裂缝。剪跨比是影响斜截面受剪承载力的重

要因素。

(2) 有腹筋梁斜截面受剪破坏形态主要有三种：斜拉破坏、斜压破坏和剪压破坏。它们都是脆性破坏，破坏特征分别类似于正截面受弯的三种破坏形态：少筋破坏、超筋破坏和适筋破坏。由剪压破坏建立斜截面受剪承载力计算公式，由防止斜拉破坏和斜压破坏的措施给出计算公式的适用条件。

① 斜拉破坏是由于没有足够有效的腹筋穿过斜截面阻止斜裂缝开展而发生的，一裂即坏。防止斜拉破坏的措施是规定箍筋的最小配箍率。

② 斜压破坏是由于在穿过斜裂缝的腹筋屈服前，混凝土斜向受压棱柱体先被压坏，说明混凝土的受压承载力不足。防止斜压破坏的措施是截面尺寸限制条件。

③ 剪压破坏是剪压区混凝土破坏前，与斜裂缝相交的腹筋先屈服。可以通过设计计算配置相应的腹筋（箍筋、弯起钢筋）来防止剪压破坏。

(3) 根据计算确定梁中每根纵向受拉钢筋所能抵抗的弯矩，绘制抵抗弯矩图，通过抵抗弯矩图是否包住设计弯矩图判断正截面受弯承载力是否足够。充分利用截面和不需要截面是抵抗弯矩图上确定钢筋弯起和截断位置的重要截面。

(4) 纵筋的弯起、截断、锚固、连接都要符合一定的构造要求。纵向受力钢筋在抵抗弯矩图上应伸过其充分利用截面至少 $0.5h_0$ 才能弯起。梁支座截面承受负弯矩的纵向受拉钢筋分批截断要满足伸出长度和延伸长度的要求。纵向钢筋在支座处的锚固要满足锚固长度的要求，不能满足的要采取机械锚固措施。钢筋的连接有绑扎连接、机械连接和焊接，绑扎搭接长度、位于同一连接区段纵向受力钢筋的接头面积百分率都要符合一定的要求。

第 5 章拓展知识和特别提示

视频：第 5 章小结讲解

5-1 填空题

1. 为保证钢筋混凝土受弯构件斜截面承载力，除应有合适的截面尺寸和材料强度外，还应配置_____和（或）_____钢筋抗剪。

2. 抗剪钢筋也称作腹筋，腹筋的形式可以采用_____和_____。

3. 无腹筋梁中典型的斜裂缝主要有_____裂缝和_____裂缝。

4. 无腹筋梁斜截面受剪的主要破坏形态有_____破坏、_____破坏和_____破坏。

5. 无腹筋梁斜截面受剪有三种主要破坏形态，就其受剪承载力而言，对同样的构件_____破坏最低，_____破坏较高，_____破坏最高；但就其破坏性质而言，均属于_____破坏。

6. 有腹筋梁斜截面破坏形态有_____、_____和_____三种。

7. 影响有腹筋梁斜截面受剪承载力的主要因素有_____、_____、_____和_____等。

8. 梁发生剪切破坏时,斜截面承受的剪力设计值 V_u 由_____、_____和_____三部分组成。

9. 斜拉破坏大多发生在腹筋配置太少,且其剪跨比_____的情况。

10. 梁内纵向受拉钢筋的弯起点应设在正截面抗弯计算时该钢筋强度全部发挥作用的截面以外至少_____处,以保证梁斜截面的_____承载力。

11. 腹筋配置数量适当的有腹筋梁,大部分发生_____破坏,破坏时与临界斜裂缝相交的腹筋一般均能达到_____,最后斜裂缝两端的混凝土_____。

12. 受弯构件斜截面受剪承载力的公式是以_____破坏为特征推导的,公式的限制条件 $V \leqslant 0.25\beta_c f_c bh_0$ 是为了防止发生_____破坏。

5-2 选择题

1. 进行受弯构件斜截面受剪承载力计算时,若所配箍筋不能满足抗剪要求(即 $V > V_{cs}$)时,采取哪种解决办法较好?(　　)
   A. 将纵向钢筋弯起为斜筋或加焊斜筋　　B. 将箍筋加密或加粗
   C. 增大构件截面尺寸　　　　　　　　　D. 提高混凝土强度等级

2. 承受集中荷载的矩形截面简支梁配置箍筋后,若所需的弯起钢筋数量或间距不满足要求,应增设(　　)以抗剪。
   A. 纵筋　　　　B. 鸭筋　　　　C. 浮筋　　　　D. 架立钢筋

3. 《混凝土结构设计标准》(GB/T 50010—2010)中规定,钢筋的基本锚固长度是指(　　)。
   A. 受拉锚固长度　　　　　　　　　　B. 受压锚固长度
   C. 搭接锚固长度　　　　　　　　　　D. 延伸锚固长度

4. 钢筋混凝土梁出现斜裂缝的原因是(　　)。
   A. 箍筋配置不足　　　　　　　　　　B. 没有配置弯起钢筋
   C. 主拉应力超过混凝土轴心抗拉强度设计值

5. 当 $V > 0.25\beta_c f_c bh_0$ 时,应采取的措施是(　　)。
   A. 加大箍筋直径或减小箍筋间距　　　B. 增大截面尺寸
   C. 提高箍筋抗拉强度设计值　　　　　D. 加配弯起钢筋

6. 梁内弯起多排钢筋时,相邻弯起筋上下弯点间距 $\leqslant s_{max}$,其目的是保证(　　)。
   A. 斜截面受弯能力　　　　　　　　　B. 正截面受弯能力
   C. 斜截面受剪能力　　　　　　　　　D. 正截面受剪能力

7. 提高梁的斜截面受剪承载力最有效的措施是(　　)。
   A. 提高混凝土强度等级　　　　　　　B. 加大截面高度
   C. 加大截面宽度　　　　　　　　　　D. 增加抗剪钢筋

8. 下列叙述(　　)是错误的。
   A. 梁的剪跨比减小时,受剪承载力减小
   B. 简支梁纵向受拉钢筋伸入支座的锚固长度与剪力有关,因为钢筋应力随剪力增大而增加
   C. 薄腹梁容易发生腹板压碎的斜压破坏,所以截面尺寸限制条件比较严格

D. 在受剪承载力计算中未反映翼缘的作用,是因为其对有腹筋梁的作用小,可略去

E. 一般板不作抗剪计算是因为其受剪承载力大于受弯承载力

9. 只要按斜截面受剪承载力公式计算并配置梁内箍筋后( )。

A. 肯定不会发生剪切破坏   B. 斜裂缝宽度能满足要求

C. 不发生纵筋锚固破坏   D. 只可能发生受弯破坏

10. 钢筋混凝土简支梁支座截面剪力设计值 $V=250\text{kN}$,截面尺寸为 $200\text{mm}\times500\text{mm}$ ($h_0=465\text{mm}$),C25 混凝土,经斜截面受剪承载力计算,该梁腹筋配置是( )。

A. 按构造配置   B. 按计算配置

C. 截面尺寸太小   D. 条件不足,无法判断

5-3 名词解释

1. 腹剪斜裂缝	2. 弯剪斜裂缝	3. 腹筋	4. 无腹筋梁
5. 有腹筋梁	6. 剪跨	7. 剪跨比	8. 广义剪跨比
9. 斜压破坏	10. 剪压破坏	11. 斜拉破坏	12. 配箍率
13. 箍筋肢数	14. 最小配箍率	15. 箍筋最大间距	16. 斜截面受弯承载力
17. 抵抗弯矩图(材料图)	18. 充分利用截面	19. 不需要截面	20. 钢筋绑扎搭接接头连接区段
21. 同一连接区段内纵向受拉钢筋搭接接头面积百分率	22. 机械连接	23. 纵向构造钢筋	24. 拉结筋

5-4 思考题

1. 钢筋混凝土梁为什么会出现斜裂缝?斜裂缝一般沿着怎样的途径发展?

2. 有腹筋梁三种破坏形态的发生条件是什么?简述其破坏特征。

3. 简述有腹筋梁斜截面受剪承载力的计算思路。

4. 为什么当受弯构件的受剪截面 $h_w/b\leqslant4$ 时,必须符合 $V\leqslant0.25\beta_cf_cbh_0$ 的条件?这个条件的实质是什么?

5. 为什么受弯构件要控制箍筋的最小配箍率?

6. 有腹筋梁斜截面受剪承载力的计算截面是如何确定的?

7. 为什么要绘制抵抗弯矩图?

8. 为什么要将梁的纵向受力钢筋弯起?纵向受力钢筋的弯起位置应满足什么要求?

9. 纵向受拉钢筋的基本锚固长度是如何确定的?当纵向钢筋直径较大或采用机械锚固措施,其锚固长度有何变化?

10. 确定梁支座负弯矩钢筋的截断位置需要考虑哪几个方面的问题?

11. 采用绑扎搭接时,梁内纵向受拉钢筋的搭接长度是如何确定的?

12. 梁内纵向构造钢筋有哪些构造方面的规定?

5-5 计算题

1. 某均布荷载作用下钢筋混凝土矩形截面简支梁,截面尺寸为 $b\times h=200\text{mm}\times400\text{mm}$,$a_s=35\text{mm}$,混凝土强度等级为 C25,箍筋为 HPB300,承受的剪力设计值为 $V=150\text{kN}$。确定箍筋的肢数和直径,求箍筋的间距,并验算配箍率。

2. 某矩形截面简支梁环境类别为一类,结构安全等级为二级,截面尺寸为 $b\times h=$

$200\text{mm} \times 500\text{mm}$,$a_s = 40\text{mm}$,两端支承在砖墙上,净跨 $l_n = 5.7\text{m}$,承受均布荷载设计值 $g + q = 60\text{kN/m}$(包括梁自重),混凝土强度等级为C25,箍筋采用HRB400。若梁只配置箍筋,不配置弯起钢筋,确定箍筋的肢数和直径,求箍筋的间距,并验算配箍率。

3. 某钢筋混凝土梁如图5-47所示,截面尺寸为 $b \times h = 200\text{mm} \times 500\text{mm}$,$a_s = 35\text{mm}$,混凝土强度等级为C30,箍筋为HRB400,承受的均布荷载设计值 $g + q = 40\text{kN/m}$(包括梁自重),求 A、$B_{左}$、$B_{右}$ 截面的抗剪钢筋,并验算配箍率。

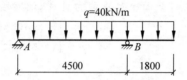

图5-47 承受均布荷载的悬挑梁

4. 某钢筋混凝土简支梁的截面尺寸为 $b \times h = 250\text{mm} \times 600\text{mm}$,计算跨度 $l_0 = 5\text{m}$,$a_s = 35\text{mm}$,梁下部配有 4⎯25 的通长纵向受拉钢筋,箍筋为HPB300。承受均布荷载设计值 $g + q = 100\text{kN/m}$(包括梁自重),混凝土强度等级为C30。(1)不设弯起钢筋,求受剪箍筋;(2)以现有纵筋为弯起钢筋,求所需箍筋,并验算配箍率。

5. 某钢筋混凝土简支梁如图5-48所示,环境类别为一类,梁截面尺寸为 $b \times h = 250\text{mm} \times 500\text{mm}$,混凝土强度等级为C25,纵向受力钢筋、箍筋均为HRB400。跨中两个集中荷载设计值 $F = 90\text{kN}$(不计梁自重),求受剪箍筋(无弯起钢筋),并验算配箍率。

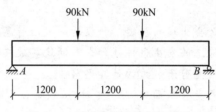

图5-48 承受集中荷载的简支梁

6. 某钢筋混凝土简支梁如图5-49所示,承受均布荷载设计值(包括梁自重)$g + q = 90\text{kN/m}$,混凝土强度等级为C25,纵向受力钢筋为HRB400,箍筋为HPB300,复核此梁是否安全。

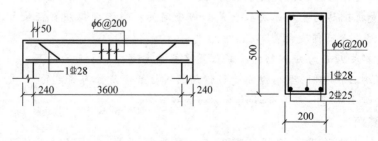

图5-49 承受均布荷载的简支梁

受压构件的截面承载力

Section bearing capacity of compression members

教学目标：

1. 理解轴心受压短柱和长柱的受力特点，掌握轴心受压构件正截面受压承载力的计算；

2. 理解偏心受压构件正截面破坏形态、矩形截面受压承载力的计算简图和基本计算公式，熟悉大、小偏心受压破坏的判别方法；

3. 掌握矩形截面大偏心受压构件的受压承载力计算，会进行小偏心受压构件受压承载力计算；

4. 了解柱的类型，能够正确选用柱的截面形式、截面尺寸和材料，领会受压构件中纵向钢筋和箍筋的构造要求。

导读：

受压构件是房屋建筑、水工建筑物、桥梁等各种土木工程结构中主要承受轴向压力的重要承重构件，应用非常广泛，如柱、剪力墙、核心筒体墙、烟囱的筒壁、桥梁结构中的桥墩及屋架等桁架结构中的受压杆、拱、桩等构件。因此，受压构件的设计很重要，如果不能依据《混凝土结构设计标准》(GB/T 50010—2010)等正确设计受压构件，将导致受压构件承载力不足，引起土木工程结构的破坏，造成的后果很严重。

钢筋混凝土柱可以预制或现浇。预制柱施工比较方便，不受气候影响，但节点连接质量不易保证，整体性、抗震性欠佳。现浇柱整体性、抗震性好，但施工工作量大，容易受气候等影响，质量不易保证。

按照柱配筋方式的不同，可分为普通箍筋柱、螺旋箍筋柱、劲性钢筋柱和钢管混凝土柱等。

普通箍筋柱中配有纵向受压钢筋和普通箍筋，普通箍筋用来固定纵向钢筋的位置，与纵筋形成空间钢筋骨架，约束核心混凝土和纵向钢筋的横向变形，防止纵筋受力后外凸。由于构造简单，施工方便，普通箍筋适用于各种截面形状的柱，是工程中最基本、最常见的箍筋类型，截面形状多为矩形或正方形，如图6-1(a)所示。在实际工程中一般采用普通箍筋柱。

螺旋箍筋柱是采用螺旋式或焊接环式箍筋，将单个箍筋做成连环形状，如图6-1(b)所示。螺旋箍筋可分为圆形和方形，柱截面形状一般为圆形或多边形，采用螺旋箍筋可提高受压承载力,加快绑扎速度,节省工时及用工。在混凝土浇捣过程中,能确保钢筋成品经受扰动、碰撞、踩踏不走样、不变形。螺旋式箍筋一般间距较密,能够显著提高核心混凝土的抗压强度,并增大其纵向变形能力。但螺旋箍筋柱施工复杂,用钢量大,造价较高,一般用于柱承受很大的轴向压力,截面尺寸由于建筑及使用上的要求又受到限制,采用普通箍筋柱会使纵向钢筋配筋率过高,而混凝土强度等级又不宜再提高的情况。

劲性钢筋柱是在柱的内部配置型钢,承担大部分荷载,其用钢量大,但可以减小柱的截面尺寸,节约空间,提高柱的刚度。一般广泛应用于大型结构中,但是造价比较高。常见的一般有 H 型钢劲性混凝土、十字钢柱混凝土、圆钢柱混凝土等。

钢管混凝土柱是在钢管中填充混凝土并捣实,钢管和核心混凝土共同承受外荷载作用。按截面形式不同,可分为圆形钢管混凝土、方形、矩形钢管混凝土和多边形钢管混凝土等。一般把混凝土强度等级在 C50 以下的钢管混凝土称为普通钢管混凝土,混凝土强度等级在 C50 以上的钢管混凝土称为钢管高强混凝土,混凝土强度等级在 C100 以上的钢管混凝土称为钢管超高强混凝土。

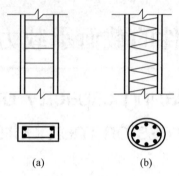

图 6-1 柱的类型
(a) 普通箍筋柱;(b) 螺旋箍筋柱
Figure 6-1 Type of column
(a) Ordinary stirrup column;
(b) Spiral stirrup column

建筑结构中柱的顶部承受楼盖中梁、板传来的永久荷载和可变荷载,柱的底部将上层楼盖和柱的全部荷载传给下层柱,底层柱和基础相连,将上部结构的荷载传给下部基础。从受力的角度,可以把受压构件分为轴心受压构件和偏心受压构件。轴心受压构件是指轴向压力作用线与构件截面形心轴线重合的构件,如图 6-2(a)所示。偏心受压构件是指轴向压力作用线与构件截面形心轴线不重合的构件。为了工程设计方便,一般不考虑混凝土材料的不均匀性和钢筋不对称布置的影响。

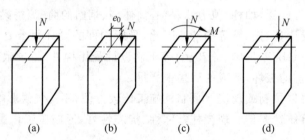

图 6-2 轴心受压构件和偏心受压构件
(a) 轴心受压构件;(b) 单向偏心受压时的轴向力和偏心距;
(c) 单向偏心受压时的轴向力和弯矩;(d) 双向偏心受压时的轴向力和偏心距
Figure 6-2 Axial compression member and eccentric compression member
(a) Axial compression member;(b) Axial force and eccentricity under uniaxial eccentric compression;
(c) Axial force and bending moment under uniaxial eccentric compression;
(d) Axial force and eccentricity under biaxial eccentric compression

根据轴向压力作用点与构件正截面形心相对位置的不同,偏心受压构件又可分为单向偏心受压构件和双向偏心受压构件。当轴向压力的作用点只对构件正截面的一个主轴有偏心距时,为单向偏心受压构件,如图 6-2(b)、(c)所示。当轴向压力的作用点对构件正截面的两个主轴都有偏心距时,为双向偏心受压构件,如图 6-2(d)所示。工程中大部分偏心受压构件都是按单向偏心受压进行截面设计的。

受压构件设计的一般步骤:选择构件配筋形式,确定截面尺寸,选择材料强度等级,确定计算简图,计算荷载和内力,计算钢筋用量,选择配置钢筋并满足构造要求,按计算结果和构造要求绘制构件结构施工图。

引例:

第 6 章引例

核心词汇:

轴心受压构件	axial compression member	轴心受压柱	axial compression column
受压承载力	compression capacity	偏心受压柱	eccentric compression column
螺旋配筋柱	spirally reinforced column	附加弯矩	additional moment
失稳破坏	unstable failure	二阶效应	second-order effect
长柱	long column	材料破坏	material failure
短柱	short column	初始偏心距	initial eccentricity
稳定系数	stability coefficient	附加偏心距	additional eccentricity
长细比	slenderness ratio	计算偏心距	calculated eccentricity
偏心荷载	eccentric load	偏心距调节系数	eccentricity adjustment coefficient
单向偏心受压	uniaxial eccentric compression		
双向偏心受压	biaxial eccentric compression	弯矩增大系数	moment increase coefficient
大偏心受压破坏	large eccentric compression failure	截面曲率修正系数	sectional curvature correction coefficient
小偏心受压破坏	small eccentric compression failure	非对称配筋	asymmetrical reinforcement
		对称配筋	symmetrical reinforcement

6.1 轴心受压构件正截面受压承载力计算

Calculation of compression bearing capacity of normal section of axial compression members

6.1.1 概述

Summary

视频 6-1

在实际工程结构中,真正的轴心受压构件几乎是不存在的。由于施工时不可避免的尺寸误差、荷载实际作用位置的偏差、混凝土材料的非均质性、纵向钢筋的非对称布置等原因,往往存在一定的初始偏心距。但为了计算方便,承受永久荷载为主的等跨多层房屋的内柱和桁架的受压腹杆等可近似按轴心受压构件计算。另外,轴心受压构件正截面承载力计算还用于偏心受压构件垂直于弯矩作用平面的承载力验算。

In actual engineering structures, true axial compression members are almost nonexistent. Due to the inevitable size error during construction, the deviation of actual action position of the load, the heterogeneity of concrete material, and the asymmetrical placement of longitudinal bars, there is often a certain initial eccentricity. However, for the convenience of calculation, the interior columns of equal-span multistory buildings and the compression web members of trusses bearing permanent loads can be calculated approximately as axial compression members. In addition, the calculation of normal section bearing capacity of axial compression member is also used to check the bearing capacity of the plane perpendicular to the bending moment acting plane of eccentric compression member.

轴心受压构件中纵向钢筋的作用是与混凝土共同承担纵向压力,提高构件的承载力,减小构件的截面尺寸,它能够抵抗因荷载偶然偏心在构件受拉边产生的拉应力,防止构件突然发生脆性破坏,改善破坏时构件的延性,减小混凝土的收缩与徐变变形。箍筋能与纵筋形成骨架,并防止纵筋受力后外凸。

The function of the longitudinal bars in the axial compression member is to bear the longitudinal compressive force together with the concrete, improve the bearing capacity of the member, reduce the cross-sectional dimensions of the member, and it can resist the tensile stress caused by the accidental eccentricity of the load on the tension side of the member, prevent sudden occurrence of brittle failure of members, improve the ductility of members during failure, and reduce the shrinkage and creep deformation of concrete. The stirrups can form a skeleton with the longitudinal bars and prevent them from protruding when stressed.

根据长细比的大小,轴心受压柱可分为短柱和长柱,当长细比 $l_c/i \leqslant 28$ 或矩形截面 $l_c/b \leqslant 8$ 或圆形截面 $l_c/d \leqslant 7$ 时为短柱,当长细比 $l_c/i > 28$ 时为长柱,其中 l_c 为柱的计算长度;i 为截面最小回转半径;b 为矩形截面短边尺寸;d 为圆形截面的直径。长柱和短柱的受压承载力和破坏形态不同。

According to the value of slenderness ratio, axial compression columns can be divided into short columns and long columns. When the slenderness ratio is $l_c/i \leqslant 28$ or the rectangular section $l_c/b \leqslant 8$ or the circular section $l_c/d \leqslant 7$, it is a short column. A column is a long column when the slenderness ratio $l_c/i > 28$, where l_c is the effective length of the column; i is the minimum radius of gyration of the cross section; b is the short side dimension of rectangular section; and d is the diameter of circular section. The compressive bearing capacity and destruction forms of long columns and short columns are different.

柱的计算长度 l_c 与柱两端的约束情况有关。在实际结构中,柱端部的连接构造比较复杂,因此,《混凝土结构设计标准》(GB/T 50010—2010)对单层厂房排架柱、框架柱等的计算长度作了具体规定。一般多层房屋中梁和柱为刚接的框架结构,各层柱的计算长度 l_c 可按表 6-1 确定。

The effective length l_c of the column is related to the constraints at both ends of the column. In the actual structure, the connection structure at the ends of the column is relatively complex. Therefore, "Standard for design of concrete structures"(GB/T 50010—2010) specifies the effective length of bent frame columns of single-storey factory and frame columns. Generally, frame structures with rigidly connected beams and columns in multistory buildings, the effective length l_c of columns on each layer can be determined according to Table 6-1.

表 6-1 框架结构各层柱的计算长度

Table 6-1　Effective length of each layer column of frame structure

楼盖类型	柱的类别	l_c
现浇楼盖	底层柱	$1.0H$
	其余各层柱	$1.25H$
装配式楼盖	底层柱	$1.25H$
	其余各层柱	$1.5H$

对底层柱,H 为基础顶面到一层楼盖顶面之间的距离;对其余各层柱,H 为上、下两层楼盖顶面之间的距离。

For the bottom column, H is the distance between the top surface of the foundation and the top surface of the first floor; for the other columns on each floor, H is the distance between the top surface of the upper and lower floors.

6.1.2　轴心受压普通箍筋柱的正截面受力分析和破坏形态
Normal section force analysis and destruction form of ordinary stirrup column under axial compression

1. 短柱的正截面受力分析和破坏形态

Normal section force analysis and destruction form of short columns

试验表明,在轴心压力作用下,配有纵筋和普通箍筋的短柱整个截面上的受压应变基本上是均匀分布的。由于钢筋与混凝土之间存在黏结作用,使两者的压应变相同。当轴心压力较小时,混凝土和钢筋的受力都处于弹性阶段,柱的压缩变形随荷载的增大成比例增加,纵筋和混凝土中压应力的增加也与荷载的增大成正比。当荷载较大时,随着混凝土塑性变形的发展,在相同的荷载增量下,钢筋的压应力比混凝土的压应力增加得更快,混凝土与钢筋之间出现了应力重分布。随着荷载的继续增加,柱中开始出现纵向微细裂缝,当轴向压力增加到破坏荷载的 90% 左右时,柱四周出现明显的纵向裂缝及压坏痕迹,混凝土保护层剥落,箍筋间的纵筋在荷载作用下压屈,向外凸出,外层混凝土剥落,柱随即破坏。短柱的破坏形态如图 6-3 所示。

试验表明,素混凝土棱柱体受压构件达到最大压应力值时的峰值应变为 0.0015~

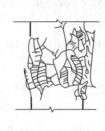

图 6-3 短柱的破坏
Figure 6-3 Failure of short column

0.002,而钢筋混凝土短柱达到最大承载力时的压应变一般为 0.0025～0.0035,甚至更大,主要原因是纵向钢筋的配置能改变混凝土的应力,更好地发挥混凝土的塑性,改善受压破坏的脆性。一般是纵筋先达到屈服强度,再继续增加荷载,最后混凝土压应变达到极限压应变而破坏。当纵向钢筋的屈服强度较高时,可能会出现钢筋未达到屈服强度而混凝土达到极限压应变的情况。

对普通混凝土构件,计算时以构件的压应变达到 0.002 为控制条件,认为此时混凝土达到了棱柱体抗压强度 f_c,相应的纵筋应力为 $\sigma'_s = E_s \varepsilon_s \approx 2 \times 10^5 \times 0.002 \text{N/mm}^2 \approx 400 \text{N/mm}^2$。对于 HRB400、HRBF400 和 RRB400 热轧钢筋,此钢筋应力 σ'_s 值已达到其屈服强度,计算时可按 f'_y 取值;当采用 HRB500、HRBF500 钢筋时,《混凝土结构设计标准》(GB/T 50010—2010)规定钢筋的抗压强度设计值 f'_y 应取 400N/mm^2。

2. 长柱的正截面受力分析和破坏形态
Normal section force analysis and destruction form of long columns

在实际工程中,较细长的长柱较多。试验表明,对于长细比较大的长柱,不能忽略偶然因素造成的初始偏心距的影响。加载后,初始偏心距产生附加弯矩,附加弯矩引起侧向挠度,侧向挠度又增大了荷载的偏心距。随着荷载增加,侧向挠度和附加弯矩不断增大,使长柱在轴力和附加弯矩的共同作用下向一侧凸出发生破坏。破坏时,首先在凹侧混凝土中出现纵向裂缝,随后混凝土被压碎,纵筋被压屈向外凸出;然后,凸侧混凝土出现垂直于纵轴方向的横向裂缝,侧向挠度急剧增大,导致柱破坏,如图 6-4 所示。

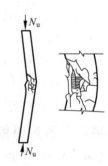

图 6-4 长柱的破坏
Figure 6-4 Failure of long column

6.1.3 轴心受压普通箍筋柱的正截面承载力
Bearing capacity of normal section of ordinary stirrup column under axial compression

1. 短柱的正截面承载力
Bearing capacity of normal section of short columns

根据以上分析可知,配有纵向受力钢筋和普通箍筋的轴心受压短柱破坏时,纵向受力钢筋先达到屈服强度 f'_y,然后混凝土达到轴心抗压强度设计值 f_c 而破坏。破坏时,轴心受压短柱正截面计算应力图形如图 6-5 所示。

According to the above analysis, when the axial compression short column with longitudinally stressed reinforcement and ordinary stirrups fails, the longitudinally stressed reinforcement first reaches the yield strength f'_y, and then the concrete reaches the design value of axial compressive strength f_c and fails. In case of failure, the calculated stress diagram of normal section of short column under axial compression is shown in Fig. 6-5.

轴心受压短柱正截面受压承载力计算公式为

The calculation formula of the compressive bearing capacity of the normal section of the axial compression short column is:

$$N_{us} = f_c A + f'_y A'_s \qquad (6-1)$$

式中，N_{us} 为轴心受压短柱正截面受压承载力设计值，N；f_c 为混凝土轴心抗压强度设计值，N/mm²；A 为柱截面面积，$A=bh$，mm²；f'_y 为纵向受力钢筋的抗压强度设计值，N/mm²；A'_s 为全部纵向受力钢筋的截面面积，mm²。

Where, N_{us} is the design value of the compressive bearing capacity of normal section of short column under axial compression, N; f_c is the design value of concrete axial compressive strength, N/mm²; A is the total cross-sectional area of the column, $A=bh$, mm²; f'_y is the design value of compressive strength of longitudinally stressed reinforcement, N/mm²; A'_s is the total cross-sectional area of all longitudinally stressed reinforcement, mm².

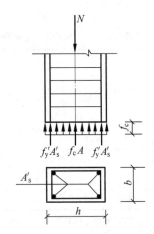

图 6-5 轴心受压短柱正截面受力示意图

Figure 6-5 Stress diagram of normal section of axial compression short column

2. 长柱的正截面承载力
Bearing capacity of normal section of long columns

试验表明，长柱的承载力 N_{ul} 低于相同条件下的短柱承载力 N_{us}，且长细比越大，长柱承载力降低越多。因为长细比越大，各种偶然因素产生的初始偏心距也越大，所以产生的附加弯矩和侧向挠度越大。长细比较大的受压构件还有可能在材料发生破坏前由于失稳而丧失承载力。在长期荷载作用下，混凝土徐变使侧向挠度增大更多，长期荷载在全部荷载中所占的比例越大，承载力降低越多。

考虑构件长细比增大的附加效应使构件承载力降低，《混凝土结构设计标准》(GB/T 50010—2010)采用稳定系数 φ 表示长柱承载力的降低程度，即

$$\varphi = \frac{N_{ul}}{N_{us}} \leqslant 1 \qquad (6-2)$$

式中，φ 为稳定系数；N_{ul} 为轴心受压长柱的受压承载力，kN；N_{us} 为轴心受压短柱的受压承载力，kN。

 特别提示 6-1

根据试验研究结果并考虑到过去的使用经验，《混凝土结构设计标准》(GB/T 50010—2010)采用的稳定系数 φ 值见表 6-2。对于长细比 l_c/b 较大的构件，考虑到初始偏心和长期

荷载作用对构件承载力的不利影响较大,为保证安全,稳定系数 φ 的取值比按试验研究得到的经验公式计算得到的 φ 值要低一些。对于长细比 $l_c/b<20$ 的构件,φ 的取值稍微提高一些。

表 6-2 钢筋混凝土轴心受压构件的稳定系数 φ

Table 6-2 Stability coefficient of reinforced concrete axial compression members φ

l_c/b	≤8	10	12	14	16	18	20	22	24	26	28
l_c/d	≤7	8.5	10.5	12	14	15.5	17	19	21	22.5	24
l_c/i	≤28	35	42	48	55	62	69	76	83	90	97
φ	1.0	0.98	0.95	0.92	0.87	0.81	0.75	0.70	0.65	0.60	0.56
l_c/b	30	32	34	36	38	40	42	44	46	48	50
l_c/d	26	28	29.5	31	33	34.5	36.5	38	40	41.5	43
l_c/i	104	111	118	125	132	139	146	153	160	167	174
φ	0.52	0.48	0.44	0.40	0.36	0.32	0.29	0.26	0.23	0.21	0.19

注:表中 l_c 为构件计算长度,b 为矩形截面的短边尺寸,d 为圆形截面的直径,i 为截面最小回转半径。

在承载能力极限状态时,轴心受压构件截面的应力如图 6-5 所示。根据构件截面竖向力的平衡条件,考虑构件可靠度调整和长柱承载力降低等因素,《混凝土结构设计标准》(GB/T 50010—2010)6.2.15 条给出了轴心受压构件正截面承载力的设计表达式,见式(6-3),其中可靠度是指考虑初始偏心的影响和主要承受永久荷载作用的轴心受压构件的可靠度。

$$N \leqslant N_u = 0.9\varphi(f_c A + f'_y A'_s) \quad (6-3)$$

式中,N 为轴向压力设计值,N;N_u 为轴心受压构件承载力设计值,N;0.9 为可靠度调整系数;φ 为稳定系数,按表 6-2 采用;f_c 为混凝土轴心抗压强度设计值,N/mm²;A 为构件截面面积,mm²;f'_y 为纵向钢筋的抗压强度设计值,N/mm²;A'_s 为全部纵向钢筋的截面面积,mm²。

当构件截面边长 $b \leqslant 300$mm 或直径 $d \leqslant 300$mm 时,式(6-3)中的 f_c 取 $0.8f_c$。当纵筋配筋率 $\rho' > 3\%$ 时,式(6-3)中的构件截面面积 A 用混凝土净截面面积 A_c 代替,即截面总面积减去钢筋所占面积:$A_c = A - A'_s$。

6.1.4 轴心受压构件的截面设计和截面复核
Section design and section review of axial compression members

1. 截面设计

Section design

已知:轴心受压构件截面尺寸 $b \times h$,混凝土轴心抗压强度设计值 f_c,纵向受压钢筋抗压强度设计值 f'_y,构件计算长度 l_c,轴向压力设计值 N。求纵向受压钢筋截面面积 A'_s。

Known: cross-sectional dimensions $b \times h$ of the axial compression member, concrete axial compressive strength design value f_c, longitudinal compression reinforcement compressive strength design value f'_y, effective length of member l_c, axial compression design value N. Calculate the cross-sectional area of longitudinal compression reinforcement A'_s.

1) 计算纵向受压钢筋截面面积 A'_s

1) Calculate the cross-sectional area of longitudinal compression reinforcement A'_s

截面设计时,取 $N=N_u$,由轴心受压构件正截面承载力计算公式(6-3),得

When designing the cross section, taking $N=N_u$, from the calculation formula Eq. (6-3) of bearing capacity of normal section of axial compression member, one obtains

$$A'_s = \frac{\dfrac{N}{0.9\varphi} - f_c A}{f'_y} \tag{6-4}$$

根据式(6-4)计算所得的纵向受压钢筋截面面积 A'_s,查附表 2-1 确定构件中纵向受压钢筋的直径和根数。选择钢筋直径和根数时,应满足《混凝土结构设计标准》(GB/T 50010—2010)对受压构件相邻钢筋之间净距和混凝土保护层厚度的要求,以保证钢筋和混凝土之间具有足够的黏结力。

According to the cross-sectional area A'_s of longitudinal compression reinforcement calculated from Eq. (6-4), refer to the Attached table 3-1 to determine the diameter and number of longitudinal compression bars in the member. When choosing the bar diameter and number, the requirements of "Standard for design of concrete structures" (GB/T 50010—2010) on the clear distance between adjacent reinforcement of compression members and the thickness of concrete cover should be met so as to ensure sufficient bond force between steel and concrete.

拓展知识 6-1

2) 验算配筋率

2) Check the reinforcement ratio

除满足计算要求外,为提高构件的延性,保证受压承载力,《混凝土结构设计标准》(GB/T 50010—2010)规定受压构件纵向受力钢筋的截面面积还应满足最小配筋率的要求。采用式(6-5)验算矩形截面受压构件纵筋的最小配筋率:

In addition to meeting the calculation requirements, in order to improve the ductility of the member and ensure the compression bearing capacity, the section area of longitudinally stressed reinforcement of compression member should meet the requirements of the minimum reinforcement ratio specified in "Standard for design of concrete structures" (GB/T 50010—2010). The minimum reinforcement ratio of longitudinal reinforcement of rectangular section compression members is checked by Eq. (6-5):

$$\rho' = \frac{A'_s}{bh} \geqslant \rho'_{\min} \tag{6-5}$$

式中,ρ' 为纵向受压钢筋的配筋率;A'_s 为纵向受压钢筋的截面积,mm^2;b 为构件截面短边尺寸,mm;h 为构件截面长边尺寸,mm;$\rho'_{\min}$ 为纵向受压钢筋的最小配筋率。

Where, ρ' is the reinforcement ratio of longitudinal compression reinforcement; A'_s is the cross-sectional area of longitudinal compression reinforcement, mm^2; b is the short side size of member section, mm; h is the long side size of member section, mm; $\rho'_{\min}$ is the minimum reinforcement ratio of longitudinal compression reinforcement.

《混凝土结构通用规范》(GB 55008—2021)4.4.6 条规定：对强度等级为 300MPa 的钢筋，纵向受力普通钢筋最小配筋率 ρ'_{min} 为 0.6%，对强度等级为 400MPa 的钢筋为 0.55%，对强度等级为 500MPa 的钢筋为 0.5%；同时一侧纵筋的配筋率不应小于 0.2%。从经济和施工方便角度考虑，全部纵向钢筋的配筋率不宜大于 5%，通常纵向受压钢筋的配筋率不超过 3%。

Article 4.4.6 of "General code for concrete structures" (GB 55008—2021) stipulates: for reinforcement with strength grade of 300MPa, the minimum reinforcement ratio ρ'_{min} of longitudinally stressed ordinary reinforcement is 0.6%, and for reinforcement with strength grade of 400MPa, it is 0.55%, 0.5% is for reinforcement with strength grade of 500MPa; and the reinforcement ratio of one side longitudinal bars should not be less than 0.2%. From the perspective of economy and construction convenience, the reinforcement ratio of all longitudinal bars should not exceed 5%, and usually the reinforcement ratio of longitudinal compression bars should not exceed 3%.

拓展知识 6-2

2. 截面复核

Section review

已知轴心受压构件的截面尺寸 $b \times h$，混凝土轴心抗压强度设计值 f_c，纵向受压钢筋抗压强度设计值 f'_y，构件计算长度 l_c，纵向受压钢筋截面面积 A'_s，求轴心受压构件承载力设计值 N_u。

先由长细比 l_c/b 查表 6-2 得到出稳定系数 φ 值，再将其他已知条件代入式(6-6)即可求出该截面所能承受的轴心压力设计值 N_u：

$$N_u = 0.9\varphi(f_c A + f'_y A'_s) \quad (6-6)$$

式(6-6)中各参数的含义同式(6-3)。

当轴心压力设计值 $N \leqslant N_u$ 时，截面承载力满足要求，构件安全；否则构件不安全。

【例题 6-1】 某教学楼为二层现浇钢筋混凝土框架结构，层高为 3.6m，平面尺寸为 45m×16m。建筑抗震设防烈度为 7 度，抗震设防类别为乙类，采取二级抗震构造措施。由结构计算软件计算得出柱底轴向压力设计值 $N = 1500$kN，柱计算长度 l_c 为 4.5m，近似按轴心受压构件设计该柱。

【Example 6-1】 A teaching building is a two-story cast-in-place reinforced concrete frame structure with floor height of 3.6m and plane size of 45m × 16m. The seismic fortification intensity of the building is 7 degrees, the seismic fortification category is Class B, and secondary seismic structural measures are adopted. Calculated by structural calculation software, the design value of axial load at the bottom of the column is $N = 1500$kN, and the effective length l_c of the column is 4.5m. The column is approximately designed as axial compression member.

解：(1) 选择柱截面尺寸

对于有抗震设防要求的框架结构，要求框架柱的截面宽度、高度不宜小于 300mm，取截

面宽度 $b=350$mm，截面高度 $h=350$mm。

(2) 选择柱材料

根据工程实际情况，选用 HRB400 钢筋和 C25 混凝土。查附表 1-3、1-4 和 1-6 得混凝土轴心抗压强度设计值 $f_c=11.9\text{N/mm}^2$，混凝土轴心抗拉强度设计值 $f_t=1.27\text{N/mm}^2$，纵向受压钢筋抗压强度设计值 $f'_y=360\text{N/mm}^2$。

(3) 确定柱轴向压力设计值

考虑附加弯矩的影响，将柱轴向压力乘以 1.2 的放大系数，则 $N=1.2\times1500\text{kN}=1800\text{kN}$（含柱自重）。

(4) 进行柱截面设计

长细比
$$\frac{l_c}{b}=\frac{4500}{350}=12.9$$

根据表 6-2，采用线性插值法，得稳定系数
$$\varphi=0.92+\frac{14-12.9}{14-12}\times(0.95-0.92)=0.92+0.0165=0.9365$$

由式(6-4)，计算纵向受压钢筋截面面积：
$$A'_s=\frac{1}{f'_y}\left(\frac{N}{0.9\varphi}-f_cA\right)=\frac{1}{360}\times\left(\frac{1800\times10^3}{0.9\times0.9365}-11.9\times350\times350\right)\text{mm}^2=1883\text{mm}^2$$

查附表 2-1 选配钢筋 4Φ25（实配钢筋截面面积 $A'_s=1964\text{mm}^2<1.05\times1883\text{mm}^2=1977\text{mm}^2$）。

按构造要求选用双肢箍筋ϕ10@200，柱端箍筋加密区箍筋为ϕ10@100。

验算纵向受压钢筋配筋率：
$$\rho'=\frac{A'_s}{A}=\frac{A'_s}{bh}=\frac{1964}{350\times350}=1.6\%>\rho'_{\min}=0.55\%$$

$\rho'<\rho'_{\max}=5\%$，且 $\rho'<3\%$，所以上述构件截面面积 A 的计算中没有减去 A'_s 是正确的。

截面每一侧纵筋的配筋率 $0.5\rho'=0.5\times1.6\%=0.8\%>\rho'_{\min}=0.2\%$（单侧纵向钢筋的最小配筋率）。所以，配筋率满足要求。

【例题 6-2】 某现浇钢筋混凝土柱的截面尺寸为 $b\times h=300\text{mm}\times300\text{mm}$，由柱两端约束情况确定其计算长度 $l_c=3\text{m}$，柱内配有 4Φ20 纵向受压钢筋，混凝土强度等级为 C30，柱的轴向压力设计值 $N=1600\text{kN}$，问：柱截面是否安全？

【Example 6-2】 The section size of a cast-in-place reinforced concrete column is $b\times h=300\text{mm}\times300\text{mm}$, the effective length $l_c=3\text{m}$ is determined by the constraints at both ends of the column. The column is reinforced by 4Φ20 longitudinal compression bars, concrete grade strength is C30, and the axial compression design value of the column is $N=1600\text{kN}$. Question: whether the column section is safe?

解：长细比 $l_c/b=3000/300=10$，查表 6-2 得稳定系数 $\varphi=0.98$。

查附表 2-1 得 4Φ20 纵向受压钢筋的截面面积 $A'_s=1256\text{mm}^2$。

根据式(6-3)得轴心受压构件承载力设计值为
$$N_u=0.9\varphi(f_cA+f'_yA'_s)=0.9\times0.98(14.3\times300\times300+360\times1256)$$
$$=1533.9\times10^3\text{N}=1533.9\text{kN}<N=1600\text{kN}$$

所以，柱截面不安全。

6.2 偏心受压构件正截面受压破坏形态
Compression destruction form of normal section of eccentric compression members

偏心受压构件的破坏是由于混凝土被压碎造成的，破坏形态与偏心距的大小以及纵向钢筋配筋率有关。试验表明，根据受力情况和破坏形态的不同，偏心受压构件的破坏可分为大偏心受压破坏和小偏心受压破坏。通常在与偏心轴方向垂直的两边布置纵向钢筋，离轴向压力较近一侧的纵向钢筋为受压钢筋，其截面面积用 A'_s 表示；远离轴向力一侧的纵向钢筋则根据轴向压力偏心距的大小，可能是大偏心受压，也可能是小偏心受压，钢筋可能受拉也可能受压，其截面面积都用 A_s 表示。

The failure of eccentric compression members is caused by the crushing of concrete, and the destruction form is related to the value of eccentricity and reinforcement ratio of longitudinal bars. The test shows that according to the different stress conditions and destruction forms, the failure of eccentric compression members can be divided into large eccentric compression failure and small eccentric compression failure. Usually, longitudinal bars are arranged on both sides perpendicular to the direction of the eccentric axis. The longitudinal bar on the side closer to the axial pressure is the compression reinforcement, and its cross-sectional area is expressed by A'_s, and the longitudinal bar away from the axial pressure side may be in tension or compression according to the eccentricity of the axial pressure, it may be a large eccentric compression or a small eccentric compression and its cross-sectional area is expressed by A_s.

6.2.1 偏心受压短柱的破坏形态
Destruction form of eccentric compression short columns

1. 大偏心受压破坏
Large eccentric compression failure

视频 6-2

大偏心受压破坏常发生在偏心压力 N 的相对偏心距 e_0/h_0 较大，且受拉钢筋配置不过多的情况下。在偏心压力 N 的作用下，构件靠近偏心压力的一侧截面受压，较远一侧截面受拉。随着偏心压力的增加，截面受拉侧混凝土首先产生水平裂缝并不断发展，受拉钢筋的应力随荷载增加发展较快，当偏心压力 N 接近破坏荷载时，受拉钢筋的应力首先达到屈服强度 f_y，进入流幅阶段，受拉变形比受压变形大，中和轴向受压区移动，使混凝土受压区高度减小，混凝土压应变增大，受压区混凝土也出现了纵向裂缝，最后受压区边缘混凝土达到极限压应变，受压区混凝土被压碎而破坏，如图 6-6(a) 所示。这时受压钢筋一般都能达到其受压屈服强度 f'_y。

The large eccentric compression failure often occurs when the relative eccentricity $e_0/$

h_0 of the eccentric pressure N is large, and the tensile reinforcement is not enough. Under the action of the eccentric pressure N, the section of the member close to the eccentric pressure is compressed, and the far side section is tensioned. With the increase of the eccentric pressure, the concrete on the tension side of the section first produces horizontal cracks and develops continuously. The stress of tensile reinforcement develops rapidly with the increase of the load. When the eccentric pressure N approaches the failure load, the stress of tensile reinforcement first reaches the yield strength f_y, entering the flow amplitude stage, the tensile deformation is larger than the compressive deformation, and the neutral axis moves to the compression zone, so that the depth of

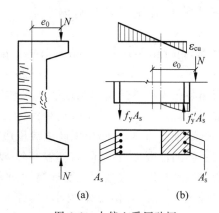

图 6-6 大偏心受压破坏
（a）破坏形态；（b）破坏时的截面应力和应变
Figure 6-6 Large eccentric compression failure
(a) Destruction form; (b) Section stress and strain at failure

concrete compression zone decreases, the compressive strain of concrete increases, and longitudinal cracks also appear in the concrete in the compression zone. Finally, the concrete at the edge of the compression zone reaches the ultimate compressive strain, and the concrete in the compression zone is crushed and destroyed, as shown in Fig. 6-6(a). At this time, the compressive reinforcement can generally reach its compressive yield strength f'_y.

大偏心受压破坏的主要特征是：破坏从受拉区开始，受拉钢筋首先屈服，然后受压区混凝土被压碎。大偏心受压破坏具有明显预兆，变形能力较大，属于延性破坏，破坏特征与双筋截面适筋梁相似。

The main characteristics of large eccentric compression failure are: the failure starts from the tension zone, the tensile reinforcement yields first, and then the concrete in the compression zone is crushed. The large eccentric compression failure has obvious advance warning, and the deformation capability is large, which belongs to ductile failure. The failure characteristics are similar to those of doubly reinforced section underreinforced beams.

大偏心受压破坏时正截面上的应力和应变如图 6-6(b)所示。

The stress and strain on the normal section during large eccentric compression failure are shown is Fig. 6-6(b).

2. 小偏心受压破坏

Small eccentric compression failure

根据偏心压力 N 的相对偏心距 e_0/h_0 的大小，小偏心受压破坏分为以下两种情况。

视频 6-3

According to the relative eccentricity e_0/h_0 of eccentric pressure N, there are two cases of small eccentric compression failure as follows.

1) 当偏心压力 N 的相对偏心距 e_0/h_0 很小或较小时，构件截面全部受压或大部分受压

1) When the relative eccentricity e_0/h_0 of eccentric pressure N is very small or small, all or most of the member section is under compression

当偏心压力从零开始加载逐渐增大时，截面受拉边缘出现水平裂缝，但进展比较缓慢，未形成明显的主裂缝，而受压区边缘混凝土的压应变增长较快，临近破坏时受压边出现纵向裂缝。靠近轴向力 N 一侧受压区边缘压应变首先达到混凝土极限压应变。破坏时，受压应力较大一侧的混凝土被压坏，破坏较突然，无明显预兆，压碎区段较长。同侧受压钢筋的应力一般能达到抗压屈服强度。而远离轴向力 N 一侧钢筋可能受拉也可能受压，一般均达不到屈服强度。只有当偏心距很小（对矩形截面 $e_0 \leqslant 0.15h_0$）而轴向力 N 又较大（$N > \alpha_1 f_c bh_0$）时，远离轴向力 N 一侧钢筋才可能受压屈服。构件破坏时截面的应力、应变情况如图 6-7(a)、(b) 所示。

When the eccentric pressure is gradually increased from zero, horizontal cracks appear on the tension side of the section, but the development is relatively slow, and no obvious main cracks are formed. However, the compressive strain of concrete at the edge of compression zone increases rapidly, and longitudinal cracks appear on the compression side near failure. The compressive strain on the side of the compression zone on the side near the axial force N first reaches the ultimate compressive strain of concrete. During failure,

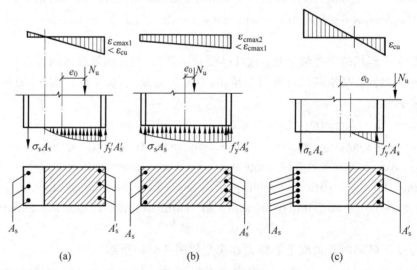

图 6-7　小偏心受压破坏

(a) 相对偏心距 e_0/h_0 较小，截面大部分受压；(b) 相对偏心距 e_0/h_0 很小，截面全部受压；
(c) 相对偏心距 e_0/h_0 较大，A_s 配筋太多

Figure 6-7　Small eccentric compression failure

(a) The relative eccentricity e_0/h_0 is small, most of the section is compressed;
(b) The relative eccentricity e_0/h_0 is very small, full section is compressed;
(c) The relative eccentricity e_0/h_0 is large, the A_s reinforcement is too much

the concrete on the side with greater compressive stress is crushed, the destruction is sudden without obvious advance warning, and the crushing section is long. The stress of compressive reinforcement on the same side can generally reach the compressive yield strength, while the reinforcement on the side away from the axial force N may be under tension or compression, generally not reaching the yield strength. Only when the eccentricity is small ($e_0 \leq 0.15h_0$ for rectangular section) and the axial force N is large ($N > \alpha_1 f_c bh_0$), the reinforcement on the side away from the axial force N may yield under compression. The stress and strain of the section when the member is destroyed are shown in Fig. 6-7(a) and Fig. 6-7(b).

当相对偏心距很小时,由于截面的实际形心和构件的几何中心不重合,若纵向受压钢筋比纵向受拉钢筋多得多,可能发生远离轴向力一侧的混凝土先被压坏的情况,称为"反向破坏"。

When the relative eccentricity is very small, since the actual centroid of the section does not coincide with the geometric center of the member, if the longitudinal compression reinforcement is much more than the longitudinal tensile reinforcement, it may occur that the concrete on the side away from the axial force is crushed first, which is called "reverse failure".

2) 虽然轴向压力 N 的相对偏心距 e_0/h_0 较大,但远离轴向力一侧的纵向钢筋 A_s 配置过多

2) Although the relative eccentricity e_0/h_0 of the axial pressure N is large, the longitudinal bars A_s on the side away from the axial force are configured excessively

破坏时,受压区边缘混凝土达到极限压应变,混凝土被压碎,受压钢筋应力达到抗压屈服强度,远离轴向力一侧的钢筋可能受拉也可能受压,因为配筋过多,受拉时始终不屈服,受压时可能屈服也可能不屈服,其截面上的应力情况如图 6-7(c)所示。破坏无明显预兆,压碎区段较长,属于脆性破坏,如图 6-8 所示,混凝土强度越高,破坏越突然。

During failure, the concrete at the edge of the compression zone reaches the ultimate compressive strain, the concrete is crushed, and the stress of the compressive reinforcement reaches the compressive yield strength. The steel bar on the side away from the axial force may be tensioned or compressed. Because of excessive reinforcement, it never yields under tension, and it may or may not yield under compression. The stress on its section is shown in Fig. 6-7(c). There is no obvious advance warning of failure,

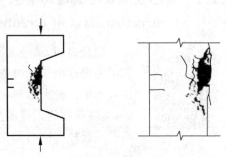

图 6-8 小偏心受压破坏形态

Figure 6-8 Small eccentric compression destruction form

and the crushing section is long, which belongs to brittle failure, as shown in Fig. 6-8, the higher the concrete strength, the more sudden the failure.

特别提示 6-2

3. 界限破坏

Balanced failure

在大、小偏心受压破坏之间必定存在着一种界限破坏状态。当构件处于界限破坏状态时,首先受拉区混凝土出现明显横向裂缝,然后受拉钢筋达到抗拉屈服强度,同时受压区边缘混凝土刚好达到极限压应变,受压钢筋也达到其抗压屈服强度。

Between the large and small eccentric compression failure, there must be a balanced failure state. When the member is in a state of balanced failure, the concrete in the tension zone first has obvious transverse cracks, and then the tensile reinforcement reaches the tensile yield strength, and the concrete at the edge of the compression zone just reaches the ultimate compressive strain, and the compression reinforcement also reaches its compressive yield strength.

4. 大、小偏心受压的判别

Discrimination of large and small eccentric compression

两类偏心受压构件的界限破坏特征与受弯构件中适筋梁与超筋梁的界限破坏特征完全相同,因此相对受压区高度 ξ 和界限相对受压区高度 ξ_b 的表达式也相同。

当 $\xi \leqslant \xi_b$ 时,属于大偏心受压破坏;当 $\xi > \xi_b$ 时,属于小偏心受压破坏。

The balanced failure characteristics of the two types of eccentric compression members are exactly the same as those of underreinforced beams and overreinforced beams in the flexural members, therefore the expressions of relative compression zone depth ξ and boundary relative compression zone depth ξ_b are also the same.

When $\xi \leqslant \xi_b$, it belongs to a large eccentric compression failure; when $\xi > \xi_b$, it belongs to a small eccentric compression failure.

6.2.2 偏心受压长柱的破坏形态

Destruction form of eccentric compression long columns

试验表明,承受偏心压力的柱在荷载作用下会产生侧向弯曲。长细比较小的短柱由于侧向弯曲很小,设计时一般可忽略不计。而长细比较大的柱会产生比较大的侧向弯曲,使柱产生不可忽略的附加弯矩或称二阶弯矩,会降低柱的正截面受压承载力,设计时必须考虑侧向弯曲的影响。偏心受压长柱在侧向弯曲影响下,可能发生材料破坏和失稳破坏两种破坏形态。

1. 材料破坏

Material failure

当偏心受压柱的长细比在一定范围内时,虽然在承受偏心受压荷载后,偏心距由 e_i 增加到 $e_i + f$(f 为侧向挠度),使柱的承载能力比相同截面的短柱小,但也和短柱一样属于材料破坏,即因截面材料强度耗尽而产生破坏。

2. 失稳破坏
Unstable failure

当偏心受压构件的长细比很大时,偏心距随偏心压力的增大而不断非线性增加,构件的破坏不是由材料引起的,而是由于偏心压力的微小增量 ΔN 引起构件不收敛的弯矩 M 增加,导致构件侧向弯曲过大而失去平衡,如图 6-9 所示,称为"失稳破坏"。失稳破坏时某些截面的承载力已达最大,但此时截面内的钢筋应力并未达到屈服强度,混凝土也未达到极限压应变。构件能够承受的偏心压力远远小于短柱时的受压承载力 N_u,达到构件最大承载能力时,控制截面上钢筋和混凝土的应力均未达到材料破坏时的强度。

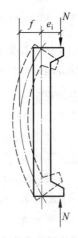

图 6-9 长柱侧向弯曲和侧向挠度 f(失稳破坏)

Figure 6-9 Lateral bending and lateral deflection f of long column(unstable failure)

6.3 偏心受压构件的二阶效应
Second-order effect of eccentric compression members

偏心受压构件的二阶效应是指轴向压力导致侧移和挠曲从而产生附加弯矩和附加曲率。对无侧移的框架结构,二阶效应指轴向压力在产生了挠曲变形的柱段中引起的附加弯矩和附加曲率,一般称为 $P\text{-}\delta$ 效应。在力学分析中,求的是构件两端截面的弯矩和轴力,考虑二阶效应后,在构件的某个其他截面,其弯矩可能会大于端部截面的弯矩,设计时应取弯矩最大的截面进行计算。对于有侧移的框架结构,二阶效应主要是指轴向压力在产生了侧移的框架中引起的附加弯矩和附加曲率,一般称为 $P\text{-}\Delta$ 效应。

The second-order effect of eccentric compression members refers to the additional moment and additional curvature produced by the axial pressure on sideway and deflection. For the frame structure with no sideway, the second-order effect refers to the additional moment and additional curvature caused by the axial pressure in the column segment that produces the deflection deformation, which is generally called the $P\text{-}\delta$ effect. In the mechanical analysis, the bending moment and axial force of the section at both ends of the member are obtained. After considering the second-order effect, in some other section of the member, the bending moment may be greater than the bending moment of the end section. The section with the maximum bending moment should be calculated during design. For the frame structure with sideway, the second-order effect mainly refers to the additional moment and additional curvature caused by the axial pressure in the frame with lateral displacement, which is generally called the $P\text{-}\Delta$ effect.

6.3.1 结构无侧移时由受压构件自身挠曲产生的 $P\text{-}\delta$ 二阶效应
$P\text{-}\delta$ second-order effect caused by the self deflection of the compression member when the structure has no sideway

1. 构件两端弯矩同号单曲率弯曲时的 $P\text{-}\delta$ 二阶效应
$P\text{-}\delta$ second-order effect of single curvature bending of bending moment at both ends of the member with the same signs

1) 控制截面的转移

1) Transfer of control section

大偏心受压或小偏心受压中的弯矩对配筋都是不利的。当轴向压力相差不多时，弯矩越大对配筋越不利，弯矩大的截面就是控制整个构件配筋的控制截面。偏心受压构件在两端同号弯矩 M_1、M_2（$M_2 > M_1$）和轴向压力 P 的共同作用下，将产生单曲率弯曲。两端同号弯矩指构件上端弯矩 M_1 和下端弯矩 M_2 使构件朝一个方向弯曲，即产生单曲率弯曲，如图 6-10(a)中的虚线所示。

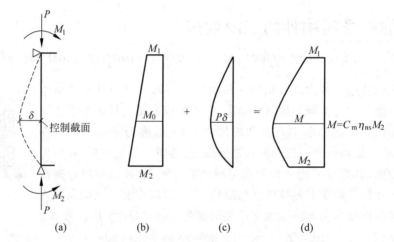

图 6-10 构件两端弯矩同号时的二阶效应（$P\text{-}\delta$ 效应）
(a) 构件单曲率弯曲；(b) 一阶弯矩图；(c) 附加弯矩图；(d) 叠加弯矩图

Figure 6-10 Second-order effect ($P\text{-}\delta$ effect) when the bending moments at both ends of the member are of the same sign
(a) Single curvature bending of the member; (b) First-order moment diagram; (c) Additional moment diagram;
(d) Superimposed moment diagram

不考虑二阶效应时,构件的弯矩图为一阶弯矩图,如图 6-10(b)所示,构件下端截面的弯矩 M_2 最大,此截面为构件的控制截面,用来进行整个构件的截面承载力计算。考虑二阶效应后,轴向压力 P 对构件中部任一截面产生附加弯矩 $P\delta$(δ 为任一截面的侧向挠度),如图 6-10(c)所示,与该截面处的一阶弯矩 M_0 叠加后,如图 6-10(d)所示,得

$$M = M_0 + P\delta \tag{6-7}$$

在构件中部总有一个截面,它的弯矩 M 是最大的。如果附加弯矩 $P\delta$ 比较大,且 M_1、M_2 比较接近,就有可能出现 $M > M_2$ 的情况,控制截面就由构件下端截面转移到构件中部弯矩最大的那个截面,这时就要考虑二阶效应。

2) 考虑 P-δ 二阶效应的条件

2) Conditions considering the P-δ second-order effect

两端弯矩同号时,发生控制截面转移的情况是不普遍的。为了减少计算工作量,《混凝土结构设计标准》(GB/T 50010—2010)6.2.3 条规定:弯矩作用平面内截面对称的偏心受压构件,当同一主轴方向的两端弯矩比 $M_1/M_2 \leqslant 0.9$ 且设计轴压比 $N/(f_c A) \leqslant 0.9$ 时,若构件的长细比满足式(6-8)的要求,可不考虑轴向压力在该方向挠曲构件中产生的附加弯矩影响,取 $\eta_{ns} = 1$。当不满足要求时,附加弯矩的影响不可忽略,需按截面的两个主轴方向分别考虑轴向压力在挠曲构件中产生的附加弯矩影响。

$$l_c/i \leqslant 34 - 12(M_1/M_2) \tag{6-8}$$

即只要满足式(6-9)三个条件中的任一条件时,就要考虑 P-δ 二阶效应的影响:

$$M_1/M_2 > 0.9 \text{ 或 } N/(f_c A) > 0.9 \text{ 或 } l_c/i > 34 - 12(M_1/M_2) \tag{6-9}$$

式中,M_1、M_2 分别为已考虑侧移影响的偏心受压构件两端截面按结构弹性分析确定的对同一主轴的组合弯矩设计值,绝对值较大端为 M_2,绝对值较小端为 M_1,当构件按单曲率弯曲时,M_1/M_2 取正值,否则取负值,N·mm;N 为轴向压力设计值,N;f_c 为混凝土轴心抗压强度设计值,N/mm^2;A 为偏心受压构件的截面面积,mm^2;l_c 为构件的计算长度,可近似取偏心受压构件相应主轴方向上下支撑点之间的距离,mm;i 为偏心方向的截面回转半径,mm。

2. 构件两端弯矩异号双曲率弯曲时的 P-δ 二阶效应

P-δ second-order effect of double curvature bending of bending moment at both ends of the member with opposite signs

构件两端弯矩异号时构件按双曲率弯曲,构件长度中部有反弯点,如图 6-11(a)所示,最典型的是框架柱。构件两端弯矩异号指构件上端弯矩 M_1 和下端弯矩 M_2 转动方向相同,使构件朝两个不同方向弯曲,即双曲率弯曲。反弯点指弯矩图中弯矩为零的点,如图 6-11(b)中的 a 点。虽然因为构件侧向弯曲,轴向压力对构件中部截面将产生附加弯矩,增大其弯矩值,如图 6-11(c)所示,但弯矩增大后仍不如端部截面的弯矩值大,即不会发生控制截面转移的情况,如图 6-11(d)所示;故不必考虑二阶效应。

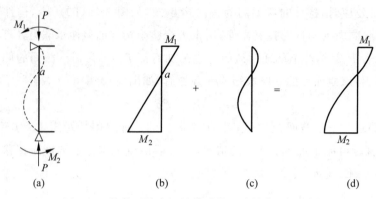

图 6-11 构件两端弯矩异号时的二阶效应（P-δ 效应）
(a) 构件双曲率弯曲；(b) 一阶弯矩图；(c) 附加弯矩图；(d) 叠加弯矩图

Figure 6-11　P-δ second-order effect of the bending moment at both ends of the member with opposite signs (P-δ effect)
(a) Double curvature bending of the member; (b) First-order moment diagram; (c) Additional moment diagram; (d) Superimposed moment diagram

6.3.2　考虑 P-δ 二阶效应后控制截面的设计弯矩
Design moment of control section considering P-δ second-order effect

1. 附加偏心距和初始偏心距

Additional eccentricity and initial eccentricity

当截面上作用的弯矩设计值为 M、轴向压力设计值为 N 时，其计算偏心距为 $e_0 = M/N$。工程中实际存在着荷载作用位置不定、混凝土质量不均匀、配筋不对称、偶然的施工偏差及其他不可预见的因素，这些因素都可能产生附加偏心距 e_a。当计算偏心距 e_0 较小时，附加偏心距 e_a 的影响较显著，随着计算偏心距 e_0 的增大，附加偏心距 e_a 对构件承载力的影响逐渐减小。参考以往工程经验和国外规范，《混凝土结构设计标准》(GB/T 50010—2010) 6.2.5 条规定：在偏心受压构件的正截面承载力计算中，应计入轴向压力在偏心方向存在的附加偏心距 e_a，为了计算方便，其值取 20mm 与 $h/30$ 中的较大值，即

When the design value of the bending moment acting on the section is M and the design value of the axial pressure is N, the calculated eccentricity is $e_0 = M/N$. The additional eccentricity e_a may be generated due to the fact that there are factors such as the uncertainty of the load action position, non-uniformity of concrete quality, asymmetry of reinforcement, accidental construction misalignment and other unforeseen factors. When the calculated eccentricity e_0 is relatively small, the influence of the additional eccentricity e_a is more significant. With the increase of calculated eccentricity e_0, the influence of additional eccentricity e_a on the bearing capacity of the member gradually decreases. Referring to previous engineering experience and foreign codes, Article 6.2.5 of "Standard for design of concrete structures" (GB/T 50010—2010) stipulates that in the calculation of the normal section bearing capacity of eccentric compression member, the additional eccentricity e_a existing in the eccentric direction of the axial pressure should be taken into account. For the convenience of calculation, its value is taken as the larger of 20mm and $h/30$, that is

$$e_a = \max\{20, h/30\} \tag{6-10}$$

式中，e_a 为附加偏心距，mm；h 为偏心方向截面尺寸，mm。

Where, e_a is the additional eccentricity, mm; h is the section size in the eccentric direction, mm.

初始偏心距 e_i 取计算偏心距 e_0 与附加偏心距 e_a 之和，即

The initial eccentricity e_i is taken as the sum of the calculated eccentricity e_0 and the additional eccentricity e_a, that is

$$e_i = e_0 + e_a \tag{6-11}$$

2. 偏心距调节系数和弯矩增大系数

Eccentricity adjustment coefficient and moment increase coefficient

实际工程中长柱比短柱更常见，在确定控制截面的弯矩设计值时，需考虑长柱在偏心压力作用下侧向弯曲引起的挠度对弯矩的影响。在工程设计中常采用增大系数法，即偏心受压柱的弯矩设计值为原柱端最大弯矩乘以偏心距调节系数和弯矩增大系数（考虑了附加弯矩的影响）。

In practical engineering long columns are more common than short columns. When determining the design value of bending moment of the control section, it is necessary to consider the influence of the deflection caused by the lateral bending of the long column under the action of eccentric pressure on the bending moment. In engineering design, the increase coefficient method is often used, that is, the design value of bending moment of eccentric compression column is the maximum bending moment at the original column end multiplied by the eccentricity adjustment coefficient and moment increase coefficient (considering the influence of additional moment).

1) 构件两端截面偏心距调节系数 C_m

1) Eccentricity adjustment coefficient C_m of the section at both ends of the member

对于弯矩作用平面内截面对称的偏心受压柱，同一主轴方向柱两端的弯矩大多不同，但也存在单曲率弯曲（M_1/M_2 为正）时二者大小接近的情况，即 $M_1/M_2 > 0.9$，此时，该柱在柱两端相同方向、几乎相同大小的弯矩作用下将产生最大的偏心距，使该柱处于最不利的受力状态，需考虑偏心距调节系数，它是考虑二阶效应影响的系数，为挠曲后的最大偏心距与初始偏心距的比值。偏心距调节系数 C_m 按式（6-12）计算：

For eccentric compression column with symmetrical section in the plane of bending moment action, the bending moments at both ends of the column in the same principal axis direction are mostly different, but there is also a situation when bending with a single curvature (M_1/M_2 is positive), the two values are similar (M_1/M_2 is positive), that is, the ratio $M_1/M_2 > 0.9$, at this time, the column will produce the maximum eccentricity under the action of bending moment in the same direction and almost the same magnitude at both ends of the column, so that the column is in the most unfavorable stress state, and the eccentricity adjustment coefficient needs to be considered. It is the coefficient considering the influence of second-order effect, and is the ratio of the maximum eccentricity after bending to the initial eccentricity. The eccentricity adjustment coefficient C_m is obtained from Eq. (6-12):

$$C_m = 0.7 + 0.3 M_1/M_2 \geqslant 0.7 \tag{6-12}$$

偏心距调节系数 C_m 考虑了构件两端截面弯矩差异的影响，当 C_m 小于 0.7 时，取 0.7。

The eccentricity adjustment coefficient C_m takes into account the influence of the difference in bending moment between the sections at both ends of the member. When the C_m is less than 0.7, it is taken as 0.7.

2) 弯矩增大系数 η_{ns}

2) Moment increase coefficient η_{ns}

弯矩增大系数是考虑侧向挠度的影响而引入的系数，考虑柱产生侧向挠度 f 后，柱中截面弯矩可表示为

The moment increase coefficient is the coefficient introduced considering the influence of lateral deflection. After considering the lateral deflection of the column, the bending moment of the section in the column can be expressed as

$$M = N(e_0 + f) = N\eta_{ns} e_0 \tag{6-13}$$

$$\eta_{ns} = \frac{e_0 + f}{e_0} = 1 + \frac{f}{e_0} \tag{6-14}$$

式中，e_0 为计算偏心距，mm；f 为柱的侧向挠度，mm；η_{ns} 为弯矩增大系数。

Where, e_0 is the calculated eccentricity, mm; f is the lateral deflection of the column, mm; η_{ns} is the moment increase coefficient.

根据大量试验结果和理论分析，《混凝土结构设计标准》(GB/T 50010—2010) 6.2.4 条给出了矩形、T 形、I 形、环形和圆形截面偏心受压构件弯矩增大系数 η_{ns} 的计算公式：

Based on a large number of experimental results and theoretical analysis, Article 6.2.4 of the "Standard for design of concrete structures" (GB/T 50010—2010) gives the calculation formula for the moment increase coefficient η_{ns} of rectangular, T shaped, I-shaped, ring and circular section eccentric compression members:

$$\eta_{ns} = 1 + \frac{1}{1300(M_2/N + e_a)/h_0} \left(\frac{l_c}{h}\right)^2 \zeta_c \tag{6-15}$$

式中，M_2 为偏心受压构件两端截面按结构分析确定的弯矩设计值中绝对值较大的弯矩设计值，N·mm；N 为与弯矩设计值 M_2 相应的轴向压力设计值，N；e_a 为附加偏心距，mm；h_0 为截面有效高度，对环形截面取 $h_0 = r_2 + r_s$，对圆形截面取 $h_0 = r + r_s$，r、r_2 和 r_s 按《混凝土结构设计标准》(GB/T 50010—2010) 的规定取值，mm；l_c 为构件计算长度，可近似取偏心受压构件相应主轴方向上下支承点之间的距离，mm；h 为截面高度，对环形截面取外直径，对圆形截面取直径，mm；ζ_c 为截面曲率修正系数，参考国外规范和试验分析结果，得

Where, M_2 is the design value of bending moment with larger absolute value in design value of bending moment determined by structural analysis for section at both ends of eccentric compression member, N·mm; N is the axial pressure design value corresponding to the bending moment design value M_2, N; e_a is the additional eccentricity, mm; h_0 is the effective depth of the section, for ring section, take $h_0 = r_2 + r_s$, for circular section, take $h_0 = r + r_s$, r, r_2 and r_s are taken according to the provisions of "Standard for design of concrete structures" (GB/T 50010—2010), mm; l_c is the effective length of the member, which can be approximated as the distance between the upper and lower support points in the direction of the corresponding principal axis of eccentric compression member, mm; h is the sectional height, for ring section take the outside diameter, for circular section take the diameter, mm; ζ_c is the sectional curvature correction coefficient, referring to foreign codes and test analysis results, one obtains

$$\zeta_c = \frac{0.5 f_c A}{N} \leqslant 1 \tag{6-16}$$

式中，f_c 为混凝土轴心抗压强度设计值，N/mm²；A 为构件截面面积，mm²。

Where, f_c is the design value of concrete axial compressive strength, N/mm²; A is the cross-sectional area of the member, mm².

当截面曲率修正系数 $\zeta_c > 1.0$ 时，取 $\zeta_c = 1.0$。

When the sectional curvature correction coefficient $\zeta_c > 1.0$, take $\zeta_c = 1.0$.

 拓展知识 6-3

3. 考虑 P-δ 二阶效应后控制截面的弯矩设计值

Design value of bending moment of control section after considering the P-δ second-order effect

1）非排架结构柱考虑二阶效应的弯矩设计值

1) Design value of bending moment of non-bent structural column considering second-order effect

《混凝土结构设计标准》(GB/T 50010—2010)6.2.4 条规定，除排架结构柱外，其他偏心受压构件考虑轴向压力在挠曲杆件中产生的二阶效应后控制截面的弯矩设计值应按式(6-17)计算：

Article 6.2.4 of "Standard for design of concrete structures" (GB/T 50010—2010) stipulates that except for bent structure columns, the design value of bending moment of control section of other eccentric compression member considering the second-order effect of axial pressure in flexural member is obtained from Eq. (6-17):

$$M = C_m \eta_{ns} M_2 \tag{6-17}$$

当 $C_m \eta_{ns} < 1.0$ 时，取 1.0。对剪力墙及核心筒墙，因为截面面积大，刚度比较大，侧向弯曲引起的 P-δ 效应不明显，可取 $C_m \eta_{ns} = 1.0$。

Take 1.0 when $C_m \eta_{ns} < 1.0$. For shear wall and core tube wall, due to its large cross-sectional area and high stiffness, the P-δ effect caused by lateral bending is not obvious, so take $C_m \eta_{ns} = 1.0$.

2）排架结构柱考虑二阶效应的弯矩设计值

2) Design value of bending moment of bent structure column considering the second-

order effect

排架结构柱考虑二阶效应的弯矩设计值可按式(6-18)、式(6-19)计算：

The design value of bending moment considering second-order effect of bent structure column is obtained from Eq. (6-18) and Eq. (6-19):

$$M = \eta_s M_0 \quad (6\text{-}18)$$

$$\eta_s = 1 + \frac{1}{1500(M_0/N + e_a)/h_0}\left(\frac{l_c}{h}\right)^2 \zeta_c \quad (6\text{-}19)$$

式中，M_0 为按一阶弹性分析确定的柱端弯矩设计值，N·mm；l_c 为排架柱的计算长度，按《混凝土结构设计标准》(GB/T 50010—2010)表 6.2.20-1 确定，mm。

Where, M_0 is the design value of bending moment at the end of the column determined by first-order elastic analysis, N·mm; l_c is the effective length of the bent structure column, determined by Table 6.2.20-1 of "Standard for design of concrete structures"(GB/T 50010—2010), mm.

拓展知识 6-4

6.3.3 结构有侧移时偏心受压构件的 P-Δ 二阶效应
P-Δ second-order effect of eccentric compression members with sideway

当框架结构上作用有水平荷载，或虽无水平荷载，但结构或荷载不对称，或两者均不对称时，结构会产生侧移，使偏心受压构件的挠曲和二阶弯矩分布发生变化。对于有侧移的框架结构，二阶效应主要指竖向荷载引起的附加弯矩，通常称 P-Δ 效应。P-Δ 效应引起的附加弯矩将增大框架柱截面的弯矩设计值，所以在框架柱的内力计算中应考虑 P-Δ 效应。但是，由 P-Δ 效应产生的弯矩增大属于结构分析中考虑几何非线性的内力计算问题，在偏心受压构件截面计算时给出的内力设计值中已经包含了 P-Δ 效应，故不必在截面承载力计算中再重复考虑。

When the frame structure is subjected to a horizontal load, or although there is no horizontal load, the structure or the load is asymmetrical, or both the structure and the load are asymmetrical, the structure will produce sideway, causing the deflection and second-order bending moment distribution of eccentric compression member change. For the frame structure with sideway, the second-order effect mainly refers to the additional moment caused by the vertical load, which is usually called the P-Δ effect. The additional moment caused by the P-Δ effect will increase the design value of bending moment of frame column section, so the P-Δ effect should be considered in the internal force calculation of frame column. However, the increase of bending moment caused by P-Δ effect belongs to the problem of internal force calculation considering geometric nonlinearity in structural analysis. The internal force design value given in the calculation of eccentric compression member section already includes the P-Δ effect. So there is no need to repeatedly consider it in the calculation of section bearing capacity.

6.4 矩形截面非对称配筋偏心受压构件正截面受压承载力计算
Calculation of normal section compression bearing capacity of rectangular section asymmetric reinforcement eccentric compression members

6.4.1 基本计算公式及适用条件
Basic calculation formulas and applicable conditions

1. 大偏心受压构件
Large eccentric compression member

试验研究表明,大偏心受压构件的破坏与适筋梁相似,纵向受拉钢筋的应力取抗拉强度设计值 f_y,纵向受压钢筋的应力取抗压强度设计值 f'_y。为简化计算,采用与受弯构件正截面承载力计算相同的基本假定和分析方法,对受压区混凝土的曲线应力图形也同样采用等效矩形应力图形代替,如图 6-12 所示。等效前混凝土实际受压区高度为 x_c,等效后混凝土受压区高度为 x,等效后的混凝土压应力取值为 $\alpha_1 f_c$。

The experimental research shows that the failure of large eccentric compression members is similar to that of underreinforced beams. The stress of longitudinal tensile reinforcement is taken as the design value of tensile strength f_y, and the stress of longitudinal compression reinforcement is taken as the design value of compressive strength f'_y. In order to simplify the calculation, the same basic assumptions and analysis methods are used to calculate the bearing capacity of normal section of flexural members.

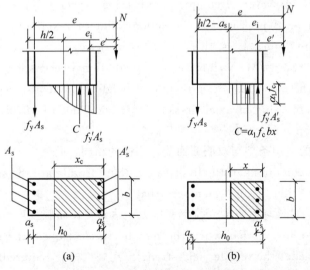

图 6-12 大偏心受压构件计算简图
(a) 实际应力图形;(b) 等效矩形应力图形

Figure 6-12 Calculation diagram of large eccentric compression member
(a) Actual stress block;(b) Equivalent rectangular stress block

The curve stress block of concrete in the compression zone is also replaced by the equivalent rectangular stress block, as shown in Fig. 6-12. The actual depth of concrete compression zone before equivalent is x_c, the depth of concrete compression zone after equivalent is x, and the compressive stress value of concrete after equivalent is $\alpha_1 f_c$.

如图 6-12(b)所示，根据大偏心受压构件截面力的平衡条件，以及截面上各力对纵向受拉钢筋合力作用点取矩的力矩平衡条件，得到式(6-20)、式(6-21)两个基本计算公式。

As shown in Fig. 6-12(b), according to the force equilibrium condition of large eccentric compression member, and the moment equilibrium condition of each force on section taking moment about resultant force action point of longitudinal tensile reinforcement, gives two basic calculation formulas Eq. (6-20) and Eq. (6-21).

力的平衡方程：

Force equilibrium equation：

$$N \leqslant N_u = \alpha_1 f_c bx + f'_y A'_s - f_y A_s \tag{6-20}$$

力矩平衡方程：

Moment equilibrium equation：

$$Ne \leqslant N_u e = \alpha_1 f_c bx \left(h_0 - \frac{x}{2}\right) + f'_y A'_s (h_0 - a'_s) \tag{6-21}$$

$$e = e_i + 0.5h - a_s \tag{6-22}$$

式中，N_u 为受压承载力设计值，N；e 为轴向压力作用点至纵向受拉钢筋合力作用点的距离，mm；e_i 为初始偏心距，$e_i = e_0 + e_a$，mm；e_0 为计算偏心距，$e_0 = M/N$，mm；e_a 为附加偏心距，取 20mm 和 $h/30$ 中的大者，mm；M 为控制截面弯矩设计值，N·mm；N 为与 M 相应的轴向压力设计值，N；x 为混凝土受压区高度，mm。

Where, N_u is the design value of compressive bearing capacity, N; e is the distance from the action point of axial pressure to the action point of resultant force of longitudinal tensile reinforcement, mm; e_i is the initial eccentricity, $e_i = e_0 + e_a$, mm; e_0 is the calculated eccentricity, $e_0 = M/N$, mm; e_a is the additional eccentricity, take the larger of 20mm and $h/30$, mm; M is the design value of bending moment of control section, N·mm; N is the design value of axial pressure corresponding to M, N; x is the depth of concrete compression zone, mm.

基本计算公式的适用条件与双筋截面受弯构件的相同，即：

The applicable conditions of the basic calculation formula are the same as those of the doubly reinforced section flexural members, that is：

(1) 为了保证构件破坏时纵向受拉钢筋应力先达到受拉屈服强度 f_y，要求 $\xi \leqslant \xi_b$。

(1) In order to ensure that when the member is destroyed, the longitudinal tensile steel stress first reaches the tensile yield strength f_y, $\xi \leqslant \xi_b$ is required.

(2) 为了保证构件破坏时纵向受压钢筋应力能达到受压屈服强度 f'_y，与双筋受弯构件一样，要求 $x \geqslant 2a'_s$（a'_s 为纵向受压钢筋合力作用点至混凝土受压区边缘的距离）。

(2) In order to ensure that the longitudinal compression steel stress can reach the compressive yield strength f'_y when the member is destroyed, as the same with the doubly

reinforced flexural member, $x \geqslant 2a'_s$ is required (a'_s is the distance from action point of resultant force of longitudinal compressive reinforcement to the edge of the compression zone).

2. 小偏心受压构件
Small eccentric compression member

试验研究表明,小偏心受压破坏时受压区混凝土已被压碎,该侧纵向受压钢筋 A'_s 的应力可以达到受压屈服强度 f'_y,而远离轴向力一侧钢筋 A_s 可能受拉或受压,可能屈服也可能不屈服。

小偏心受压可分为三种情况：

(1) 当 $\xi_b < \xi < \xi_{cy}$ 时, A_s 受拉或受压,但都不屈服,如图 6-13(a) 所示；

(2) 当 $\xi_{cy} \leqslant \xi < h/h_0$ 时, A_s 受压屈服,但混凝土受压区高度 $x < h$,如图 6-13(b)所示；

(3) 当 $\xi > \xi_{cy}$,且 $\xi \geqslant h/h_0$ 时, A_s 受压屈服,且全截面受压,如图 6-13(c)所示。

ξ_{cy} 为 A_s 受压屈服时的混凝土相对受压区高度。

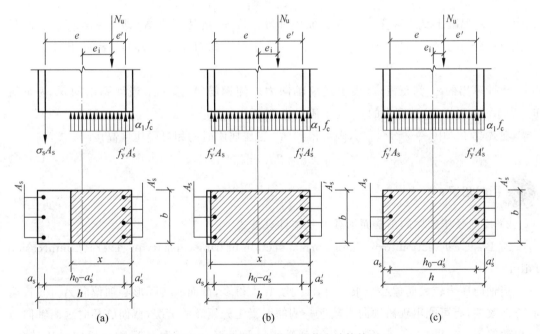

图 6-13 小偏心受压截面承载力计算简图
(a) 当 $\xi_b < \xi < \xi_{cy}$ 时, A_s 受拉或受压,但都不屈服； (b) 当 $\xi_{cy} \leqslant \xi < h/h_0$ 时, A_s 受压屈服,但 $x < h$；
(c) 当 $\xi > \xi_{cy}$,且 $\xi \geqslant h/h_0$ 时, A_s 受压屈服,且全截面受压

Figure 6-13 Calculation diagram of section bearing capacity of small eccentric compression
(a) When $\xi_b < \xi < \xi_{cy}$, A_s is tensioned or compressed, but neither yield;
(b) When $\xi_{cy} \leqslant \xi < h/h_0$, A_s yield under compression, but $x < h$;
(c) When $\xi > \xi_{cy}$, and $\xi \geqslant h/h_0$, A_s yield under compression, and the whole section is compressed

1) 假定 A_s 受拉，但未达到受拉屈服强度 f_y 时的基本计算公式

1) Basic calculation formula assuming that A_s is under tension but does not reach the tensile yield strength f_y

根据截面上力的平衡条件和力矩平衡条件，如图 6-13(a)所示，可得力的平衡方程：

$$N \leqslant N_u = \alpha_1 f_c bx + f'_y A'_s - \sigma_s A_s \qquad (6\text{-}23)$$

对远离轴向力一侧钢筋 A_s 的合力作用点取力矩：

$$Ne \leqslant N_u e = \alpha_1 f_c bx\left(h_0 - \frac{x}{2}\right) + f'_y A'_s (h_0 - a'_s) \qquad (6\text{-}24)$$

或对纵向受压钢筋 A'_s 的合力作用点取力矩：

$$Ne' \leqslant N_u e' = \alpha_1 f_c bx\left(\frac{x}{2} - a'_s\right) - \sigma_s A_s (h_0 - a'_s) \qquad (6\text{-}25)$$

$$e = 0.5h - a_s + e_i \qquad (6\text{-}26)$$

$$e' = 0.5h - a'_s - e_i \qquad (6\text{-}27)$$

式中，x 为混凝土受压区高度，当 $x > h$ 时，取 $x = h$，mm；σ_s 为远离轴向力一侧钢筋 A_s 的应力，可根据截面应变保持平面的假定计算，也可近似取

$$\sigma_s = \frac{\xi - \beta_1}{\xi_b - \beta_1} f_y \qquad (6\text{-}28)$$

式中，β_1 为混凝土受压区高度 x 与截面中和轴高度 x_c 的比值系数（即 $x = \beta_1 x_c$），当混凝土强度等级小于等于 C50 时，$\beta_1 = 0.8$；ξ、ξ_b 分别为相对受压区高度、界限相对受压区高度。

当计算出的 σ_s 为正值时，表示远离轴向力一侧钢筋 A_s 受拉；当计算出的 σ_s 为负值时，表示 A_s 受压，且计算出的 σ_s 要求满足 $-f'_y \leqslant \sigma_s \leqslant f_y$。

在式(6-28)中，令 $\sigma_s = -f_y$，则可得到 A_s 受压屈服时的相对受压区高度

$$\xi_{cy} = 2\beta_1 - \xi_b \qquad (6\text{-}29)$$

拓展知识 6-5

2) 反向受压破坏的正截面承载力基本计算公式

2) Basic calculation formula of normal section bearing capacity of reverse compression failure

当轴向压力较大而偏心距很小时，A_s 比 A'_s 大得多，截面的实际形心轴偏向 A'_s，导致偏心方向改变，有可能出现离轴向力较远一侧的混凝土先压坏的情况，截面破坏时远离轴向力一侧钢筋 A_s 有可能处于受压状态并达到受压屈服强度 f'_y，这种情况称为反向受压破坏，这时的正截面承载力计算简图如图 6-14 所示。

这时，附加偏心距 e_a 反向了，使 e_0 减小，则初始偏心距

$$e_i = e_0 - e_a \qquad (6\text{-}30)$$

$$e' = 0.5h - a'_s - (e_0 - e_a) \qquad (6\text{-}31)$$

对 A'_s 合力作用点取矩：

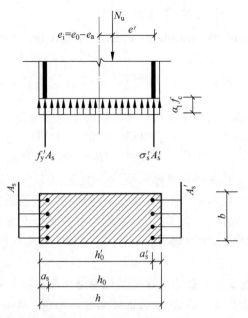

图 6-14 小偏心受压构件反向受压破坏的正截面承载力计算简图

Figure 6-14 Calculation diagram of normal section bearing capacity of reverse compression failure of small eccentric compression members

$$Ne' \leqslant N_u e' = f_c bh\left(\frac{h}{2} - a'_s\right) + f'_y A_s (h_0 - a'_s) \tag{6-32}$$

式中，e' 为轴向压力作用点至纵向钢筋 A'_s 合力作用点的距离，mm。

特别提示 6-3

3. 界限偏心距
Balanced eccentricity

对于普通热轧钢筋和常用的混凝土强度等级，相对界限偏心距 e_{ib}/h_0 的最小值在 0.3 附近变化，对常用材料，取 $e_{ib}=0.3h_0$ 作为大、小偏心受压的界限偏心距比较合适。当 $e_i > 0.3h_0$ 时，可能为大偏心受压，也可能为小偏心受压，可先按大偏心受压设计，等计算结果出来后进行校验，若不符合再按小偏心受压进行设计；当 $e_i \leqslant 0.3h_0$ 时，按小偏心受压设计。

For ordinary hot-rolled reinforcement and commonly used concrete strength grade, the minimum relative balanced eccentricity e_{ib}/h_0 is about 0.3. For common material, it is more appropriate to take $e_{ib}=0.3h_0$ as the balanced eccentricity of large and small eccentric compression. When $e_i > 0.3h_0$, it may be subjected to large eccentric compression or small eccentric compression. It can be designed according to the large eccentric compression first, and then verified after the calculation results are out, if not, design according to the small eccentric compression; when $e_i \leqslant 0.3h_0$, design according to the small eccentric compression.

6.4.2 非对称配筋偏心受压构件截面设计
Section design of asymmetric reinforcement eccentric compression members

1. 概述
Summary

为增强偏心受压构件抵抗轴向压力和弯矩的能力,一般在截面弯矩作用方向的两侧同时配置纵向受力钢筋 A_s 和 A'_s(A_s 为远离轴向力一侧钢筋,A'_s 为近轴向力一侧钢筋,一般为受压钢筋)。纵向受力钢筋的配置方式有非对称配筋和对称配筋,非对称配筋是指在截面弯矩作用方向两侧配置不同的纵向受力钢筋,受拉钢筋和受压钢筋的根数、直径或强度级别不相同,截面面积 A_s 和 A'_s 不相等。如果对称配置相同的纵向受力钢筋,则称为对称配筋。非对称配筋用钢量较省,但施工易出错;对称配筋构造简单,施工方便,不易出错,但用钢量较大。

偏心受压构件正截面受压承载力计算包括截面设计和截面复核两种情况。截面设计时首先判别是否需要考虑挠曲的二阶效应,若需要则计算考虑二阶效应后的弯矩设计值,再根据偏心距的大小初步判别构件的偏心类型。当 $e_i > 0.3h_0$ 时,可能为大偏心受压,也可能为小偏心受压,可先按大偏心受压设计,再进行校验,若不符合再按小偏心受压进行设计;当 $e_i \leqslant 0.3h_0$ 时,按小偏心受压设计。根据大、小偏心受压基本计算公式进行截面设计时,取 $N = N_u$。

不论大、小偏心受压,都要按轴心受压构件的式(6-3)验算垂直于弯矩作用平面的受压承载力 N_u,当其不小于轴向压力设计值 N 时为满足,否则要重新设计。式(6-3)中的 A'_s 应取截面上全部纵向钢筋的截面面积,包括受压钢筋的 A'_s 和远离轴向力一侧钢筋的 A_s。计算长度 l_c 应按垂直于弯矩作用平面方向确定,稳定系数 φ 应按该方向的长细比查表 6-2 确定,如矩形截面在该方向上的长细比为 l_c/b(b 为截面短边尺寸)。

2. 非对称配筋大偏心受压构件截面设计
Section design of asymmetric reinforcement large eccentric compression members

截面设计分为纵向钢筋截面面积 A_s、A'_s 均未知及 A'_s 已知、A_s 未知两种情况。

Section design can be divided into two situations: cross-sectional area A_s and A'_s of the longitudinal reinforcement are unknown, and A'_s is known, A_s is unknown.

1) 情况 1:A_s 和 A'_s 均未知,求 A_s 和 A'_s

1) Case 1: Both A_s and A'_s are unknown, calculate A_s and A'_s

已知:某构件截面尺寸($b \times h$)、构件计算长度 l_c、混凝土强度等级和钢筋级别(f_c, f_y, f'_y)、轴向压力设计值 N 和弯矩设计值 M,求纵向钢筋截面面积 A_s 和 A'_s。

Known: cross-sectional dimensions of the member ($b \times h$), effective length l_c of the member, concrete strength grade and reinforcement grade (f_c, f_y, f'_y), design value of axial pressure N and design value of bending moment M, calculate the cross-sectional area A_s and A'_s of longitudinal reinforcement.

式(6-20)、式(6-21)这两个平衡方程中有 A_s、A'_s 和 x 三个未知数,无唯一解。与双筋截面受弯构件相似，考虑配筋的经济性，为使总配筋面积（$A_s+A'_s$）最小，充分发挥受压区混凝土的抗压能力，且为简化计算，取 $\xi=\xi_b$（ξ_b 为界限相对受压区高度），即 $x=x_b=\xi_b h_0$（x_b 为界限破坏时的混凝土受压区计算高度），代入式(6-21)中，得纵向受压钢筋截面面积 A'_s：

The two equilibrium equations of Eq. (6-20) and Eq. (6-21) have three unknowns, A_s、A'_s and x with no unique solution. Similar to the flexural member with doubly reinforced section, considering the economy of reinforcement, in order to minimize the total reinforcement area ($A_s+A'_s$) to fully utilize the compressive capacity of concrete in the compression zone, for simplified calculation, take $\xi=\xi_b$ (ξ_b is the balanced relative compression zone depth), that is $x=x_b=\xi_b h_0$ (x_b is the calculated depth of concrete compression zone in the case of balanced failure), substituting it into Eq. (6-21), and gives the cross-sectional area of longitudinal compressive reinforcement A'_s:

$$A'_s=\frac{Ne-\alpha_1 f_c bh_0^2 \xi_b(1-0.5\xi_b)}{f'_y(h_0-a'_s)} \geq \rho'_{min}bh \qquad (6-33)$$

若 $A'_s/bh<\rho'_{min}$（受压钢筋配筋率小于受压构件单侧钢筋的最小配筋率），应取 $A'_s=\rho'_{min}bh$，然后按第二种情况（A'_s 已知）计算 A_s。

If $A'_s/bh<\rho'_{min}$ (the reinforcement ratio of compressive reinforcement is less than the minimum reinforcement ratio of reinforcement on one side of the compression member), $A'_s=\rho'_{min}bh$ should be taken, and then calculate A_s according to the second case (A'_s is known).

将求得的 A'_s 和 $x=\xi_b h_0$ 代入式(6-20)中，得纵向受拉钢筋截面面积：

Substituting the obtained A'_s and $x=\xi_b h_0$ into Eq. (6-20), gives cross-sectional area of longitudinal tensile reinforcement：

$$A_s=\frac{\alpha_1 f_c bh_0 \xi_b + f'_y A'_s - N}{f_y} \geq \rho'_{min}bh \qquad (6-34)$$

若 $A_s<\rho'_{min}bh$，应取 $A_s=\rho'_{min}bh$。

If $A_s<\rho'_{min}bh$, gives $A_s=\rho'_{min}bh$.

【例题 6-3】 某钢筋混凝土偏心受压柱承受轴向压力设计值 $N=600$kN，柱两端弯矩设计值分别为 $M_1=360$kN·m，$M_2=380$kN·m，柱截面尺寸为 $b\times h=400$mm$\times 500$mm，计算长度 $l_c=5$m，采用 C30 混凝土和 HRB400 钢筋，计算过程中取 $a_s=a'_s=40$mm，要求按非对称配筋进行截面设计。

【Example 6-3】 Reinforced concrete eccentric compression column is subjected to the design value of axial pressure $N=600$kN, the design values of bending moment at both ends of the column are $M_1=360$kN·m, $M_2=380$kN·m respectively, and cross-sectional dimensions of the column are $b\times h=400$mm$\times 500$mm, effective length is $l_c=5$m, C30 concrete and HRB400 reinforcement are used, during the calculation process, $a_s=a'_s=40$mm is taken, and it is required to design the section according to asymmetrical reinforcement.

解：(1) 确定设计参数

C30 混凝土，$\alpha_1=1.0$，$\beta_1=0.8$，$f_c=14.3$N/mm^2。HRB400 钢筋，$f_y=f'_y=360$N/mm^2，

$\xi_b = 0.518$。

$$I = bh^3/12 = 400 \times 500^3/12 \, \text{mm}^4 = 41667 \times 10^5 \, \text{mm}^4$$

$$b \times h = 400\text{mm} \times 500\text{mm} = 2 \times 10^5 \, \text{mm}^2$$

$$i = \sqrt{\frac{I}{A}} = \sqrt{\frac{41667 \times 10^5}{2 \times 10^5}} \, \text{mm} = 144\text{mm}, a_s = a'_s = 40\text{mm}$$

$$h_0 = h - a_s = (500 - 40)\text{mm} = 460\text{mm}$$

(2) 求设计弯矩

由于 $M_2/M_1 = 380/360 = 1.06 > 0.9$,$i = 144\text{mm}$,则 $l_c/i = 5000/144 = 34.72 > 34 - 12M_2/M_1 = 34 - 12 \times 1.06 = 21.28$,因此需要考虑二阶效应附加弯矩的影响。

截面曲率修正系数:
$$\xi_c = \frac{0.5 f_c A}{N} = \frac{0.5 \times 14.3 \times 2 \times 10^5}{600 \times 10^3} = 2.38 > 1$$

取 $\xi_c = 1.0$。

偏心距调节系数:
$$C_m = 0.7 + 0.3 M_1/M_2 = 0.7 + 0.3 \times 360/380 = 0.984 > 0.7$$

取 $C_m = 0.984$。

附加偏心距:
$$e_a = h/30 = 500/30 = 16.7\text{mm} < 20\text{mm}$$

取 20mm。

弯矩增大系数:
$$\eta_{ns} = 1 + \frac{1}{1300(M_2/N + e_a)/h_0} \left(\frac{l_c}{h}\right)^2 \zeta_c$$

$$= 1 + \frac{1}{1300(380 \times 10^6/600 \times 10^3 + 20)/460} \times \left(\frac{5000}{500}\right)^2 \times 1 = 1.05$$

$C_m \eta_{ns} = 0.984 \times 1.05 = 1.03 > 1$,取 $C_m \eta_{ns} = 1.03$。

考虑附加弯矩后的柱设计弯矩为
$$M = C_m \eta_{ns} M_2 = 1.03 \times 380 \text{kN} \cdot \text{m} = 391.4 \text{kN} \cdot \text{m}$$

(3) 计算纵向受拉钢筋截面面积 A_s 和纵向受压钢筋截面面积 A'_s

计算偏心距:
$$e_0 = M/N = 391.4 \times 10^6/(600 \times 10^3) \text{mm} = 652\text{mm}$$

初始偏心距:
$$e_i = e_0 + e_a = (652 + 20)\text{mm} = 672\text{mm} > 0.3h_0 = 0.3 \times 460\text{mm} = 138\text{mm}$$

先按大偏心受压计算。
$$e = e_i + 0.5h - a_s = (672 + 0.5 \times 500 - 40)\text{mm} = 882\text{mm}$$

由式(6-33)得纵向受压钢筋截面面积 A'_s:
$$A'_s = \frac{Ne - \alpha_1 f_c b h_0^2 \xi_b (1 - 0.5\xi_b)}{f'_y (h_0 - a'_s)}$$

$$= \frac{600 \times 10^3 \times 882 - 1.0 \times 14.3 \times 400 \times 460^2 \times 0.518 \times (1 - 0.5 \times 0.518)}{360 \times (460 - 40)} \text{mm}^2$$

$$= 427 \text{mm}^2 > \rho'_{min} bh = 0.2\% bh = 0.002 \times 400 \times 500 \text{mm}^2 = 400 \text{mm}^2$$

由式(6-34)得纵向受拉钢筋截面面积 A_s：

$$A_s = \frac{\alpha_1 f_c b h_0 \xi_b + f'_y A'_s - N}{f_y}$$

$$= \frac{1.0 \times 14.3 \times 400 \times 460 \times 0.518 + 360 \times 427 - 600 \times 10^3}{360} \text{mm}^2$$

$$= 2190 \text{mm}^2 \geqslant \rho'_{\min} bh = 0.2\% bh = 0.002 \times 400 \times 500 \text{mm}^2 = 400 \text{mm}^2$$

（4）选配钢筋并验算配筋率

根据计算的纵向受力钢筋截面面积 A_s、A'_s，柱截面受压区选配 3⏀14，$A'_s=461\text{mm}^2$，柱截面受拉区选配 2⏀25+3⏀22，$A_s=(982+1140)\text{mm}^2=2214\text{mm}^2$。

全部纵筋配筋率为

$$\rho' = (A_s + A'_s)/bh = (2214 + 461)/(400 \times 500) = 1.34\% > \rho'_{\min}$$

$$= 0.55\% (\rho'_{\min} 可查附表 3-3)$$

且 $\rho' < \rho'_{\max} = 5\%$（最大配筋率），所以配筋率满足要求。

（5）判别大、小偏心受压

根据非对称配筋大偏心受压构件力的平衡方程式(6-20)，得混凝土受压区高度 x：

$$x = \frac{N - f'_y A'_s + f_y A_s}{\alpha_1 f_c b} = \frac{600 \times 10^3 - 360 \times 461 + 360 \times 2214}{1.0 \times 14.3 \times 400} \text{mm}$$

$$= 215.2 \text{mm} < \xi_b h_0 = 0.518 \times 460 \text{mm} = 238.28 \text{mm}$$

属于大偏心受压，且 $x > 2a'_s = 2 \times 40 \text{mm} = 80 \text{mm}$。

2) 情况 2：纵向受压钢筋截面面积 A'_s 为已知，求纵向受拉钢筋截面面积 A_s

2) Case 2：Cross-sectional area A'_s of longitudinally compressed steel bars is known, calculate the cross-sectional area A_s of longitudinally tensioned steel bars

已知：构件截面尺寸($b \times h$)、构件计算长度 l_c、混凝土强度等级和钢筋级别(f_c, f_y, f'_y)、轴向压力设计值 N 和弯矩设计值 M、纵向受压钢筋截面面积 A'_s，求纵向受拉钢筋截面面积 A_s。

Known：cross-sectional dimensions ($b \times h$) of the member, effective length l_c of the member, concrete strength grade and reinforcement grade (f_c, f_y, f'_y), axial pressure design value N and bending moment design value M, cross-sectional area A'_s of the longitudinally compressive reinforcement, calculate the cross-sectional area A_s of the longitudinal tensile reinforcement.

令 $N=N_u$，$M=Ne_0$，则式(6-20)和式(6-21)中仅有 x，A_s 两个未知数，完全可以通过式(6-20)和式(6-21)联立方程直接求解 A_s，但要求解 x 的二次方程，比较麻烦。可参考双筋截面受弯构件已知 A'_s 时的情况，如图 6-12(b)所示，将混凝土压应力的合力对纵向受拉钢筋合力作用点取矩，得

Let $N=N_u$, $M=Ne_0$, then there are only two unknowns x and A_s in Eq.(6-20) and Eq.(6-21), which is completely possible through the simultaneous equations of Eq.(6-20) and Eq.(6-21) to solve the A_s directly, but it is quite troublesome to solve the quadratic equation of x. Reference can be made to the situation when the A'_s is known for flexural member of doubly reinforced section, as shown in Fig. 6-12(b), taking the moment of the resultant force of concrete compressive stress about the resultant force action point of longitudinal tensile reinforcement, one obtains

$$M_{u2} = \alpha_1 f_c bx(h_0 - x/2) \tag{6-35}$$

由式(6-21)得

From Eq. (6-21), one obtains

$$M_{u2} = Ne - f'_y A'_s (h_0 - a'_s) \tag{6-36}$$

则截面抵抗矩系数 α_s 和相对受压区高度 ξ 分别为

Then the sectional resistance moment coefficient α_s and the relative compression zone depth ξ are respectively

$$\alpha_s = \frac{M_{u2}}{\alpha_1 f_c b h_0^2} \tag{6-37}$$

$$2a'_s/h_0 \leqslant \xi = 1 - \sqrt{1 - 2\alpha_s} \leqslant \xi_b \tag{6-38}$$

若 $\xi > \xi_b$，说明纵向受压钢筋数量不足，则：①改用小偏心受压重新计算；②若仍按大偏心受压计算，则应加大截面尺寸或提高混凝土强度等级，应增加 A'_s 的数量；③按 A_s、A'_s 均未知的第一种情况重新计算，使其满足 $\xi \leqslant \xi_b$ 的条件。

If $\xi > \xi_b$, it shows that the number of longitudinally compressive steel bars is insufficient, then, the treatment is：① Recalculate using small eccentric compression instead；② If the calculation is still based on large eccentric compression, the cross-sectional dimensions or concrete strength grade should be increased, and A'_s should be increased；③Recalculate as the first situation where both A_s、A'_s are unknown, so that it satisfies the condition $\xi \leqslant \xi_b$.

若混凝土受压区高度 $x = \xi h_0 < 2a'_s$，则参照双筋截面受弯构件的计算，取 $x = 2a'_s$，如图 6-12(b)所示，对纵向受压钢筋 A'_s 合力作用点取矩，计算 A_s，得

If the concrete compression zone depth is $x = \xi h_0 < 2a'_s$, referring to the calculation of flexural member of doubly reinforced section, taken $x = 2a'_s$ as shown in Fig. 6-12(b), taking the moment about resultant force action point of longitudinally compressive reinforcement A'_s, calculating A_s, one obtains

$$A_s = \frac{N(e_i - h/2 + a'_s)}{f_y(h_0 - a'_s)} \tag{6-39}$$

另外，再按不考虑受压钢筋 A'_s，即取 $A'_s = 0$，利用式(6-20)、式(6-21)计算 A_s 值，然后与式(6-39)求得的 A_s 值比较，取其中较小值进行配筋。

In addition, the A_s value is calculated by using Eq. (6-20) and Eq. (6-21) without considering the compression reinforcement A'_s, that is, take $A'_s = 0$, and then compare it with the A_s value obtained from Eq. (6-39), take the smaller value for reinforcement.

3. 非对称配筋小偏心受压构件截面设计
Section design of asymmetric reinforcement small eccentric compression members

此时式(6-23)和式(6-24)中的未知数有 x、A_s 和 A'_s 三个，而独立的平衡方程只有两个，故必须补充一个条件才能求解。若仍以 $(A_s + A'_s)$ 总量最小作为补充条件，则计算过程非常复杂。可以按以下两个步骤进行截面设计。

1) 确定 A_s 作为补充条件

1) Determine A_s as a supplementary condition

当 $\xi_b < \xi < \xi_{cy}$ 时,试验研究表明,小偏心受压时远离轴向力一侧钢筋 A_s 无论受拉还是受压,一般都不能达到屈服强度,为了经济,不需要配置较多的 A_s,实用上可按单侧钢筋最小配筋率初步拟定 A_s 值,取 $A_s = \rho'_{min} bh = 0.2\% bh$。

为了防止出现反向受压破坏时 A_s 数量不足的情况,当 $N > f_c bh$ 时,A_s 应由反向受压破坏的计算公式(6-32)求得。若求得的 $A_s < 0.2\% bh$,取 $A_s = 0.2\% bh$,并应符合钢筋的构造要求。

2) 求出相对受压区高度 ξ,再按 ξ 的三种情况求出纵向受压钢筋 A'_s

2) Calculate the relative depth of compression zone ξ, and then calculate the longitudinal compressive reinforcement A'_s according to the three cases of ξ

将步骤1)实际选配的 A_s 代入力的平衡方程式(6-23)和力矩平衡方程式(6-25)中,并利用 σ_s 的近似公式(6-28)求解 A'_s。

首先求解 ξ,对式(6-23)、式(6-25)进行适当处理,得到关于 ξ 的一元二次方程 $\xi^2 + 2B\xi + 2C = 0$,求解此方程,得

$$\xi = -B + \sqrt{B^2 - 2C}$$

$$B = \frac{f_y A_s (h_0 - a'_s)}{\alpha_1 f_c b h_0^2 (\beta_1 - \xi_b)} - \frac{a'_s}{h_0}$$

$$C = \frac{N(e - h_0 + a'_s)(\beta_1 - \xi_b) - \beta_1 f_y A_s (h_0 - a'_s)}{\alpha_1 f_c b h_0^2 (\beta_1 - \xi_b)} \quad (6\text{-}40)$$

3) 得到相对受压区高度 ξ 值后,如图 6-13 所示,按小偏心受压 ξ 的三种情况求出 A'_s

3) After obtaining the value of relative depth of compression zone ξ, as shown in Fig. 6-13, calculate A'_s according to three cases of small eccentric compression ξ

(1) 当 $\xi_b < \xi \leqslant \xi_{cy}$ 时,其中 ξ_{cy} 按式(6-29)求得,将 $x = \xi h_0$ 代入力的平衡方程式(6-23)或力矩平衡方程式(6-24)中,即可求出 A'_s。

(2) 当 $\xi_{cy} \leqslant \xi < h/h_0$ 时,取 $\sigma_s = -f'_y$,说明混凝土受压区高度 x 未超出截面高度 h,A_s 的应力 σ_s 已达到受压屈服强度 f'_y,按第2)步计算的 ξ 值无效,按下式重新计算 ξ:

$$\xi = \frac{a'_s}{h_0} + \sqrt{\left(\frac{a'_s}{h_0}\right)^2 + 2\left[\frac{Ne'}{\alpha_1 f_c b h_0^2} - \frac{A_s}{bh_0}\frac{f_y}{\alpha_1 f_c}\left(1 - \frac{a'_s}{h_0}\right)\right]} \quad (6\text{-}41)$$

再将 $x = \xi h_0$ 代入力的平衡方程式(6-23)求出 A'_s。

(3) 当 $\xi > \xi_{cy}$,且 $\xi \geqslant h/h_0$ 时,混凝土全截面受压,取 $x = h$,$\xi = h/h_0$,$\sigma_s = -f'_y$,$\alpha_1 = 1$,式(6-23)、式(6-24)中的未知数为 A_s、A'_s,由式(6-24)计算 A'_s,再代入式(6-23)求出 A_s,与第1)步求出的 A_s 比较,取大值。

如果以上求得的 A'_s 小于 $0.2\% bh$,取 $A'_s = 0.2\% bh$。

4) 按轴心受压构件验算垂直于弯矩作用平面的受压承载力,如果不满足要求,应重新计算。

【例题 6-4】 某钢筋混凝土偏心受压柱承受轴向压力设计值 $N = 5000$kN,柱两端弯矩设计值分别为 $M_1 = 350$kN·m,$M_2 = 390$kN·m,柱截面尺寸为 $b \times h = 500$mm $\times 700$mm,

弯矩作用平面内和垂直于弯矩作用平面方向的柱计算长度 $l_c=6m$,采用 C30 混凝土和 HRB400 钢筋,计算过程中取 $a_s=a'_s=40mm$,要求按非对称配筋进行截面设计。

【Example 6-4】 The design value of axial compression of a reinforced concrete eccentric compression column is $N=5000kN$, the bending moment design values at both ends of the column are $M_1=350kN·m$, $M_2=390kN·m$ respectively, and the cross-sectional dimensions of the column are $b×h=500mm×700mm$, the effective length of the column in the action plane of bending moment and perpendicular to the action plane of bending moment is $l_c=6m$, C30 concrete and HRB400 reinforcement are used, in the calculation process, $a_s=a'_s=40mm$ is taken, and the section design is required to be carried out based on asymmetric reinforcement.

解：(1) 确定设计参数

C30 混凝土，$\alpha_1=1.0, \beta_1=0.8, f_c=14.3N/mm^2$。HRB400 钢筋，$f_y=f'_y=360N/mm^2$，$\xi_b=0.518$。

$$a_s=a'_s=40mm, \quad h_0=h-a_s=(700-40)mm=660mm$$

(2) 求设计弯矩

由于 $M_2/M_1=390/350=1.11>0.9$，因此需要考虑二阶效应的影响。

截面曲率修正系数：

$$\xi_c=\frac{0.5f_cA}{N}=\frac{0.5×14.3×500×700}{5000×10^3}=0.5<1$$

取 $\xi_c=0.5$。

偏心距调节系数：

$$C_m=0.7+0.3M_1/M_2=0.7+0.3×350/390=0.969>0.7$$

取 $C_m=0.969$。

附加偏心距：

$$e_a=h/30=700/30=23.3mm>20mm$$

取 23.3mm。

弯矩增大系数：

$$\eta_{ns}=1+\frac{1}{1300(M_2/N+e_a)/h_0}\left(\frac{l_c}{h}\right)^2\zeta_c$$

$$=1+\frac{1}{1300(390×10^6/5000×10^3+23.3)/460}×\left(\frac{6000}{700}\right)^2×0.5=0.255$$

$C_m\eta_{ns}=0.969×0.255=0.247<1$，取 $C_m\eta_{ns}=1$。

考虑附加弯矩后的柱设计弯矩为

$$M=C_m\eta_{ns}M_2=1×390kN·m=390kN·m$$

(3) 判别偏心受压类型

计算偏心距：

$$e_0=M/N=390×10^6/(5000×10^3)mm=78mm$$

初始偏心距：

$$e_i=e_0+e_a=(78+23.3)mm=101.3mm<0.3h_0=0.3×660mm=198mm$$

先按小偏心受压计算。

$$e = e_i + 0.5h - a_s = (101.3 + 0.5 \times 700 - 40)\text{mm} = 411.3\text{mm}$$

(4) 初步计算纵向钢筋截面面积 A_s

$$A_s = \rho_{min}bh = 0.2\%bh = 0.002 \times 500 \times 700\text{mm}^2 = 700\text{mm}^2$$

$$f_cbh = 14.3 \times 500 \times 700\text{N} = 5005 \times 10^3\text{N} = 5005\text{kN} > N = 5000\text{kN}$$

所以,可以不进行反向受压破坏的验算,取 $A_s = 700\text{mm}^2$,选配 $1 \underline{\Phi} 20 + 1 \underline{\Phi} 22$, $A_s = 694.3\text{mm}^2$。

(5) 计算相对受压区高度 ξ

由式(6-40)计算相对受压区高度 ξ:

$$B = \frac{f_y A_s(h_0 - a'_s)}{\alpha_1 f_c b h_0^2(\beta_1 - \xi_b)} - \frac{a'_s}{h_0}$$

$$= \frac{360 \times 700 \times (660 - 40)}{1.0 \times 14.3 \times 500 \times 660^2 \times (0.8 - 0.518)} - \frac{40}{660} = 0.117$$

$$C = \frac{N(e - h_0 + a'_s)(\beta_1 - \xi_b) - \beta_1 f_y A_s(h_0 - a'_s)}{\alpha_1 f_c b h_0^2(\beta_1 - \xi_b)}$$

$$= \frac{5000 \times 10^3 \times (411.3 - 660 + 40) \times (0.8 - 0.518) - 0.8 \times 360 \times 700 \times (660 - 40)}{1.0 \times 14.3 \times 500 \times 660^2 \times (0.8 - 0.518)}$$

$$= -0.477$$

$$\xi = -B + \sqrt{B^2 - 2C} = -0.117 + \sqrt{0.117^2 - 2 \times (-0.477)} = 0.861 > \xi_b = 0.518$$

$$\xi < \xi_{cy} = 2\beta_1 - \xi_b = 2 \times 0.8 - 0.518 = 1.082$$

按图 6-13(a)小偏心受压 ξ 的第一种情况计算。

(6) 计算纵向钢筋截面面积 A'_s

将 ξ 代入式(6-28)得远离轴向力一侧纵向钢筋的应力:

$$\sigma_s = \frac{\xi - \beta_1}{\xi_b - \beta_1}f_y = \frac{0.861 - 0.8}{0.518 - 0.8} \times 360\text{N/mm}^2 = -78\text{N/mm}^2 < f_y = 360\text{N/mm}^2$$

说明 A_s 受压但未达到屈服强度,由式(6-24)得

$$A'_s = \frac{Ne - \alpha_1 f_c b h_0^2 \xi(1 - 0.5\xi)}{f'_y(h_0 - a'_s)}$$

$$= \frac{5000 \times 10^3 \times 411.3 - 1.0 \times 14.3 \times 500 \times 660^2 \times 0.861 \times (1 - 0.5 \times 0.861)}{360 \times (660 - 40)}\text{mm}^2$$

$$= 2371\text{mm}^2 \geqslant \rho'_{min}bh = 0.2\% \times 500 \times 700\text{mm}^2 = 700\text{mm}^2$$

选配钢筋 $2 \underline{\Phi} 28 + 3 \underline{\Phi} 22$, $A'_s = (1232 + 1140)\text{mm}^2 = 2372\text{mm}^2$

全部纵筋配筋率为

$$\rho' = (A_s + A'_s)/bh = (694.3 + 2372)/(500 \times 700) = 0.88\% > \rho'_{min} = 0.55\%$$

且 $\rho' < \rho'_{max} = 5\%$,所以配筋率满足要求。

(7) 验算垂直于弯矩作用平面的受压承载力

垂直于弯矩作用方向的长细比为

$$\frac{l_c}{b} = \frac{6000}{500} = 12$$

查表 6-2 得稳定系数 $\varphi=0.95$。由式(6-6),得

$$N_u = 0.9\varphi(f_c A + f'_y A'_s)$$
$$= 0.9 \times 0.95 \times [14.3 \times 500 \times 700 + 360 \times (694.3 + 2372)]N$$
$$= 5223 \times 10^3 N = 5223 kN > N = 5000 kN$$

垂直于弯矩作用平面的受压承载力满足要求。

6.4.3 非对称配筋偏心受压构件截面承载力复核
Review of section bearing capacity of asymmetric reinforcement eccentric compression members

在实际工程中,有时需要对已有的偏心受压构件进行截面承载力复核。此时,一般已知构件截面尺寸($b \times h$)、构件计算长度 l_c、混凝土强度等级和钢筋级别(f_c, f_y, f'_y)、轴向压力设计值 N、弯矩设计值 M、钢筋截面面积 A_s 及 A'_s,要求判断截面承载力是否满足要求或确定截面能承受的轴向压力设计值 N_u。

分为两种情况:一种是已知轴向力设计值 N,求偏心距 e_0,即验算截面能承受的弯矩设计值 M;另一种是已知偏心距 e_0,求轴向压力设计值 N。不论哪一种情况,都要进行垂直于弯矩作用平面的承载力复核,此时应考虑稳定系数 φ,并按 l_c/b 计算长细比。

6.5 矩形截面对称配筋偏心受压构件正截面受压承载力计算
Calculation of normal section compression bearing capacity of rectangular section symmetrical reinforcement eccentric compression members

实际工程中,受压构件经常承受变号弯矩的作用。在不同荷载(如风荷载、竖向荷载)组合下,偏心受压构件在同一截面内可能承受相反方向的弯矩,即在某一种内力组合作用下受拉的部位,在另一种内力组合作用下可能变为受压。当两种不同符号的弯矩绝对值相差不大或者相差较大,但对称配筋计算所得钢筋总量与非对称配筋计算所得钢筋总量相比相差不多时,为了设计、施工方便,宜采用对称配筋,如图 6-15 所示。采用对称配筋不易在施工中发生差错,为了使装配式构件吊装不容易出错,一般也采用对称配筋。对称配筋是指在构件截面两侧配置相同数量、相同直径、相同种类、相同位置的钢筋,即 $A_s = A'_s, f_y = f'_y, a_s = a'_s$。

In practical engineering, the compression member is often subjected to the action of sign changing bending moment. Under the combination of different loads (such as wind load, vertical load), the eccentric compression member may be subjected to bending moments in the opposite direction in the same section, that is, a part that is in tension under the action of one certain combination of internal forces may become compressed under the action of another combination of internal forces. When the difference between the absolute value of bending moments of two different symbols is not large or the difference is large but the total amount of steel bars calculated by symmetrical reinforcement is similar to the total amount of steel bars calculated by asymmetrical reinforcement, for the convenience of design and construction, symmetrical reinforcement should be adopted, as shown in Fig. 6-15. Adopting symmetrical reinforcement is not prone to errors during construction. In order to make the hoisting of prefabricated members not easy to make mistakes, symmetrical reinforcement is generally adopted. Symmetrical reinforcement refers to the configurations of reinforcement with the same number, diameter, type and position on both sides of the member section, that is, $A_s = A'_s$, $f_y = f'_y$, $a_s = a'_s$.

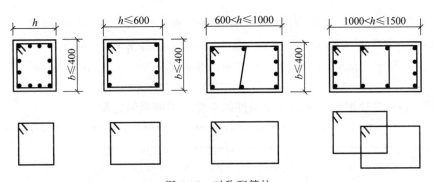

图 6-15 对称配筋柱

Figure 6-15 Symmetrical reinforcement column

6.5.1 对称配筋偏心受压构件截面设计
Section design of symmetrical reinforcement eccentric compression members

1. 对称配筋大偏心受压构件

Large eccentric compression member with symmetrical reinforcement

由非对称配筋大偏心受压构件截面力的平衡方程式(6-20),令 $N=N_u$,得

From the force equilibrium equation Eq. (6-20) of the large eccentric compression member with asymmetrical reinforcement, letting $N=N_u$, one obtains

$$N = \alpha_1 f_c b x + f'_y A'_s - f_y A_s \tag{6-42}$$

对称配筋时 $A_s = A'_s$, $f_y = f'_y$,代入式(6-42)得

In the case of symmetrical reinforcement, substituting $A_s = A'_s$, $f_y = f'_y$ into Eq. (6-42), gives

$$N = \alpha_1 f_c b x \tag{6-43}$$

则

$$x = \frac{N}{\alpha_1 f_c b} \tag{6-44}$$

对称配筋时,可根据相对受压区高度 ξ 进行大、小偏心受压构件的判别:

In case of symmetrical reinforcement, according to the relative depth of compression zone ξ, the large and small eccentric compression members can be identified as follows

$$\xi = \frac{x}{h_0} = \frac{N}{\alpha_1 f_c b h_0} \tag{6-45}$$

当 $\xi \leqslant \xi_b$ 时，为大偏心受压构件；当 $\xi > \xi_b$ 时，说明纵向受拉钢筋应力达不到受拉屈服强度 f_y，为小偏心受压构件。

When $\xi \leqslant \xi_b$, it is a large eccentric compression member; when $\xi > \xi_b$, it means that the longitudinal tensile steel stress cannot reach the tensile yield strength f_y, and it is a small eccentric compression member.

特别提示 6-4

由非对称配筋大偏心受压构件截面力矩平衡方程式(6-21)，令 $N = N_u$，得

According to the section moment equilibrium equation Eq. (6-21) of asymmetrical reinforcement large eccentric compression members, let $N = N_u$, one obtains

$$Ne = \alpha_1 f_c bx(h_0 - x/2) + f'_y A'_s (h_0 - a'_s) \tag{6-46}$$

由式(6-46)，对称配筋大偏心受压构件纵向受力钢筋截面面积为

From Eq. (6-46), the cross-sectional area of longitudinally stressed reinforcement of symmetrical reinforcement large eccentric compression members is

$$A_s = A'_s = \frac{Ne - \alpha_1 f_c bx(h_0 - x/2)}{f'_y(h_0 - a'_s)} = \frac{Ne - N(h_0 - x/2)}{f'_y(h_0 - a'_s)} \geqslant \rho'_{min} bh \tag{6-47}$$

若 $x < 2a'_s$，可近似取 $x = 2a'_s$，如图 6-12(b)所示对纵向受压钢筋 A'_s 合力作用点取矩，得

If $x < 2a'_s$, taking $x = 2a'_s$ approximated as shown in Fig. 6-12(b), and taking moment about the resultant force action point of longitudinal compression reinforcement A'_s, one obtains

$$A_s = A'_s = \frac{Ne'}{f_y(h_0 - a'_s)} \tag{6-48}$$

$$e' = e_i - 0.5h + a'_s \tag{6-49}$$

若 $A'_s < \rho'_{min} bh$（ρ'_{min} 为受压构件单侧钢筋的最小配筋率），取 $A'_s = \rho'_{min} bh$。

If $A'_s < \rho'_{min} bh$ (ρ'_{min} is the minimum reinforcement ratio of reinforcement on one side of the compression member), gives $A'_s = \rho'_{min} bh$.

当 $A_s + A'_s > \rho'_{max} bh = 5\% bh$，超过纵向受力钢筋最大配筋率时，说明截面尺寸过小，宜加大构件截面尺寸重新计算，并且求得的 A_s、A'_s 均应满足最小配筋率 ρ'_{min} 的要求，然后再验算垂直于弯矩作用平面的受压承载力。

When $A_s + A'_s > \rho'_{max} bh = 5\% bh$, the maximum reinforcement ratio of longitudinally stressed steel bars is exceeded, it indicates that the cross-sectional dimensions are too small, and the cross-sectional dimensions of the member should be increased for recalculation, and the obtained A_s、A'_s should all meet the requirements of the minimum reinforcement ratio. Then check and calculate the compressive bearing capacity of the plane perpendicular to the bending moment action plane.

【例题 6-5】 某钢筋混凝土偏心受压柱承受轴向压力设计值 $N = 600$kN，柱两端弯矩设计值分别为 $M_1 = 360$kN·m，$M_2 = 380$kN·m，柱截面尺寸为 $b \times h = 400$mm $\times 500$mm，计

算长度 $l_c = 5$m，采用 C30 混凝土和 HRB400 钢筋，计算过程中取 $a_s = a'_s = 40$mm，要求按对称配筋进行截面设计。

【Example 6-5】 Design value of axial pressure of reinforced concrete eccentric compression column is $N = 600$kN, the design values of bending moment at both ends of the column are $M_1 = 360$kN·m, $M_2 = 380$kN·m respectively, and the cross-sectional dimensions of the column are $b \times h = 400$mm $\times 500$mm, the effective length is $l_c = 5$m, C30 concrete and HRB400 steel bar are used, in the calculation process, $a_s = a'_s = 40$mm is taken, and it is required to design the section according to symmetrical reinforcement.

解：(1) 确定设计参数

C30 混凝土，$\alpha_1 = 1.0$，$\beta_1 = 0.8$，$f_c = 14.3$N/mm^2。HRB400 钢筋，$f_y = f'_y = 360$N/mm^2，$\xi_b = 0.518$。

$$a_s = a'_s = 40\text{mm}, \quad h_0 = h - a_s = (500-40)\text{mm} = 460\text{mm}$$

截面惯性矩：
$$I = bh^3/12 = 400 \times 500^3/12 \text{mm}^4 = 41667 \times 10^5 \text{mm}^4$$

截面面积：
$$b \times h = 400 \times 500 \text{mm}^2 = 2 \times 10^5 \text{mm}^2$$

截面回转半径：
$$i = \sqrt{\frac{I}{A}} = \sqrt{\frac{41667 \times 10^5}{2 \times 10^5}} \text{mm} = 144 \text{mm}$$

(2) 求设计弯矩

由于 $M_2/M_1 = 380/360 = 1.06 > 0.9$，$i = 144$mm，则 $l_c/i = 5000/144 = 34.72 > 34 - 12 M_2/M_1 = 34 - 12 \times 1.06 = 21.28$，因此需要考虑二阶效应附加弯矩的影响。

截面曲率系数：
$$\xi_c = \frac{0.5 f_c A}{N} = \frac{0.5 \times 14.3 \times 2 \times 10^5}{600 \times 10^3} = 2.38 > 1$$

取 $\xi_c = 1.0$。

偏心距调节系数：
$$C_m = 0.7 + 0.3 M_1/M_2 = 0.7 + 0.3 \times 360/380 = 0.984 > 0.7$$

取 $C_m = 0.984$。

附加偏心距：
$$e_a = h/30 = 500/30 \text{mm} = 16.7 \text{mm} < 20 \text{mm}$$

取 e_a 为 20mm。

弯矩增大系数：
$$\eta_{ns} = 1 + \frac{1}{1300(M_2/N + e_a)/h_0} \left(\frac{l_c}{h}\right)^2 \zeta_c$$

$$= 1 + \frac{1}{1300(380 \times 10^6/600 \times 10^3 + 20)/460} \times \left(\frac{5000}{500}\right)^2 \times 1 = 1.05$$

$C_m \eta_{ns} = 0.984 \times 1.05 = 1.03 > 1$，取 $C_m \eta_{ns} = 1.03$。

考虑附加弯矩后的柱设计弯矩为
$$M = C_m \eta_{ns} M_2 = 1.03 \times 380 \text{kN·m} = 391.4 \text{kN·m}$$

(3) 判别大、小偏心受压

混凝土受压区高度：

$$x = \frac{N}{\alpha_1 f_c b} = \frac{600 \times 10^3}{1.0 \times 14.3 \times 400} \text{mm} = 104.9 \text{mm} < \xi_b h_0$$

$$= 0.518 \times 460 \text{mm} = 238.28 \text{mm}$$

属于大偏心受压，且 $x > 2a'_s = 2 \times 40 \text{mm} = 80 \text{mm}$。

(4) 计算纵向受拉钢筋截面面积 A_s 和纵向受压钢筋截面面积 A'_s

计算偏心距：

$$e_0 = M/N = 391.4 \times 10^6/(600 \times 10^3) \text{mm} = 652 \text{mm}$$

初始偏心距：

$$e_i = e_0 + e_a = (652 + 20) \text{mm} = 672 \text{mm}$$

则

$$e = e_i + 0.5h - a_s = (672 + 0.5 \times 500 - 40) \text{mm} = 882 \text{mm}$$

对称配筋，由式(6-47)得纵向受拉钢筋截面面积 A_s 和纵向受压钢筋截面面积 A'_s 为

$$A_s = A'_s = \frac{Ne - N(h_0 - x/2)}{f'_y(h_0 - a'_s)}$$

$$= \frac{600 \times 10^3 \times 882 - 600 \times 10^3 \times (460 - 104.9/2)}{360 \times (460 - 40)} = 1883 \text{mm}^2 > \rho'_{\min} bh$$

$$= 0.2\% bh = 0.002 \times 400 \times 500 \text{mm}^2 = 400 \text{mm}^2$$

(5) 选配钢筋并验算配筋率

根据计算的纵向受力钢筋截面面积 A_s、A'_s，柱截面受拉区和受压区分别选配钢筋 4⌀25，$A_s = A'_s = 1964 \text{mm}^2$。

全部纵筋配筋率为

$$\rho' = (A_s + A'_s)/bh = (1964 + 1964)/(400 \times 500) = 1.96\% > \rho'_{\min}$$

$$= 0.55\% (\rho'_{\min} 可查附表 3-3)$$

且 $\rho' < \rho'_{\max} = 5\%$，所以配筋率满足要求。

与例题 6-3 比较可知，同样情况下，对称配筋的配筋量比非对称配筋要多一些。

2. 对称配筋小偏心受压构件

Small eccentric compression member with symmetrical reinforcement

英文翻译 6-10

对于对称配筋，取 $A_s = A'_s$，$f_y = f'_y$，根据式(6-23)、式(6-24)或式(6-25)直接计算混凝土受压区高度 x。取 $x = \xi h_0$，$N = N_u$，由式(6-23)得

$$N = \alpha_1 f_c b h_0 \xi + (f'_y - \sigma_s) A'_s \quad (6-50)$$

将式(6-28)代入式(6-50)，得

$$f'_y A'_s = \frac{N - \alpha_1 f_c b h_0 \xi}{\dfrac{\xi_b - \xi}{\xi_b - \beta_1}} \quad (6-51)$$

将式(6-51)和 $x = \xi h_0$ 代入式(6-24)，得

$$Ne = \alpha_1 f_c bh_0^2 \xi(1-0.5\xi) + \frac{N - \alpha_1 f_c bh_0 \xi}{\dfrac{\xi_b - \xi}{\xi_b - \beta_1}}(h_0 - a'_s) \qquad (6\text{-}52)$$

即

$$Ne \frac{\xi_b - \xi}{\xi_b - \beta_1} = \alpha_1 f_c bh_0^2 \xi(1-0.5\xi)\frac{\xi_b - \xi}{\xi_b - \beta_1} + (N - \alpha_1 f_c bh_0 \xi)(h_0 - a'_s) \qquad (6\text{-}53)$$

由式(6-53)可知，求 ξ 需要求解三次方程，手算不方便，可采用降阶简化方法，令

$$\bar{y} = \xi(1-0.5\xi)\frac{\xi - \xi_b}{\beta_1 - \xi_b} \qquad (6\text{-}54)$$

将式(6-54)代入式(6-53)中，得

$$\frac{Ne}{\alpha_1 f_c bh_0^2} \cdot \frac{\xi_b - \xi}{\xi_b - \beta_1} - \left(\frac{N}{\alpha_1 f_c bh_0^2} - \frac{\xi}{h_0}\right)(h_0 - a'_s) = \bar{y} \qquad (6\text{-}55)$$

对于给定的混凝土强度等级和钢筋级别，ξ_b、β_1 为确定值，则由式(6-55)可知，在小偏心受压 $\xi_b < \xi < \xi_{cy}$ 的区段内，$\bar{y}$-ξ 接近直线关系。为了简化计算，《混凝土结构设计标准》(GB/T 50010—2010)规定对 HPB300、HRB400(或 RRB400)级钢筋，$\bar{y}$-ξ 的线性方程近似取

$$\bar{y} = 0.43\frac{\xi - \xi_b}{\beta_1 - \xi_b} \qquad (6\text{-}56)$$

将式(6-56)代入式(6-55)，整理后就将求解 ξ 的方程降为一次方程，得到《混凝土结构设计标准》(GB/T 50010—2010)6.2.17 条给出的 ξ 的近似公式

$$\xi = \frac{N - \xi_b \alpha_1 f_c bh_0}{\dfrac{Ne - 0.43\alpha_1 f_c bh_0^2}{(\beta_1 - \xi_b)(h_0 - a'_s)} + \alpha_1 f_c bh_0} + \xi_b \qquad (6\text{-}57)$$

将式(6-57)代入式(6-51)即可求得钢筋截面面积 A'_s。

设计对称配筋小偏心受压构件时，除了上述将求解 ξ 的三次方程作降阶处理的近似方法外，还可采用迭代法求解 ξ 和 A'_s，在此不再详述。

6.5.2 对称配筋偏心受压构件截面承载力复核

Review of section bearing capacity of symmetrical reinforcement eccentric compression members

对称配筋截面承载力复核方法与非对称配筋的情况基本相同，取 $A_s = A'_s, f_y = f'_y$。计算控制截面所能承受的轴向压力设计值 N_u 时，无论大偏心受压还是小偏心受压，其未知量均为两个(N_u 和 x 或 ξ)，可由两个基本计算公式直接求解 x 或 ξ 和 N_u。对于对称配筋小偏心受压构件，由于 $A_s = A'_s$，因此不必再进行反向受压破坏的验算。

The review method of the bearing capacity of symmetrical reinforcement section is basically the same as that of asymmetrical reinforcement, taking $A_s = A'_s, f_y = f'_y$. When calculating the design value of axial pressure N_u that the control section can withstand, whether it is subjected to large or small eccentric compression, there are two unknowns

(N_u and x or ξ), x or ξ and N_u can be directly solved by two basic calculation formulas. For symmetrical reinforcement small eccentric compression members, since $A_s = A'_s$, it is not necessary to perform reverse compression failure verification.

6.6 I 形截面对称配筋偏心受压构件正截面受压承载力计算
Calculation of normal section compression bearing capacity of I-shaped section symmetrical reinforcement eccentric compression members

在单层工业厂房中，当柱截面尺寸较大时，为了节省混凝土、减轻柱自重，截面高度 h 大于 600mm 的柱一般采用 I 形截面。I 形截面柱的翼缘高度一般不小于 120mm，腹板宽度一般不小于 100mm。I 形截面偏心受压构件的受力性能、破坏形态、计算原理与矩形截面偏心受压构件类似，区别只在于增加了受压翼缘参与工作，由于受压区截面形状不同而使基本计算公式稍有差别。实际工程中，I 形截面柱一般都采用对称配筋，分为大偏心受压和小偏心受压两种情况。

In a single-storey industrial factory, when the cross-sectional dimensions of the column are large, in order to save concrete and reduce the weight of the column, the I-shaped section is generally used for the column with a section depth h greater than 600mm. The flange thickness of I-shaped section column is generally not less than 120mm, and the web width is generally not less than 100mm. The mechanical performance, destruction form and calculation principle of eccentric compression member of I-shaped section are similar to those of eccentric compression member of rectangular section. The only difference is that the compressive flange is added to participate in the work. The basic calculation formulas are slightly different due to the different cross-sectional shapes of the compression zone. In practical engineering, symmetrical reinforcement is generally used in I-section columns, which are divided into two types: large eccentric compression and small eccentric compression.

6.6.1 I 形截面对称配筋大偏心受压
Large eccentric compression of I-shaped section with symmetrical reinforcement

1. 基本计算公式
Basic calculation formula

(1) 当 $x \leqslant h'_f$ (h'_f 为受压翼缘高度)时，如图 6-16(a) 所示，中和轴在受压翼缘内，受力情况等同于宽度为 b'_f 的矩形截面，由力的平衡条件和力矩平衡条件(对纵向受拉钢筋合力作用点取力矩)得

(1) When $x \leqslant h'_f$ (h'_f is the depth of compressive flange), as shown in Fig. 6-16(a), the

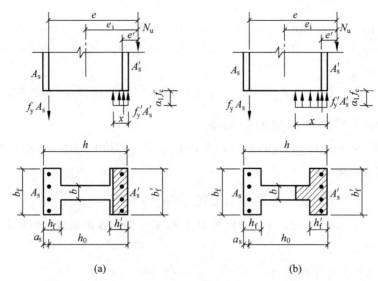

图 6-16 I 形截面大偏心受压计算简图
(a) 受压区在翼缘内 ($x<h'_f$); (b) 受压区在腹板内 ($x>h'_f$)

Fig. 6-16 Calculation diagram of I-shaped section large eccentric compression
(a) Compression zone in the flange ($x<h'_f$); (b) Compression zone in the web ($x>h'_f$)

neutral axis is within the compressive flange, and the stress situation is equivalent to a rectangular section with a width of b'_f. From the force equilibrium condition and the moment equilibrium condition (taking the moment about the action point of the resultant force of the longitudinal tensile steel bars), one obtains

$$N \leqslant N_u = \alpha_1 f_c b'_f x \tag{6-58}$$

$$Ne \leqslant N_u e = \alpha_1 f_c b'_f x \left(h_0 - \frac{x}{2}\right) + f'_y A'_s (h_0 - a'_s) \tag{6-59}$$

(2) 当 $x>h'_f$,且 $\xi \leqslant \xi_b$ 时,如图 6-16(b)所示,中和轴在腹板内,受压区为 T 形,由力的平衡条件和力矩平衡条件得

(2) When $x>h'_f$, and $\xi \leqslant \xi_b$, as shown in Fig. 6-16(b), the neutral axis is within the web, the compression zone is T-shaped, and from the force equilibrium condition and moment equilibrium condition, one obtains

$$N \leqslant N_u = \alpha_1 f_c b x + \alpha_1 f_c (b'_f - b) h'_f \tag{6-60}$$

$$Ne \leqslant N_u e = \alpha_1 f_c b x \left(h_0 - \frac{x}{2}\right) + \alpha_1 f_c (b'_f - b) h'_f \left(h_0 - \frac{h'_f}{2}\right) + f'_y A'_s (h_0 - a'_s) \tag{6-61}$$

式中,b'_f 为 I 形截面受压翼缘的计算宽度,mm;h'_f 为 I 形截面受压翼缘的高度,mm。
Where, b'_f is the effective width of compressive flange of I-shaped section, mm; h'_f is the depth of compressive flange of I-shaped section, mm.

2. 适用条件

Applicable conditions

为了保证纵向受拉钢筋 A_s 和纵向受压钢筋 A'_s 都能达到屈服强度,必须满足以下条件：

In order to ensure that both the longitudinal tensile reinforcement A_s and the longitudinal compressive reinforcement A'_s can reach the yield strength, the following conditions must be met：

$$\xi \leqslant \xi_b \text{ 或 } x \leqslant \xi_b h_0, \quad x \geqslant 2a'_s \text{（对 } x > h'_f \text{ 的情况，一般自然满足）} \tag{6-62}$$

6.6.2 I 形截面对称配筋小偏心受压
Small eccentric compression of I-shaped section with symmetrical reinforcement

对于小偏心受压 I 形截面，一般不会出现 $x < h'_f$ 的情况。

1. 当 $\xi_b h_0 \leqslant x \leqslant h - h_f$（$h_f$ 为离轴向压力 N 较远一侧翼缘的高度）时，中和轴在腹板内（图 6-17(a)）

由力的平衡和力矩平衡条件，可得

$$N \leqslant N_u = \alpha_1 f_c b x + \alpha_1 f_c (b'_f - b) h'_f + f'_y A'_s - \sigma_s A_s \tag{6-63}$$

$$Ne \leqslant N_u e = \alpha_1 f_c b x \left(h_0 - \frac{x}{2}\right) + \alpha_1 f_c (b'_f - b) h'_f \left(h_0 - \frac{h'_f}{2}\right) + f'_y A'_s (h_0 - a'_s) \tag{6-64}$$

$$\sigma_s = \frac{\xi - \beta_1}{\xi_b - \beta_1} f_y \tag{6-65}$$

2. 当 $x > h - h_f$ 时，中和轴进入离轴向压力 N 较远一侧的翼缘内（图 6-17(b)），计算中应考虑翼缘 h_f 的作用

由力的平衡和力矩平衡条件，可得

$$N \leqslant N_u = \alpha_1 f_c b x + \alpha_1 f_c (b'_f - b) h'_f + \alpha_1 f_c (b_f - b)(h_f + x - h) + f'_y A'_s - \sigma_s A_s \tag{6-66}$$

$$Ne \leqslant N_u e = \alpha_1 f_c b x \left(h_0 - \frac{x}{2}\right) + \alpha_1 f_c (b'_f - b) h'_f \left(h_0 - \frac{h'_f}{2}\right) +$$
$$\alpha_1 f_c (b_f - b)(h_f + x - h)\left(h_f - \frac{h_f + x - h}{2} - a_s\right) +$$
$$f'_y A'_s (h_0 - a'_s) \tag{6-67}$$

式中，$x > h$ 时，取 $x = h$ 计算。

3. 当 $x = h$ 时，全截面受压

如图 6-17(c)所示，对于小偏心受压构件，尚应满足以下要求：

$$N_u \left[\frac{h}{2} - a'_s - (e_0 - e_a)\right] \leqslant \alpha_1 f_c \left[bh\left(h'_0 - \frac{h}{2}\right) + (b_f - b) h_f \left(h'_0 - \frac{h_f}{2}\right) + (b'_f - b) h'_f \left(\frac{h'_f}{2} - a'_s\right)\right] + f'_y A_s (h'_0 - a_s) \tag{6-68}$$

式中，h'_0 为钢筋 A'_s 合力点至离轴向力 N 较远一侧边缘的距离，即 $h'_0 = h - a'_s$。

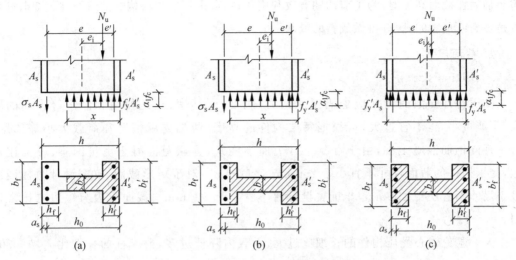

图 6-17 I 形截面小偏心受压计算简图
(a) 受压区在腹板内($\xi_b h_0 \leqslant x \leqslant h - h_f$); (b) 受压区在翼缘高度 h_f 内($x > h - h_f$); (c) 全截面受压($x = h$)

Figure 6-17 Calculation diagram of I-shaped section small eccentric compression
(a) Compression zone in the web($\xi_b h_0 \leqslant x \leqslant h - h_f$); (b) Compression zone in the flange height $h_f(x > h - h_f)$;
(c) Full section compression($x = h$)

6.7 受压构件的一般构造要求
General structural requirements of compression members

6.7.1 截面形式和尺寸
Section form and size

1. 截面形式
Section form

受压构件的截面形式有方形、矩形、圆形、I 形、多边形、环形等,主要根据工程性质和使用要求确定,还要考虑受力合理、节约模板、模板制作方便和保证结构的刚性等要求。为了施工和制作方便起见,通常采用方形或矩形截面,在特殊情况下才采用圆形或多边形截面。从受力合理考虑,轴心受压构件和在两个方向偏心距大小接近的双向偏心受压构件宜采用正方形截面,而单向偏心受压构件和双向偏心受压构件则宜采用矩形截面(较大弯矩方向通常采用较长边)。偏心受压构件除了满足轴心受压构件的截面尺寸要求外,当截面尺寸较大时,为了节约混凝土,减轻结构自重,同时满足强度和刚度要求,装配式受压构件也常采用 I 形截面或双肢截面。采用离心法制作的桩、电杆、烟囱、水塔支筒等常采用环形截面。

方形截面和矩形截面的受压构件最省模板,且制作方便,因此使用广泛。方形截面构件适用于近似轴心受压的情况,矩形截面是偏心受压构件截面的基本形式。当单层厂房柱的弯矩较大时,常采用薄壁 I 形截面的预制柱。当厂房的吊车吨位较大,根据吊车定位尺寸,需要加大柱截面高度时,为了节约和有效利用材料,可采用空腹格构式的双肢柱。双肢柱可以是现浇或预制的,腹杆可做成斜的或水平的。

2. 截面尺寸
Section size

受压构件的截面尺寸应能满足承载力、刚度、配筋率、建筑使用和经济等方面的要求,不能过小,也不宜过大,可根据每层构件的高度、两端支承情况和荷载大小等选用。对于现浇钢筋混凝土柱,由于混凝土自上而下浇筑,为避免混凝土浇筑困难、长细比过大,受压构件的截面尺寸不宜小于 250mm×250mm。对于矩形截面,抗震设计时柱截面最小尺寸不宜小于 250mm,非抗震设计时不宜小于 300mm。截面长边与短边之比常选用 1.5～3。

为了避免轴心受压构件的长细比过大,承载力降低过多,要求柱的长细比 $l_c/b \leqslant 30$, $l_c/h \leqslant 25$, $l_c/d \leqslant 25$ (其中 l_c 为柱的计算长度,b 为矩形截面短边,h 为矩形截面长边,d 为圆形截面直径)。考虑到模板的规格,为施工支模方便,柱截面尺寸宜符合模数,当柱边长不大于 800mm 时,取 50mm 的倍数;当柱边长大于 800mm 时,可取 100mm 的倍数。

对于 I 形截面,翼缘高度不宜小于 120mm,因为翼缘太薄,会使构件过早出现裂缝,靠近柱底处的混凝土容易碰坏,影响柱的承载力和工作年限。腹板厚度不宜小于 100mm,地震区采用 I 形截面柱时,其腹板宜再加厚些。

框架柱可取截面高度 $h = (1/15 \sim 1/20)H$,H 为层高,柱截面高度与宽度的比值不宜大于 3,矩形截面框架柱的截面宽度不应小于 300mm,圆形截面柱的直径不应小于 350mm。为避免发生剪切破坏,柱净高与截面长边之比宜大于 4。框架柱的剪跨比宜大于 2,否则按短柱进行相应的计算。柱的轴压比是抗震设计时确定柱截面尺寸的重要依据。

6.7.2 材料强度要求
Material strength requirements

1. 混凝土的选用
Selection of concrete

合理选用材料是结构设计的基础。混凝土强度等级对受压构件正截面受压承载力影响较大,为了减小构件的截面尺寸,节省钢材,宜采用较高强度等级的混凝土,一般常用 C25、C30、C35、C40,多层及高层建筑的下层柱和重要结构中,必要时应选择强度等级更高的混凝土。

Reasonable selection of materials is the basis of structural design. The concrete strength grade has a great influence on the normal section compressive capacity of the compression member. In order to reduce the cross-sectional dimensions of the member and save reinforcement, concrete with higher strength grade should be used. C25, C30, C35 and

C40 are commonly used. In the lower columns and important structures of multistory and high-rise buildings, concrete with higher strength grade should be selected if necessary.

2. 钢筋的选用

Selection of reinforcement

受压构件中纵向受力钢筋的主要作用是协助混凝土共同承担荷载引起的内力,防止脆性破坏,承担混凝土收缩和温度变化引起的拉应力等。钢筋与混凝土共同受压时,若钢筋强度过高,则不能充分发挥其作用。受压构件纵向受力普通钢筋宜采用 HRB400、HRB500、HRBF400、HRBF500 钢筋,箍筋宜采用 HRB400、HRB500、HRBF400、HRBF500、HPB300 钢筋。

The main function of longitudinally stressed reinforcement in compression members is to assist concrete to share the internal forces caused by loads, prevent brittle failure, bear the tensile stress caused by shrinkage and temperature changes of concrete, etc. When the reinforcement is compressed together with concrete, if the strength of the reinforcement is too high, its effect cannot be fully exerted. The ordinary longitudinally stressed steel bars of compression members should be HRB400, HRB500, HRBF400, HRBF500 steel bars, the stirrups should be HRB400, HRB500, HRBF400, HRBF500, HPB300 steel bars.

6.7.3 纵向钢筋

Longitudinal bars

对于轴心受压构件和偏心距较小、截面上不存在拉力的偏心受压构件,纵向受力钢筋主要用来协助混凝土受压,减小截面尺寸;同时,也可增加构件的延性,抵抗偶然因素所产生的拉力。对偏心较大、部分截面上产生拉力的偏心受压构件,截面受拉区的纵向受力钢筋则是用来承受拉力的。

For axial compression members and eccentric compression members with small eccentricity and no tension in the section, the longitudinally stressed reinforcement is mainly used to assist the concrete compression and reduce the cross-sectional dimensions; at the same time, it can also increase the ductility of the member and resist the tensile force generated by accidental factors. For eccentric compression members with large eccentricity and tension in part of the section, the longitudinally stressed reinforcement in the tension zone of the section is used to withstand the tension.

1. 纵筋的直径和配筋率

Diameter and reinforcement ratio of longitudinal bars

为了增大钢筋骨架的刚度,减少钢筋在施工时可能产生的纵向弯曲和受压时的局部屈曲,减少箍筋用量,纵向受力钢筋宜采用直径较大的钢筋,纵向受力钢筋的直径不宜小于 12mm,一般在 12~32mm 范围内选用。

In order to increase the stiffness of the steel skeleton, reduce the longitudinal bending during the construction and local buckling in compression, and reduce the amount of stirrups, the longitudinally stressed reinforcement should be used with larger diameter, and the diameter of longitudinally stressed reinforcement should not be less than 12mm, generally selected in the range of 12-32mm.

受压构件全部纵向钢筋的配筋率不宜大于 5%, 全部纵向钢筋配筋率不应小于附表 3-3 中给出的最小配筋率 $\rho'_{\min}$ 的要求,且截面一侧纵向钢筋配筋率不应小于 0.2%,一般配筋率控制在 1%~2%为宜。

The reinforcement ratio of all longitudinal bars of compression member should not be greater than 5%, the reinforcement ratio of all longitudinal bars should not be less than the minimum reinforcement ratio $\rho'_{\min}$ given in the Attached table 3-3, and the longitudinal reinforcement ratio on one side of the section should not be less than 0.2%, and generally the reinforcement ratio should be controlled at 1%-2%.

2. 纵筋的根数

Number of longitudinal bars

矩形截面受压柱中纵向受力钢筋根数不得少于 4 根,以便与箍筋形成刚性骨架。轴心受压构件中的纵向受力钢筋应沿构件截面周边均匀布置,钢筋根数不得少于 4 根。偏心受压构件中的纵向受力钢筋应布置在截面偏心方向的两侧。圆形截面柱中纵向钢筋不宜少于 8 根,不应少于 6 根,且宜沿截面周边均匀布置。

The number of longitudinally stressed reinforcement in the rectangular section compression column should be not less than 4 in order to form a rigid skeleton with stirrups. The longitudinally stressed reinforcement in axial compression member should be evenly arranged along the perimeter of the member section, and the number of reinforcement should not be less than 4. The longitudinally stressed reinforcement in the eccentric compression member should be arranged on both sides of the eccentric direction of the section. The longitudinal bars in the circular section column are not suitable for less than 8, and should not be less than 6, and should be evenly arranged along the perimeter of the section.

当截面高度 $h \geqslant 600$mm 时,柱截面两个侧面上(长边)应设置直径为 10~16mm 的纵向构造钢筋并相应设置多肢箍筋或拉筋,其间距不宜大于 500mm,以防止构件因温度变化和混凝土收缩而产生裂缝,如图 6-18 所示。拉筋的直径和间距可与基本箍筋相同,位置与基本箍筋错开。

When the section depth $h \geqslant 600$mm, longitudinally structural reinforcement with a diameter of 10-16mm should be provided on the two sides (long sides) of the column section, and multiple-leg stirrups or tie bars should be provided accordingly, the spacing should not be greater than 500mm to prevent cracks in members due to temperature changes and concrete shrinkage, as shown in Fig. 6-18. The diameter and spacing of the tie bars can be the same as that of the basic stirrups, and the position is staggered with the basic stirrups.

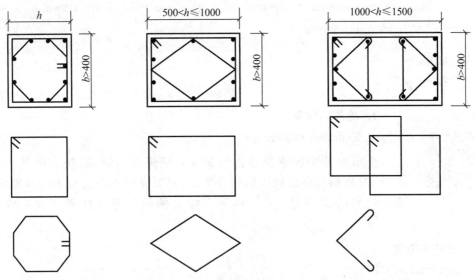

图 6-18 柱的纵向构造钢筋与多肢箍筋

Figure 6-18 Longitudinally structural bars and multiple-leg stirrups of columns

3. 纵筋的间距

Spacing of longitudinal bars

柱内纵向钢筋的净间距不应小于 50mm，且不宜大于 300mm。偏心受压构件中，垂直于弯矩作用平面的侧面上的纵向受力钢筋和轴心受压构件中各边的纵向受力钢筋的中距不宜大于 300mm。在水平位置上浇筑混凝土预制柱，其纵筋的最小净间距可按受弯构件的有关规定取用。

The clear distance of longitudinal bars in the column should not be less than 50mm, and not more than 300mm. In eccentric compression member, longitudinally stressed reinforcement on the side perpendicular to the action plane of bending moment and longitudinally stressed reinforcement on each side of axial compression members, the center distance should not be longer than 300mm. In prefabricated columns poured in a horizontal position, the minimum clear spacing of longitudinal bars can be used according to the relevant provisions of flexural members.

4. 纵筋的连接

Connection of longitudinal bars

纵筋的连接接头宜设置在受力较小处。同一根钢筋宜少设接头。钢筋的接头可采用机械连接接头，也可采用焊接接头和搭接接头。对于直径大于 25mm 的受拉钢筋和直径大于 28mm 的受压钢筋，不宜采用绑扎搭接接头。

Splice joints of longitudinal bars should be set at places with less stress. The same steel bar should have fewer joints. The joints of steel bars can be mechanical connection splices, or welded splices and lap splices. For tensile reinforcement with diameter greater than 25mm and compression reinforcement with diameter greater than 28mm, it is not advisable to use binding lap joint.

6.7.4 箍筋
Stirrup

1. 箍筋的作用

Function of stirrups

受压构件中配置箍筋的目的是：箍住纵向钢筋，防止纵筋压曲外凸，同时箍筋与纵筋构成钢筋骨架，一起形成对核心混凝土的围箍约束；保证纵筋的正确位置；一些剪力较大的偏心受压柱需要利用箍筋来抗剪。

2. 箍筋的构造

Structure of stirrups

1）箍筋的基本要求

1) Basic requirements of stirrups

受压构件中的周边箍筋应做成封闭式，纵筋至少每隔一根放置于箍筋转弯处。对于截面复杂的构件，不可采用含有内折角的箍筋，如图 6-19 所示，避免内折角处箍筋产生向外的拉力，可能使折角处混凝土崩裂、破损。可将复杂截面划分成若干简单截面，再分别配置分离式箍筋。

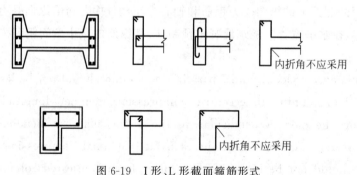

图 6-19 I 形、L 形截面箍筋形式

Figure 6-19 I-shaped and L-shaped section stirrup forms

2）箍筋的形式

2) Form of stirrups

受压构件的箍筋根据形状分为矩形箍和螺旋箍，根据肢数不同分为普通箍和多肢箍。普通矩形箍指单个矩形箍筋，多肢矩形箍指单个矩形箍筋内附加有矩形、多边形、圆形箍筋或拉筋。普通螺旋箍指单个螺旋箍筋，仅用于圆形柱；复合螺旋箍指由螺旋箍筋与矩形、多边形、圆形箍筋或拉筋组成的箍筋。

当柱截面的短边尺寸大于 400mm 且每边纵向钢筋多于 3 根时，或当柱截面的短边尺寸不大于 400mm，但每边纵向钢筋多于 4 根时，应设置多肢箍筋，其布置要求是使纵向钢筋至少每隔一根位于箍筋转角处，如图 6-20 所示。

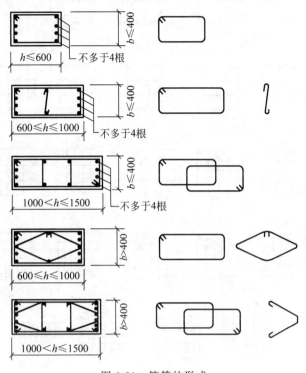

图 6-20 箍筋的形式

Figure 6-20 The form of stirrups

多肢矩形箍用 $m \times n$ 表示，m 和 n 均为不小于 3 的自然数，其中，m 为沿柱 b 边的箍筋肢数，n 为沿柱 h 边的箍筋肢数。常见的箍筋形式如图 6-21 所示。

多肢箍筋排布应对称、均匀，箍筋转角处应有纵向钢筋。多肢箍筋应采用截面周边外封闭大箍加内封闭小箍的组合方式（大箍套小箍），内部多肢箍筋的相邻两肢形成一个内封闭小箍，当多肢箍筋的肢数为单数时，设一个单肢箍。矩形多肢箍筋的复合原则为：沿外封闭箍筋周边，箍筋局部重叠不宜多于两层。以多肢箍筋最外围的封闭箍筋为准，柱内的横向箍筋紧挨其位置在下（或在上），柱内的纵向箍筋紧挨其位置在上（或在下）。柱封闭箍筋弯钩位置应沿柱竖向按顺时针方向（或逆时针方向）顺序排布。柱内部多肢箍筋采用拉筋时，拉筋需同时钩住纵向钢筋和外封闭箍筋。为使箍筋外围局部重叠不多于两层，当拉筋设在旁边时，可沿竖向将相邻两道箍筋按其各自平面位置交错放置。抗震设防时，箍筋对纵筋应满足隔一拉一的要求。

 特别提示 6-5

3）箍筋的直径和间距

3) Diameter and spacing of stirrups

箍筋直径不应小于 $d/4$（d 为纵向钢筋的最大直径），且不应小于 6mm。

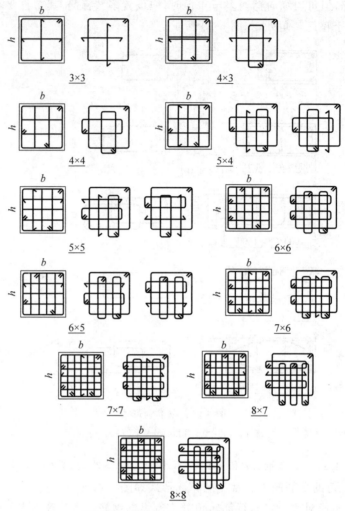

图 6-21 非焊接矩形箍筋复合方式

Figure 6-21 Composite method of non-welded rectangular stirrup

  箍筋的间距在绑扎骨架中不应大于 $15d$（d 为纵向受力钢筋的最小直径），且不应大于 400mm，也不大于构件截面的短边尺寸。当全部纵向受力钢筋的配筋率大于 3% 时，箍筋直径不应小于 8mm，间距不应大于 $10d$（d 为纵向受力钢筋的最小直径），且不应大于 200mm。箍筋末端应做成 135°弯钩，且弯钩末端平直段长度不应小于箍筋直径的 10 倍。

  一级抗震等级时，框架柱箍筋加密区的箍筋间距不宜大于 200mm；二、三级抗震等级时，不宜大于 250mm 和 20 倍箍筋直径中的较大值；四级抗震等级时，不宜大于 300mm。

  柱内纵向钢筋搭接长度范围内的箍筋间距应加密，其直径不应小于搭接钢筋较大直径的 0.25 倍。当搭接钢筋受压时，箍筋间距不应大于 $10d$，且不应大于 200mm；当搭接钢筋受拉时，箍筋间距不应大于 $5d$（d 为纵向钢筋的最小直径），且不应大于 100mm。当搭接受压钢筋直径 $d>25$mm 时，应在搭接接头两个端面外 100mm 范围内各设置两个箍筋。

第 6 章　受压构件的截面承载力

Section bearing capacity of compression members

(1) 受压构件按轴向压力作用位置的不同分为轴心受压构件和偏心受压构件。根据长细比的大小,柱可分为长柱和短柱。轴心受压长柱在加载后将产生侧向弯曲,加大了初始偏心距,产生附加弯矩,使长柱在弯矩和轴力共同作用下发生破坏。长柱的受压承载力比相应短柱的受压承载力低,降低程度用稳定系数 φ 反映。当柱的长细比更大时,还可能发生失稳破坏。

(2) 偏心受压构件正截面破坏有大偏心受压破坏和小偏心受压破坏两种破坏形态,它们的根本区别在于受压区混凝土压碎时远离轴向力一侧钢筋是否达到屈服。大偏心受压破坏与受弯构件中双筋截面适筋梁的受弯破坏类似,属于延性破坏,而小偏心受压破坏为脆性破坏。大偏心受压破坏时,纵向受拉钢筋先达到屈服强度,最后另一侧受压区混凝土被压碎,纵向受压钢筋也达到屈服强度。小偏心受压破坏时,靠近轴向力一侧的混凝土先被压碎,纵向受压钢筋也达到屈服强度,而远离轴向力一侧钢筋可能受拉也可能受压,一般均达不到屈服强度。只有当偏心距很小(对矩形截面 $e_0 \leqslant 0.15h_0$)而轴向力 N 又较大($N > \alpha_1 f_c bh_0$)时,远离轴向力一侧钢筋才可能受压屈服。非对称配筋小偏心受压构件还可能发生远离轴向力一侧的混凝土先被压坏的反向破坏。

(3) 符合特定条件的偏心受压构件应考虑侧向弯曲引起的二阶效应,采用偏心距调节系数和弯矩增大系数考虑二阶效应对弯矩的影响。计算偏心受压构件时,还需要计算附加偏心距、初始偏心距和计算偏心距等。

(4) 大、小偏心受压构件正截面受压承载力的基本计算公式都是由力的平衡条件和力矩平衡条件得到的,两者计算的根本区别在于大偏心受压破坏远离轴向力一侧的钢筋应力按钢筋抗拉屈服强度 f_y 计算,小偏心受压破坏远离轴向力一侧的钢筋受拉时未达到屈服强度,应力是未知数 σ_s,需要求解。

(5) 对于矩形截面非对称配筋偏心受压构件截面设计,当 $e_i > 0.3h_0$ 时,可先按大偏心受压进行计算,如果计算得到的 $x \leqslant x_b = \xi_b h_0$(或 $\xi \leqslant \xi_b$),说明的确是大偏心受压,否则应按小偏心受压重新计算;当 $e_i \leqslant 0.3h_0$(或 $\xi > \xi_b$)时,可直接按小偏心受压计算。

(6) 偏心受压构件的计算包括非对称配筋和对称配筋两种情况,对称配筋可以使计算得到简化。矩形截面非对称配筋大偏心受压构件的截面设计方法与双筋矩形截面受弯构件的相同。矩形截面对称配筋大偏心受压构件截面设计时可直接根据力的平衡条件求出混凝土受压区高度 x。与受弯构件一样,截面承载力复核时一定要先求出混凝土受压区高度 x。

(7) 对偏心受压构件,无论是截面设计还是截面复核,是大偏心受压还是小偏心受压,除了在弯矩作用平面内按照偏心受压计算外,都必须按轴心受压构件验算垂直于弯矩作用平面的受压承载力,对矩形截面计算稳定系数 φ 时的长细比应取截面宽度 b 计算,即长细比取 l_c/b。

(8) 偏心受压构件在承受弯矩的同时还承受较大剪力时,除应进行正截面受压承载力计算外,还应进行斜截面受剪承载力计算。

第 6 章拓展知识和特别提示

视频：第 6 章小结讲解

习 题

6-1 填空题

1. 根据受压构件的轴向压力作用点是否与构件截面形心位置重合，受压构件分为_____和_____构件。

2. 根据长细比大小的不同，轴心受压柱可分为_____和_____。

3. 在轴心受压构件的计算中，采用_____系数来考虑_____对构件承载力的影响。

4. 配置普通箍筋轴心受压长柱的承载力计算公式为_____。

5. 偏心受压构件在侧向弯曲的影响下，破坏特征有两种类型，短柱和长柱属于_____破坏，细长柱属于_____破坏。

6. 偏心受压构件正截面承载力破坏有_____和_____两种破坏形态。

7. 判别大、小偏心受压破坏的条件是：_____时为大偏心受压破坏，_____时为小偏心受压破坏。

8. 大偏心受压的破坏特征是远离轴向力一侧钢筋_____，靠近轴向力一侧钢筋_____，受压区混凝土_____。

9. 小偏心受压的破坏特征是受压区混凝土_____，靠近轴向力一侧钢筋_____，远离轴向力一侧钢筋_____。

10. 在大偏心受压构件正截面承载力计算公式中，适用条件 $\xi \leqslant \xi_b$ 是保证_____，$x \geqslant 2a_s'$ 是保证_____。

11. 偏心受压长柱计算中，由于侧向弯曲引起的附加弯矩是通过_____考虑的。

12. 柱截面尺寸为 $b \times h$，计算长度为 l_c，当按偏心受压计算时，其长细比为_____；当按平面外轴心受压验算时，其长细比为_____。

13. 在偏心受压构件正截面承载力计算中，由于实际工程中存在施工偏差等因素，应考虑轴向压力在偏心方向的附加偏心距 e_a，其值取_____和_____两者中的较大值。

14. 受压构件一侧纵筋的配筋率不应小于_____；全部纵筋的配筋率不应大于_____。

15. 偏心受压构件的计算包括_____配筋和_____配筋两种情况。

16. 受压构件配筋计算时，当_____，可先按小偏心受压构件计算；当_____，可先按大偏心受压构件计算。

6-2 选择题

1. 影响钢筋混凝土轴心受压构件稳定系数 φ 的最主要因素是（　　）。

　　A. 配筋率　　　　B. 混凝土强度　　　C. 钢筋强度　　　D. 构件的长细比

2. 轴心受压构件的箍筋是由( )确定的。
   A. 正截面受弯承载力计算
   B. 斜截面受剪承载力计算
   C. 构造要求

3. 在保持不变的长期荷载作用下,钢筋混凝土轴心受压构件中( )。
   A. 徐变使混凝土压应力减小,因为钢筋与混凝土共同变形,所以钢筋压应力也减小
   B. 根据平衡条件,徐变使混凝土压应力减小,钢筋压应力增大
   C. 由于徐变是应力不增加而变形随时间增长的现象,所以混凝土及钢筋的压应力均不变

4. 轴心受压短柱加载过程中,由于混凝土塑性变形的发展,在受压钢筋屈服之前,柱内应力的发展规律为( )。
   A. 混凝土应力增长变慢,钢筋应力增长变快
   B. 混凝土应力增长变快,钢筋应力增长变慢
   C. 混凝土应力不变,钢筋应力不断增大
   D. 混凝土应力不断增大,钢筋应力不变

5. 大偏心受压构件( )。
   A. M 不变时,N 越大越危险
   B. M 不变时,N 越小越危险
   C. N 不变时,M 越小越危险

6. 对于小偏心受压破坏特征,下列表述不正确的是( )。
   A. 远离轴向力一侧钢筋受拉时未屈服,靠近轴向力一侧钢筋受压屈服,混凝土被压碎
   B. 远离轴向力一侧钢筋受拉时屈服,靠近轴向力一侧钢筋受压屈服,混凝土被压碎
   C. 远离轴向力一侧钢筋受压时一般未屈服,靠近力一侧钢筋受压屈服,混凝土被压碎

7. 混凝土被压碎,其他条件相同,仅边界条件有下述区别的四根受压构件,其承载力最大的是( )。
   A. 两端铰支   B. 两端固定
   C. 一端铰支,一端自由   D. 一端固定,一端铰支

8. 对钢筋混凝土偏心受压构件,大、小偏心受压破坏的根本区别是( )。
   A. 截面破坏时,受压钢筋是否屈服
   B. 偏心距的大小
   C. 截面破坏时,受拉钢筋是否屈服
   D. 受压一侧混凝土是否达到极限压应变

6-3 名词解释

1. 轴心受压构件	2. 受压承载力	3. 螺旋箍筋	4. 失稳破坏
5. 长细比	6. 长柱	7. 短柱	8. 稳定系数
9. 偏心受压构件	10. 单向偏心受压	11. 双向偏心受压	12. 大偏心受压破坏
13. 小偏心受压破坏	14. 附加弯矩	15. 二阶效应	16. 材料破坏
17. 附加偏心距	18. 初始偏心距	19. 计算偏心距	20. 偏心距调节系数
21. 弯矩增大系数	22. 截面曲率修正系数	23. 非对称配筋	24. 对称配筋

6-4 思考题

1. 简述轴心受压短柱和长柱的破坏特点。为什么轴心受压长柱的承载力比短柱的小?如何确定轴心受压长柱的稳定系数?轴心受压构件设置纵向钢筋的目的是什么?
2. 什么是偏心受压构件?如何分类?列举实际工程中的几种偏心受压构件。
3. 简述大、小偏心受压破坏的破坏特征。
4. 什么是偏心受压构件的 P-δ 二阶效应?在哪些情况下要考虑 P-δ 二阶效应?
5. 计算偏心受压构件承载力时,为什么要考虑附加偏心距?如何取值?
6. 如何区分大、小偏心受压破坏的界限?
7. 绘制矩形截面非对称配筋大、小偏心受压构件正截面受压承载力的计算简图,并列出力的平衡方程和力矩平衡方程。
8. 偏心受压构件和受弯构件的正截面应力和应变分布有哪些异同?
9. 比较非对称大偏心受压构件截面设计方法与受弯构件双筋截面梁设计方法的异同。
10. 什么是对称配筋?对称配筋和非对称配筋各有什么优缺点?比较矩形截面非对称配筋与对称配筋偏心受压的截面设计方法,对比各自配筋量的差别。
11. 为什么受压构件中的纵向钢筋不宜采用强度很高的钢筋?
12. 受压构件的纵向钢筋和多肢箍筋有哪些构造要求?

6-5 计算题

1. 某多层三跨现浇框架结构的第三层内柱,柱截面尺寸为 $350mm \times 350mm$,楼层高 $H=5m$,轴向压力设计值 $N=1000kN$,采用 HRB400 钢筋,混凝土强度等级为 C25,求纵向受力钢筋截面面积,选配钢筋,并画出柱横截面配筋图。

2. 某偏心受压柱截面尺寸为 $350mm \times 500mm$,计算长度 $l_c=3m$。采用 HRB400 钢筋,混凝土强度等级为 C30,承受轴向压力设计值 $N=700kN$,柱端弯矩设计值 $M_1=0.6M_2$,$M_2=150kN \cdot m$,$a_s=a'_s=40mm$。按非对称配筋求纵向受力钢筋截面面积 A_s 及 A'_s,选配钢筋,并画出柱横截面配筋图。

3. 柱的截面尺寸为 $350mm \times 600mm$,$a_s=a'_s=40mm$,轴向压力设计值 $N=500kN$,柱端弯矩设计值 $M_1=370kN \cdot m$,$M_2=400kN \cdot m$,采用 HRB400 钢筋,混凝土强度等级为 C25,计算长度 $l_c=3m$。按非对称配筋求纵向受力钢筋截面面积 A_s 及 A'_s,选配钢筋,并画出柱横截面配筋图。

4. 柱轴向压力设计值 $N=6000kN$,柱端弯矩设计值 $M_1=0.95M_2$,$M_2=1800kN \cdot m$,截面尺寸为 $800mm \times 1000mm$,$a_s=a'_s=40mm$,采用 HRB500 钢筋,混凝土强度等级为 C35,计算长度 $l_c=5m$。按对称配筋求纵向受力钢筋截面面积 A_s、A'_s,选配钢筋,并画出柱横截面配筋图。

5. 矩形截面偏心受压柱截面尺寸为 $400mm \times 600mm$,$a_s=a'_s=40mm$,计算长度 $l_c=6m$,混凝土强度等级为 C25,纵向钢筋为 HRB400,箍筋采用 HRB400 钢筋,承受轴向压力设计值 $N=500kN$,柱端弯矩设计值 $M_1=270kN \cdot m$,$M_2=300kN \cdot m$,受压区配置 3⌀18($A'_s=763mm^2$)。按非对称配筋求纵向受力钢筋的截面面积 A_s,选配钢筋,并画出柱横截面配筋图。

6. 柱截面尺寸为 $400mm \times 600mm$,$a_s=a'_s=40mm$,承受轴向压力设计值 $N=3200kN$,考虑二阶效应后的弯矩设计值 $M=80kN \cdot m$,柱的计算长度 $l_c=5m$,采用

HRB400钢筋，混凝土强度等级为C40，采用非对称配筋，求纵向受力钢筋的截面面积A_s、A_s'，选配钢筋，并画出柱横截面配筋图。

7. 柱轴向压力设计值$N=5000\text{kN}$，考虑二阶效应后的弯矩设计值$M=1700\text{kN}\cdot\text{m}$，截面尺寸为$b\times h=800\text{mm}\times1000\text{mm}$，$a_s=a_s'=40\text{mm}$，柱的计算长度$l_c=6\text{m}$，混凝土强度等级为C35，钢筋为HRB500，采用对称配筋。求纵向受力钢筋截面面积A_s、A_s'，选配钢筋，并画出柱横截面配筋图。

8. 矩形截面偏心受压构件的截面尺寸为$b\times h=400\text{mm}\times600\text{mm}$，两个方向的计算长度均为$l_c=5\text{m}$，轴向压力组合设计值为$N=1600\text{kN}$，相应的弯矩组合设计值为$M_1=M_2=170\text{kN}\cdot\text{m}$，混凝土强度等级为C25，纵向受力钢筋HRB400，按对称配筋进行截面设计。求纵向受力钢筋截面面积A_s、A_s'，选配钢筋，并画出构件横截面配筋图。

9. 柱承受轴向压力设计值$N=3000\text{kN}$，考虑二阶效应后的弯矩设计值$M=80\text{kN}\cdot\text{m}$，截面尺寸为$b\times h=400\text{mm}\times600\text{mm}$，$a_s=a_s'=40\text{mm}$，计算长度$l_c=5\text{m}$，混凝土强度等级为C30，HRB400钢筋，3⏀16($A_s'=603\text{mm}^2$)，4⏀25($A_s=1964\text{mm}^2$)。沿长边方向的偏心距$e_0=600\text{mm}$，求该柱能承受的轴向压力设计值。

10. 矩形截面柱截面尺寸为$b\times h=400\text{mm}\times600\text{mm}$，$a_s=a_s'=40\text{mm}$，计算长度$l_c=5\text{m}$，混凝土强度等级为C30，HRB400钢筋。采用对称配筋，4⏀20($A_s=A_s'=1256\text{mm}^2$)。沿长边方向的偏心距$e_0=700\text{mm}$，求该柱能承受的轴向压力设计值。

11. 矩形截面柱截面尺寸为$b\times h=400\text{mm}\times600\text{mm}$，$a_s=a_s'=40\text{mm}$，计算长度$l_c=6\text{m}$，混凝土强度等级为C40，HRB400钢筋，$A_s$为4⏀25，$A_s'$为4⏀20。沿长边方向轴向力的偏心距$e_0=90\text{mm}$。求柱的受压承载力$N_u$。

12. 柱承受轴向压力设计值$N=3000\text{kN}$，柱端弯矩设计值$M_1=0.95M_2$，$M_2=80\text{kN}\cdot\text{m}$，柱截面尺寸为$400\text{mm}\times600\text{mm}$，$a_s=a_s'=40\text{mm}$，采用HRB400钢筋，混凝土强度等级为C35，配有4⏀25($A_s=1964\text{mm}^2$)，3⏀16($A_s'=603\text{mm}^2$)的纵向受力钢筋，柱的计算长度$l_c=5\text{m}$。复核柱截面是否安全。

第 7 章

受拉构件的截面承载力

Section bearing capacity of tension members

教学目标：
1. 了解受拉构件的分类；
2. 理解大、小偏心受拉构件的破坏特征；
3. 了解轴心受拉构件和偏心受拉构件的正截面受拉承载力计算方法。

导读：

受拉构件是指构件截面上主要作用垂直于截面的纵向拉力的作用。根据轴向拉力作用位置的不同，受拉构件可分为轴心受拉构件和偏心受拉构件，如图 7-1 所示。轴心受拉构件是指轴向拉力作用点与构件正截面形心重合，如屋架的下弦杆、圆形水池的池壁等；偏心受拉构件是指轴向拉力作用点偏离构件正截面形心或构件截面上同时承受轴向拉力 N 和弯矩 M，如承受节点之间横向荷载的屋架下弦杆、矩形水池的池壁、浅仓的仓壁、工业厂房中双肢柱的受拉肢杆等。

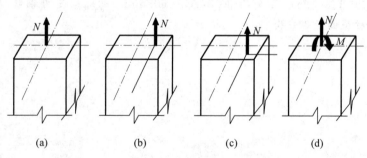

图 7-1 受拉构件
(a) 轴心受拉；(b) 单向偏心受拉；(c) 双向偏心受拉；(d) 拉力、弯矩共同作用
Figure 7-1 Tension member
(a) Axial tension; (b) Uniaxial eccentric tension; (c) Biaxial eccentric tension;
(d) Combined action of tension and bending moment

在实际工程中，理想的轴心受拉构件实际上是不存在的，但有些构件如桁架式屋架或托架的受拉弦杆和腹杆、拱的拉杆，当自重和节点约束引起的弯矩很小时，可近似按轴心受拉构件计算。轴心受拉构件的刚度是通过限制长细比来保证的。另外，单纯受水压力的管道壁、圆形水池的池壁在净水压力的作用下，池壁的垂直截面在水平方向处于环向受拉状态，

可以按轴心受拉构件计算。

偏心受拉构件是一种介于轴心受拉构件与受弯构件之间的受力构件。偏心受拉构件纵向钢筋的布置方式与偏心受压构件相同,离轴向拉力较近一侧的钢筋截面面积用 A_s(并简称该钢筋为钢筋 A_s)表示,离轴向拉力较远一侧的钢筋截面面积用 A'_s(并简称该钢筋为钢筋 A'_s)表示。偏心受拉构件按轴向拉力的作用位置不同,分为大偏心受拉构件和小偏心受拉构件。当轴向拉力 N 作用在钢筋 A_s 和钢筋 A'_s 之间时,称为小偏心受拉构件;当轴向拉力 N 作用在 A_s 和 A'_s 之外时,称为大偏心受拉构件。根据偏心距大小的不同,偏心受拉构件破坏分为大偏心受拉破坏和小偏心受拉破坏,大、小偏心受拉构件的承载力计算是截然不同的。

对受拉构件除了要进行正截面承载力或斜截面承载力计算外,根据不同的要求,还需进行抗裂或裂缝宽度验算。受拉构件的截面形状一般为方形、矩形和圆形等对称截面,为方便施工,通常采用矩形截面。

核心词汇:

拉力	tensile force	受拉承载力	tensile capacity
轴心受拉构件	axial tension member	偏心受拉构件	eccentric tension member
大偏心受拉构件	large eccentric tension member	小偏心受拉构件	small eccentric tension member
全截面受拉	full section tension	部分截面受拉	partial section tension

7.1 轴心受拉构件承载力计算
Calculation of bearing capacity of axial tension members

7.1.1 轴心受拉构件的受力特点
Force characteristics of axial tension members

试验研究表明,轴心受拉构件的受力特点与适筋梁相似,从开始加载到构件破坏,轴心受拉构件的受力和变形也可分为三个阶段。

英文翻译 7-1

第 Ⅰ 阶段为从开始加载到混凝土受拉开裂前。钢筋与混凝土在弹性范围内共同工作,轴向拉力与其平均应变基本上呈线性关系。随荷载增大,混凝土很快达到极限拉应变,构件上即将产生裂缝。对于使用阶段不允许开裂的构件,应以此受力状态作为抗裂验算的依据。

第 Ⅱ 阶段为混凝土开裂到受拉钢筋即将屈服。裂缝发展方向与主拉应力方向垂直,当荷载进一步增加时,裂缝不断发展,混凝土达到轴心抗拉强度,裂缝贯穿整个截面,裂缝截面处的混凝土逐渐停止抵抗拉力,开裂截面上的全部拉力由钢筋承受。对于使用阶段允许出现裂缝的构件,应以此阶段作为裂缝宽度验算的依据。

第 Ⅲ 阶段为受拉钢筋屈服到构件破坏。某一裂缝截面处的受拉钢筋应力首先达到屈服强度,随即裂缝迅速变宽,变形大幅度增加,最后裂缝截面处的全部钢筋达到屈服强度,可认为构件达到了破坏状态,即达到极限荷载 N_u。轴心受拉构件承载力计算是以第 Ⅲ 阶段的应力状态作为依据的。

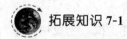 拓展知识 7-1

7.1.2 轴心受拉构件正截面承载力计算
Calculation of normal section bearing capacity of axial tension members

在实际工程中,轴心受拉构件在混凝土开裂前,混凝土与钢筋共同承受拉力,构件破坏时,裂缝截面的全部拉力由钢筋承受,钢筋应力达到抗拉强度设计值 f_y,构件达到其极限承载力,轴心受拉破坏时混凝土裂缝贯通整个截面。

In practical engineering, before the concrete of axial tension member cracks, the concrete and the reinforcement jointly bear the tensile force. When the member is damaged, all the tension of the crack section is borne by the reinforcement, the bar stress reaches the design value of tensile strength f_y, and the member reaches its ultimate bearing capacity. The concrete cracks penetrate the whole section in case of axial tension failure.

轴心受拉构件正截面受拉承载力计算公式为

The calculation formula of the normal section tensile capacity of the axial tension member is

$$N \leqslant N_u = f_y A_s \tag{7-1}$$

式中,N 为轴向拉力设计值,N;N_u 为轴心受拉承载力设计值,N;f_y 为纵向钢筋抗拉强度设计值,N/mm²;A_s 为全部纵向受拉钢筋的截面面积,mm²。

Where, N is the axial tension design value, N; N_u is the axial tensile capacity design value, N; f_y is the design value of tensile strength of longitudinal bar, N/mm²; A_s is the cross-sectional area of total longitudinal tensile reinforcement, mm².

由式(7-1)可知,轴心受拉构件的承载力仅与钢筋抗拉强度设计值和钢筋截面面积有关,与构件尺寸、混凝土强度等级等无关。因此,可尽量减小构件截面尺寸并采用较低的混凝土强度等级。

It can be seen from Eq. (7-1) that the bearing capacity of the axial tension member is only related to the design value of the tensile strength of the reinforcement and the cross-sectional area of the reinforcement, and has nothing to do with the member cross-sectional dimensions and concrete strength grade, etc. Therefore, the cross-sectional dimensions of the member can be reduced as much as possible and the lower concrete strength grade can be adopted.

7.1.3 轴心受拉构件的构造要求
Structural requirements of axial tension members

 英文翻译 7-2

轴心受拉构件中的纵向受拉钢筋在截面中应对称布置或沿截面周边均匀布置,宜选配直径小的钢筋。一侧受拉钢筋的配筋率应不小于 0.2% 和 $0.45 f_t / f_y$ 中的较大值。箍筋直径不宜小于 6mm,间距不宜大于

200mm，屋架腹杆的箍筋间距不宜大于 150mm。

纵向受拉钢筋必须采用焊接接头，仅圆形池壁或管中允许采用搭接接头，但接头位置应错开，搭接长度不应小于 $1.2l_a$ 和 300mm。在接头处左右 $35d$（或 500mm）的区段内焊接受拉钢筋截面积≤受拉钢筋总截面面积的 50%。构件端部受力钢筋应可靠地锚固在支座内。

【例题 7-1】 某钢筋混凝土屋架下弦的截面尺寸为 $b \times h = 200\text{mm} \times 150\text{mm}$，其所受的轴心拉力设计值为 280kN，混凝土强度等级为 C25，纵向钢筋采用 HRB400，求纵向钢筋截面面积并选配钢筋。

视频 7-1

【Example 7-1】 The cross-sectional dimensions of the bottom chord of reinforced concrete roof truss are $b \times h = 200\text{mm} \times 150\text{mm}$, the design value of the axial tensile force is 280kN, the concrete strength grade is C25, and the longitudinal reinforcement is HRB400. Calculate the cross-sectional area of the longitudinal reinforcement and select the reinforcement.

解：由附表 1-6 查得 $f_y = 360\text{N/mm}^2$，代入式(7-1)，得纵向钢筋截面面积

$$A_s = \frac{N}{f_y} = \frac{280 \times 10^3}{360}\text{mm}^2 = 778\text{mm}^2$$

选用 4⌀16，$A_s = 804\text{mm}^2$。

7.2 偏心受拉构件承载力计算
Calculation of bearing capacity of eccentric tension members

7.2.1 偏心受拉构件正截面的受力特点
Force characteristics of normal section of eccentric tension members

1. 大偏心受拉构件

Large eccentric tension member

大偏心受拉构件轴向拉力 N 的偏心距 e_0 较大，$e_0 \geq h/2 - a_s$，轴向拉力 N 在 A_s 合力点和 A'_s 合力点外侧，截面部分受拉、部分受压，离 N 近的一侧 A_s 受拉，离 N 远的一侧 A'_s 受压。加载开始后，随着轴向拉力的增加，裂缝首先从拉应力较大侧开始出现，离轴向拉力较远一侧还有受压区，裂缝不会贯通整个截面，否则对轴向拉力 N 作用点取矩将不满足平衡条件。

英文翻译 7-3

破坏特征与 A_s 的数量多少有关，当 A_s 的数量适当时，随荷载继续增加，受拉钢筋 A_s 先达到抗拉强度设计值 f_y，离轴向力较远一侧的受压区边缘混凝土应变达到混凝土极限压应变 ε_{cu}，受压混凝土压碎破坏，受压区混凝土强度达到 $\alpha_1 f_c$，受压钢筋应力 σ'_s 达到其抗压强度设计值 f'_y，构件达到极限承载力而破坏，如图 7-2(a)所示。与大偏心受压柱及双筋截面受弯构件的受力特点和破坏形态类似，大偏心受拉构件截面设计时，应以这种破坏特征为依据。而当 A_s 的数量过多时，受压区混凝土先被压碎，受压钢筋应力能够达到受压屈服强度 f'_y，但受拉钢筋不屈服，这种破坏特征具有脆性性质，设计时应予以避免。

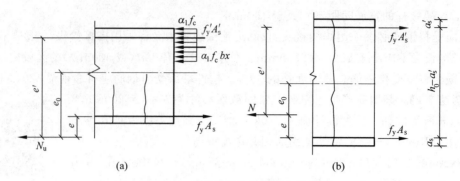

图 7-2 偏心受拉构件的受力特点
(a) 大偏心受拉构件；(b) 小偏心受拉构件

Figure 7-2 Stress characteristics of eccentric tension members
(a) Large eccentric tension member；(b) Small eccentric tension member

2. 小偏心受拉构件

Small eccentric tension member

当偏心受拉构件轴向拉力 N 的偏心距 e_0 较小，$0<e_0<h/2-a_s$，轴向拉力 N 处于 A_s 合力点与 A'_s 合力点之间时，发生小偏心受拉破坏，如图 7-2(b)所示。小偏心受拉构件全截面均承受拉应力，但 A_s 一侧拉应力较大，A'_s 一侧拉应力较小。随着轴向拉力 N 的增加，A_s 一侧截面边缘混凝土达到极限拉应变而首先开裂，裂缝很快贯通整个截面，拉力全部由钢筋承受，构件达到其极限承载力。破坏时纵向钢筋是否屈服取决于其配筋方式(非对称配筋还是对称配筋)。

7.2.2 矩形截面偏心受拉构件正截面承载力计算

Calculation of normal section bearing capacity of eccentric tension members with rectangular section

1. 大偏心受拉构件

Large eccentric tension member

1) 基本计算公式

1) Basic calculation formula

大偏心受拉构件破坏时，受拉钢筋 A_s 的应力取抗拉强度设计值 f_y，受压钢筋 A'_s 的应力取抗压强度设计值 f'_y，混凝土压应力分布仍采用等效矩形应力分布图形，破坏时受压区混凝土应力达到 $\alpha_1 f_c$，混凝土受压区高度为 x。

When the large eccentric tension member is damaged, the stress of the tensile reinforcement A_s is taken as the design value of tensile strength f_y, the stress of the compression reinforcement A'_s is taken as the design value of compressive strength f'_y, and the concrete compressive stress distribution still adopts the equivalent rectangular stress distribution block, and the concrete stress in the compression zone reaches $\alpha_1 f_c$ when the member is damaged, and the depth of concrete compression zone is x.

大偏心受拉构件的截面应力计算图形如图 7-3(a)所示,根据截面力的平衡条件和力矩平衡条件,可得大偏心受拉构件正截面承载力基本计算公式为

The section stress calculation diagram of the large eccentric tension member is shown in Fig. 7-3(a). According to the section force equilibrium condition and moment equilibrium condition, the basic calculation formula of the normal section bearing capacity of the large eccentric tension member can be obtained as follows

$$N \leqslant N_u = f_y A_s - f'_y A'_s - \alpha_1 f_c bx \tag{7-2}$$

$$Ne \leqslant N_u e = \alpha_1 f_c bx \left(h_0 - \frac{x}{2}\right) + f'_y A'_s (h_0 - a'_s) \tag{7-3}$$

式中,e 为轴向拉力作用点至受拉钢筋 A_s 合力作用点之间的距离,$e = e_0 - h/2 + a_s$,mm。

Where, e is the distance between the action point of axial tensile force and the resultant force action point of tensile reinforcement A_s, $e = e_0 - h/2 + a_s$, mm.

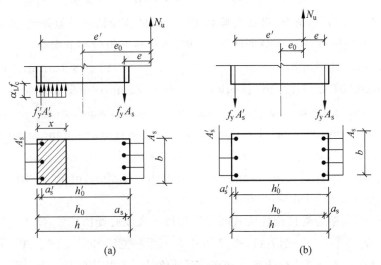

图 7-3 偏心受拉构件受拉承载力计算简图
(a) 大偏心受拉构件；(b) 小偏心受拉构件

Figure 7-3 Calculation diagram of tensile capacity of eccentric tension members
(a) Large eccentric tension member; (b) Small eccentric tension member

2) 适用条件

2) Applicable conditions

上述基本计算公式的适用条件为

$$2a'_s \leqslant x \leqslant \xi_b h_0 \tag{7-4}$$

$$A_s \geqslant \rho_{\min} bh \tag{7-5}$$

式(7-5)是为了满足最小配筋率的要求,其中 $\rho_{\min} = \max(0.45 f_t / f_y, 0.2\%)$。

为确保受拉钢筋 A_s 达到抗拉强度设计值 f_y,应满足 $\xi \leqslant \xi_b$。当 $\xi > \xi_b$ 时,受拉钢筋不屈服,这是受拉钢筋 A_s 配筋率过大导致的,类似于受弯构件中的超筋梁,应避免采用。

为确保受压钢筋 A'_s 达到抗压强度设计值 f'_y，应满足 $x \geqslant 2a'_s$。若 $x < 2a'_s$，则受压钢筋 A'_s 没有屈服，近似认为受压钢筋合力作用点与混凝土压力作用点重合，和大偏心受压构件截面设计时相同，近似且偏安全地取 $x = 2a'_s$，并对受压钢筋 A'_s 的合力作用点取矩，得

$$Ne' \leqslant N_u e' = f_y A_s (h_0 - a'_s) \tag{7-6}$$

式中，e' 为轴向拉力作用点至受压钢筋 A'_s 合力作用点的距离，$e' = e_0 + h/2 - a'_s$，mm。

3）截面设计

3) Section design

应用大、小偏心受拉构件基本计算公式进行截面设计时，取 $N = N_u$。

When the basic calculation formula of large and small eccentric tension members is used for section design, $N = N_u$ is taken.

对称配筋时，$A_s = A'_s$，$f_y = f'_y$，将其代入基本计算公式(7-2)，求得混凝土受压高度 x 为负值，即属于 $x < 2a'_s$ 的情况，可按大偏心受压构件的相应情况类似处理，即取 $x = 2a'_s$，对 A'_s 合力作用点取矩和取 $A'_s = 0$ 分别计算 A_s 值，最后按所得较小值配筋。

For symmetrical reinforcement, $A_s = A'_s$, $f_y = f'_y$, it is substituted into the basic calculation formula Eq. (7-2), the depth of concrete compression zone x is obtained as a negative value, that is, it belongs to the situation $x < 2a'_s$, which can be handled similarly according to the corresponding situation of the large eccentric compression member, that is, $x = 2a'_s$ is taken. The A_s value is calculated by taking the moment about the A'_s resultant force action point and $A'_s = 0$ respectively, and the reinforcement is finally configured according to the smaller value obtained.

其他情况下，大偏心受拉构件的截面设计类似于大偏心受压构件，所不同的是轴向力为拉力。当 A_s、A'_s 均未知时，式(7-2)、式(7-3)中有三个独立未知量 x、A_s、A'_s，但只有两个独立方程，需要补充一个条件以便求解。和双筋截面梁 A_s、A'_s 均未知的情况一样，以 $A_s + A'_s$ 总量最小作为补充条件，取 $\xi = \xi_b$，即 $x = \xi_b h_0$，剩下两个未知量 A_s、A'_s，由式(7-2)、式(7-3)可以联立求解。如果 $A'_s < \rho_{\min} bh$ 且 A'_s 与 $\rho_{\min} bh$ 数值相差较多，则取 $A'_s = \rho_{\min} bh$，改为按已知 A'_s 计算 A_s。

In other cases, the section design of large eccentric tension members is similar to that of large eccentric compression members, the difference is that the axial force is the tensile force. When both A_s and A'_s are unknown, Eq. (7-2) and Eq. (7-3) have three independent unknowns x、A_s and A'_s, but there are only two independent equations, and a condition needs to be added to solve them. As in the case where both A_s and A'_s of doubly reinforced section beam are unknown, the minimum total amount of $A_s + A'_s$ is taken as the supplementary condition, and $\xi = \xi_b$ is taken, that is, $x = \xi_b h_0$, and the remaining two unknown A_s and A'_s can be solved jointly according to Eq. (7-2) and Eq. (7-3). If $A'_s < \rho_{\min} bh$ and the numerical value of A'_s is much different from $\rho_{\min} bh$, $A'_s = \rho_{\min} bh$ is taken, and calculate A_s by known A'_s instead.

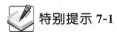 **特别提示 7-1**

4) 截面承载力复核

4) Section bearing capacity review

进行大偏心受拉构件截面复核时,截面尺寸、截面配筋 A_s 和 A_s'、混凝土强度等级、钢筋种类、轴向拉力 N 和弯矩 M 均已知,要求验算是否满足正截面受拉承载力的要求。截面复核时大偏心受拉构件基本计算公式(7-2)和式(7-3)中只有 N_u 和 x 两个未知数,联立方程求解,可消去 N_u,求出 x。

若求出的 $x > \xi_b h_0$ 即 $\xi > \xi_b$,表明受压钢筋配置不足,此时可按式 $\sigma_s = \dfrac{\xi - \beta_1}{\xi_b - \beta_1} f_y$ 计算受拉钢筋的应力 σ_s,将 σ_s 代替 f_y 代入式(7-2)中,重新联立式(7-2)和式(7-3)求解。或近似取 $x = \xi_b h_0$,由大偏心受拉构件基本计算公式(7-2)和式(7-3)各计算出一个 N_u,取两者之中的较小者作为最终的 N_u,即为该截面所能承受的轴向拉力设计值。

若求出的 $x < 2a_s'$,取 $x = 2a_s'$,由式(7-6)得

$$N_u = \frac{f_y A_s (h_0 - a_s')}{e'} \tag{7-7}$$

2. 小偏心受拉构件

Small eccentric tension member

1) 基本计算公式

如图 7-3(b)所示,根据力矩平衡条件,分别对钢筋 A_s、A_s' 的合力作用点取矩。

对钢筋 A_s 合力点取矩,得

$$Ne \leqslant N_u e = f_y A_s' (h_0 - a_s') \tag{7-8}$$

对钢筋 A_s' 合力点取矩,得

$$Ne' \leqslant N_u e' = f_y A_s (h_0 - a_s') \tag{7-9}$$

式中,e、e' 分别为轴向拉力 N 至钢筋 A_s、钢筋 A_s' 合力点的距离,计算公式如下:

$$e = h/2 - a_s - e_0 \tag{7-10}$$

$$e' = h/2 - a_s' + e_0 \tag{7-11}$$

2) 截面设计

已知:构件截面尺寸、材料强度等级和内力等,求配筋。此时式(7-8)、式(7-9)中有两个未知量 A_s 和 A_s',可直接求解,得

$$A_s' = \frac{Ne}{f_y (h_0 - a_s')} \tag{7-12}$$

$$A_s = \frac{Ne'}{f_y (h_0 - a_s')} \tag{7-13}$$

当采用对称配筋时,可按式(7-9)计算得到 A_s,并取 $A_s = A_s'$,即

$$A_s = A_s' = \frac{Ne'}{f_y (h_0 - a_s')} \tag{7-14}$$

按上述公式计算所得 A_s 和 A_s' 均应满足最小配筋率的要求。

3) 截面承载力复核

已知：小偏心受拉构件的截面尺寸、截面配筋面积 A_s 和 A_s'、混凝土强度等级、钢筋种类、偏心距 e_0，由小偏心受拉构件基本公式(7-8)和式(7-9)可以各计算出一个 N_u，取两者之中的较小者作为最终的 N_u，即为该截面所能承受的轴向拉力设计值。

本 章 小 结

（1）受拉构件是指构件截面上主要作用垂直于截面的纵向拉力的作用。根据轴向拉力作用位置的不同，受拉构件可分为轴心受拉构件和偏心受拉构件。轴心受拉构件是指轴向拉力作用点与构件正截面形心重合，偏心受拉构件是指轴向拉力作用点偏离构件正截面形心或构件截面上同时承受轴向拉力 N 和弯矩 M。

（2）偏心受拉构件纵向受力钢筋的布置方式与偏心受压构件相同，离轴向拉力较近一侧的钢筋截面面积用 A_s 表示，离轴向拉力较远一侧的钢筋截面面积用 A_s' 表示。偏心受拉构件按轴向拉力的作用位置不同，分为大偏心受拉构件和小偏心受拉构件。当轴向拉力 N 作用在钢筋 A_s 和钢筋 A_s' 之间时，称为小偏心受拉构件；当轴向拉力 N 作用在 A_s 和 A_s' 之外时，称为大偏心受拉构件。

（3）大偏心受拉构件与大偏心受压构件正截面承载力的计算公式是相似的，其计算方法也可参照大偏心受压构件进行。

（4）由于轴向拉力会降低混凝土的抗剪能力，偏心受拉构件斜截面承载力计算应考虑轴向拉力的不利影响，即在受弯构件受剪承载力公式的基础上减去一项由于轴向拉力的存在对构件受剪承载力产生的不利影响。

第 7 章拓展知识和特别提示　　　　　　视频：第 7 章小结讲解

习 题

7-1 填空题

1. 根据轴向拉力作用位置的不同，受拉构件可分为_____受拉构件和_____受拉构件。

2. 轴心受拉构件是指轴向拉力作用点与构件正截面_____重合，偏心受拉构件是指轴向拉力作用点偏离构件正截面_____或构件截面上同时承受_____和_____。

3. 偏心受拉构件按_____的作用位置不同，分为_____受拉构件和_____受拉构件。当_____作用点在钢筋 A_s 和钢筋 A_s' 之间时，称为_____受拉构件；当_____作用点在钢筋 A_s 和钢筋 A_s' 之外时，称为_____受拉构件。

4. 条件相同的钢筋混凝土轴心受拉构件和预应力轴心受拉构件，其受拉承载力

_____（填相同或不相同），抗裂度是_____受拉构件高。

7-2 选择题

1. 下列构件中属于受拉构件的是（   ）。
   A. 基础　　　　　　B. 桥墩　　　　　　C. 悬索　　　　　　D. 主梁

2. 轴心受拉构件配有两种不同面积和不同等级的钢筋,混凝土开裂后,钢筋应力增量 $\Delta\sigma_s$ 的变化为（   ）。
   A. 强度高的钢筋 $\Delta\sigma_s$ 大　　　　　　B. 两种钢筋的 $\Delta\sigma_s$ 相同
   C. A_s 大的钢筋 $\Delta\sigma_s$ 小　　　　　　D. A_s 小的钢筋 $\Delta\sigma_s$ 小

3. 下列表述中（   ）项错误。
   A. 轴心受拉构件从加载至开裂前,钢筋应力增加速度比混凝土快
   B. 轴心受压构件随荷载增加,钢筋应力的增长速度快于混凝土,是由于混凝土割线模量不断减小所致
   C. 轴心受拉构件即将开裂时,配筋率越大则钢筋应力越大
   D. 轴心受压构件需控制最大配筋率是为了避免可变荷载突然卸去时在混凝土中产生过大的拉应力

4. 在轴心受拉及小偏心受拉构件破坏时,（   ）。
   A. 轴心受拉时外荷载由混凝土承受
   B. 轴心受拉时外荷载由混凝土和钢筋承受
   C. 小偏心受拉时外荷载由混凝土和钢筋承受
   D. 两种情况外荷载均由钢筋承受

5. 关于小偏心受拉构件,以下（   ）是正确的。
   A. 若偏心距 e_0 改变,则总用钢量 $A_s + A'_s$ 不变
   B. 若偏心距 e_0 改变,则总用钢量 $A_s + A'_s$ 改变
   C. 若偏心距 e_0 增大,则总用钢量 $A_s + A'_s$ 增加
   D. 若偏心距 e_0 增大,则总用钢量 $A_s + A'_s$ 减小

6. 在钢筋混凝土双筋截面梁、大偏心受压构件和大偏心受拉构件的正截面承载力计算中要求混凝土受压区高度 $x \geq 2a'_s$,是为了（   ）。
   A. 保证受压钢筋在构件破坏时能达到其抗压强度设计值
   B. 防止受压钢筋压屈
   C. 避免混凝土保护层剥落

7-3 名词解释

1. 受拉构件	2. 轴心受拉构件	3. 偏心受拉构件	4. 大偏心受拉构件
5. 小偏心受拉构件	6. 受拉承载力		

7-4 思考题

1. 当轴心受拉构件的受拉钢筋强度不同时,如何计算其正截面承载力?
2. 如何区分大、小偏心受拉?它们的受力特点和破坏特征有何不同?
3. 大偏心受拉构件的正截面承载力计算中,混凝土受压区高度 x 为什么取与受弯构件的相同?

4. 大偏心受拉构件承载力基本计算公式的适用条件有哪些？为什么也有和受弯构件中双筋截面梁一样的适用条件？

5. 偏心受拉构件计算中为什么不考虑类似于受压构件中二阶效应的影响？

6. 从破坏形态、截面应力、计算公式及计算步骤等方面分析大偏心受拉与大偏心受压的异同点。

7-5 计算题

1. 矩形水池池壁厚 $h=200\mathrm{mm}$，$a_s=a_s'=35\mathrm{mm}$，根据水池荷载组合计算出的沿水池长度方向每米最大轴向拉力设计值 $N=80\mathrm{kN}$，弯矩设计值 $M=70\mathrm{kN\cdot m}$，混凝土强度等级为 C30，钢筋采用 HRB500，水池沿池壁高度方向取 $b=1000\mathrm{mm}$，计算池壁截面配筋。

2. 受拉构件截面尺寸为 $b\times h=350\mathrm{mm}\times 400\mathrm{mm}$，$a_s=a_s'=35\mathrm{mm}$。承受轴向拉力设计值 $N=500\mathrm{kN}$，弯矩设计值 $M=40\mathrm{kN\cdot m}$，混凝土强度等级为 C25，采用 HRB400 钢筋，求纵筋截面面积。

受扭构件的扭曲截面承载力

Twisted section bearing capacity of torsional members

教学目标：
1. 理解平衡扭转、协调扭转、开裂扭矩、截面受扭塑性抵抗矩等概念；
2. 了解纯扭构件的受力特点、破坏形态及扭曲截面承载力计算；
3. 理解变角度空间桁架模型；
4. 了解弯扭构件、剪扭构件、弯剪扭构件的扭曲截面承载力计算；
5. 了解受扭构件的构造要求。

导读：

受扭构件是截面上承受扭矩或截面所受剪应力的合力不通过截面弯曲中心的结构构件，是钢筋混凝土结构的基本构件之一。按照受扭构件截面所受内力的不同，分为纯扭构件、剪扭构件、弯扭构件、弯剪扭构件、压弯剪扭构件等。纯扭构件是只承受扭矩的构件，实际工程结构中，纯扭构件是很少的，大多数构件处于弯矩、剪力、扭矩共同作用下的复合受扭状态。如图 8-1 所示的雨篷梁、折梁、吊车梁、框架边梁等均为弯、剪、扭复合受扭构件。

工程中的扭转作用根据其形成原因分为平衡扭转和协调扭转。如图 8-1(c)所示吊车梁为静定的受扭构件，通过构件的静力平衡条件可以确定吊车横向水平制动力 H 和竖向轮压 P 对吊车梁截面产生的扭矩 T，扭矩 T 的确定与受扭构件的扭转刚度无关，称为平衡扭转。另外，雨篷梁、曲线形桥梁的大梁、承受偏心荷载的箱形梁和螺旋楼梯板等都属于平衡扭转。平衡扭转是由荷载作用直接引起的，是混凝土结构中的主要扭转。

协调扭转是指由于超静定结构构件之间的连续性，相邻构件的位移受到该构件的约束而引起的扭转，它是混凝土结构中的次要扭转。如图 8-1(d)所示框架边梁为超静定受扭构件，当次梁在荷载作用下受弯变形时，边梁对次梁梁端的转动产生约束作用，边梁承受的扭矩 T 即为作用在楼面梁支座处的负弯矩，扭矩 T 由楼面梁支座处的转角和该处边梁扭转角的变形协调条件确定，称为协调扭转。

平衡扭转的扭矩在构件中不会产生内力重分布，而协调扭转的扭矩会由于构件开裂产生内力重分布而减小。例如，当框架边梁和楼面梁开裂后，楼面梁的抗弯刚度和边梁的抗扭刚度发生显著变化，框架边梁和楼面梁均产生内力重分布，边梁的扭转角急剧增大，使作用

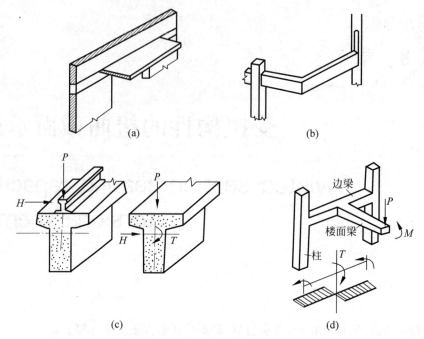

图 8-1　钢筋混凝土受扭构件示例
(a) 雨篷梁；(b) 折梁；(c) 吊车梁；(d) 框架边梁

Figure 8-1　Examples of reinforced concrete torsional members
(a) Canopy beam；(b) Folded beam；(c) Crane beam；(d) Side frame beam

于边梁的扭矩 T 迅速减小。

核心词汇：

受扭构件	torsional member	变角度空间桁架模型	variable-angle space truss model
纯扭构件	pure torsion member		
弯剪扭构件	bending-shear-torsion member	纵筋与箍筋的配筋强度比	reinforcement strength ratio of the longitudinal bars to the stirrups
平衡扭转	equilibrium torsion		
协调扭转	compatibility torsion		
全超筋破坏	full overreinforced failure	受扭承载力降低系数	torsional capacity reduction coefficient
部分超筋破坏	partial overreinforced failure		
开裂扭矩	cracking torque	截面受扭塑性抵抗矩	section torsional plastic resistance moment

8.1　纯扭构件的扭曲截面承载力计算
Calculation of twisted section bearing capacity of pure torsional members

纯扭构件的受力性能是复合受扭构件承载力计算的基础，所以研究纯扭构件的受力性能十分必要。

The mechanical performance of pure torsional members is the basis for calculating the bearing capacity of composite torsional members, so it is necessary to study the mechanical performance of pure torsional members.

8.1.1 纯扭构件的试验研究
Experimental study of pure torsional members

1. 素混凝土纯扭构件
  Plain concrete pure torsional member

如图 8-2 所示,矩形截面素混凝土棱柱体纯扭构件两端承受大小相等、方向相反的扭矩 T,在扭矩作用下,产生的扭转变形不符合平截面假定,截面之间除产生相对转动以外,截面本身还会发生翘曲变形。在扭矩作用下构件截面上产生剪应力,与构件纵轴成 45°方向产生主拉应力和主压应力,当主拉应力达到混凝土抗拉强度时构件开裂。首先在构件的一个长边中点附近出现与构件成 45°夹角的斜裂缝,此裂缝迅速沿着与构件轴线成 45°方向以螺旋形向上、向下延伸,形成三面开裂一面受压的空间扭曲面,最后受压面上的混凝土被压碎,构件断裂破坏。其破坏是突然的,无明显预兆,属于脆性破坏。

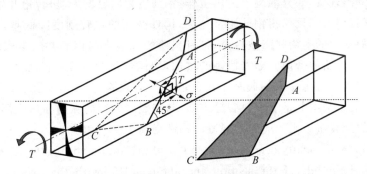

图 8-2 素混凝土纯扭构件的破坏面

Figure 8-2 Failure surface of plain concrete pure torsional members

由于素混凝土构件的受扭承载力很低且呈现出明显的脆性破坏特征,因此在混凝土受扭构件中,配置适当的受扭钢筋以改善其受力性能,当混凝土开裂后,可由钢筋继续承担拉应力,这对于提高构件的受扭承载力和变形能力有很大的作用。理论上,受扭钢筋的最佳形式是做成垂直于裂缝方向,与构件纵轴线成 45°的螺旋形钢筋,当混凝土开裂后,受扭钢筋直接承受主拉应力。但是,这种配筋方式施工复杂,并且当扭矩改变方向时会完全失去作用,试验表明,仅配置受扭纵筋很难提高受扭承载力,因为没有箍筋约束的纵向钢筋只能通过销栓作用抗扭,若沿纵筋发生劈裂破坏,则销栓作用可能失去作用。因此,实际工程中通常是沿构件纵轴方向布置封闭的受扭箍筋以及沿构件周边均匀对称地布置受扭纵筋,以共同组成抗扭钢筋骨架。受扭箍筋和受扭纵筋必须同时设置,缺一不可,并且在配筋数量上相互匹配,否则不能充分发挥其抗扭作用。

2. 钢筋混凝土纯扭构件
  Reinforced concrete pure torsional member

试验研究表明,加载初期,钢筋混凝土受扭构件的截面扭转变形很小,受力性能与素混凝土构件相似,构件的抗扭刚度相对较大。裂缝出现之前,钢筋混凝土纯扭构件的受力性能大体上符合圣维南弹性扭转理论。在扭矩较小时,其扭矩-扭转角曲线为直线,扭转刚度与按弹性理论的计算值非常接近,纵筋和箍筋的应力都很小。当扭矩增大到接近开裂扭矩T_{cr}时,扭矩-扭转角曲线会偏离原直线。

The experimental research shows that at the initial stage of loading, the torsional deformation of the section of reinforced concrete torsional member is small, the mechanical performance is similar to that of plain concrete members, and the torsional stiffness of the member is relatively large. Before cracks appear, the mechanical performance of reinforced concrete pure torsional members generally conforms to the Saint-Venant elastic torsion theory. When the torque is small, the torque-torsion angle curve is a straight line, the torsional stiffness is very close to the calculated value according to the elastic theory, and the stress of longitudinal bars and stirrups is very small. When the torque increases close to the cracking torque T_{cr}, the torque-torsion angle curve deviates from the original straight line.

初始裂缝通常发生在剪应力最大处,即截面长边的中点附近且与构件轴线约成45°。当加载至斜裂缝出现后,部分混凝土退出工作,钢筋应力明显增大,扭转角开始显著增大,构件的抗扭刚度明显降低,且受扭钢筋用量越少,构件截面的扭转刚度降低越多。此时,裂缝出现前构件截面受力的平衡状态被打破,带有裂缝的混凝土与钢筋共同组成新的受力体系以抵抗扭矩,并获得新的平衡。随着扭矩的增加,裂缝数量及宽度逐渐加大,在构件的四个表面形成大致连续的与构件纵轴成某个角度的螺旋形裂缝,如图8-3所示。斜裂缝之间的混凝土受压,压应力基本上是沿着斜裂缝方向,与斜裂缝相交的纵筋和箍筋都受拉。钢筋混凝土构件截面的开裂扭矩比相应的素混凝土构件提高10%～30%。

The initial crack usually occurs at the point of maximum shear stress, that is, near the midpoint of the long side of the section and about at 45° with the axis of the member. After loading until diagonal cracks appear, a portion of the concrete quits the work, the stress of the steel bars increases significantly, the torsion angle begins to increase significantly, and the torsional stiffness of the member decreases significantly. At this time, the equilibrium state of the force on the section of the member before cracks appear is broken, and the cracked concrete and the reinforcement together form a new force system to resist torque and achieve a new balance. With the increase of torque, the number and width of cracks gradually increase, and roughly continuous spiral cracks at a certain angle with the longitudinal axis of the member are formed on the four surfaces of the member, as shown in Fig. 8-3. The concrete between diagonal cracks is compressed, and the compressive stress is basically along the direction of diagonal cracks, and the longitudinal bars and stirrups intersecting with the diagonal cracks are all tensioned. The cracking torque of the reinforced concrete member section is about 10% to 30% higher than that of the corresponding plain concrete member.

接近极限扭矩时,构件某一长边上的斜裂缝中有一条发展为临界斜裂缝,与这条斜裂缝相交的纵筋或箍筋应力首先达到屈服强度。达到极限扭矩时,临界斜裂缝沿短边发展,与短

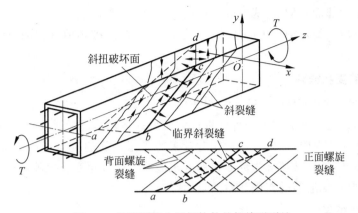

图 8-3 钢筋混凝土纯扭构件的螺旋形裂缝

Figure 8-3 Spiral cracks of reinforced concrete pure torsional members

边上斜裂缝相交的纵筋或箍筋应力也达到屈服强度,斜裂缝宽度不断增加,直到沿空间扭曲破坏面受压边混凝土压碎,构件破坏。这样的破坏称为适筋受扭破坏,属于延性破坏,只有受扭纵筋和受扭箍筋都配置适量时才可能出现。

When approaching the ultimate torque, one of the diagonal cracks on a long side of the member develops into a critical diagonal crack, and the stress of longitudinal reinforcement or stirrup intersecting with this diagonal crack first reaches the yield strength. When the ultimate torque is reached, the critical diagonal crack develops along the short side, and the stress of the longitudinal reinforcement or stirrup intersecting with the diagonal crack on the short side also reaches the yield strength, and the width of the diagonal crack increases continuously until the concrete on the compression side along the space distortion failure surface is crushed, the member is destroyed. Such failure is called underreinforced torsional failure, which belongs to the ductile failure, and can only occur when the torsional longitudinal reinforcement and the torsional stirrup are configured appropriately.

8.1.2 纯扭构件的破坏形态
Destruction forms of pure torsional members

对配置受扭纵筋和受扭箍筋的矩形截面纯扭构件,裂缝产生的过程与素混凝土纯扭构件基本相似,但其破坏形态则与受扭纵筋和受扭箍筋的配筋率有关,主要有适筋破坏、部分超筋破坏、超筋破坏和少筋破坏四种破坏形态。

1. 适筋受扭破坏
   Underreinforced torsional failure

对于受扭纵筋和受扭箍筋都配置适量的受扭构件,混凝土开裂前,受扭纵筋和受扭箍筋的应力很小。随着扭矩增加,构件的三个侧面上逐渐形成许多大致平行的连续螺旋形斜裂缝。在开裂截面,混凝土原来承受的拉应力转移给与裂缝相交的纵筋和箍筋,使纵筋和箍筋应力逐渐增大,当与某条斜裂缝相交的纵筋与箍筋先达到受拉屈服强度时,这条斜裂缝会迅速加宽并向两个相邻面延伸,形成空间扭曲破坏面,最后受压面上的混凝土被压碎,构件破坏。这种破坏类似于受弯构件中的适筋梁,属于延性破坏,破坏前有较明显的预兆,称为适筋受扭破坏。

2. 部分超筋受扭破坏

Partial overreinforced torsional failure

当受扭纵筋或受扭箍筋中的一种配置过多而另一种基本适量时,则在破坏时,配置数量相对较少的受扭钢筋屈服,而配置数量相对较多的受扭钢筋的应力直到受压边混凝土压碎仍达不到屈服强度,这种破坏称为部分超筋受扭破坏。部分超筋受扭破坏有一定的预兆,具有一定的延性,但不如适筋受扭破坏时的延性好。

3. 超筋受扭破坏

Overreinforced torsional failure

当受扭纵筋和受扭箍筋两者都配置过多时,破坏前的螺旋形裂缝多而密,直至构件破坏,这些裂缝的宽度都不大。受扭破坏是由于裂缝间的混凝土压碎而引起的,破坏时受扭纵筋和受扭箍筋应力都达不到屈服强度。这种破坏和受弯构件中的超筋梁类似,破坏前没有明显预兆,属于脆性破坏,称为超筋受扭破坏。

4. 少筋受扭破坏

Lightly reinforced torsional failure

当受扭纵筋和受扭箍筋均配置过少或其中之一配置过少时,在扭矩作用下,先在构件截面的长边最薄弱处产生一条与纵轴成 45°左右的斜裂缝。受扭纵筋与受扭箍筋不足以承担混凝土开裂后转移给钢筋的拉力,不仅会达到屈服强度,而且可能进入强化阶段,其破坏过程类似于受弯构件中的少筋梁,破坏时的扭矩与开裂时的扭矩接近。与素混凝土纯扭构件破坏类似,其破坏过程急速而突然,破坏前没有明显预兆,属于脆性破坏,称为少筋受扭破坏。

适筋受扭构件的塑性变形比较充分,部分超筋受扭构件次之,而超筋受扭构件和素混凝土受扭构件的塑性变形很小。工程中为了保证构件受扭时具有一定的塑性,应避免将受扭构件设计成超筋受扭构件和少筋受扭构件,尽可能设计成适筋受扭构件,部分超筋受扭构件在工程中也是可以采用的,但经济性不好。

8.1.3 矩形截面纯扭构件的开裂扭矩
Cracking torque of pure torsional members with rectangular section

试验研究表明,钢筋混凝土纯扭构件在裂缝出现前,受扭纵筋和受扭箍筋的应力都很小,且钢筋对开裂扭矩的影响很小。因此,研究开裂扭矩时可忽略钢筋的作用,扭矩主要由混凝土承担,素混凝土纯扭构件一裂即坏,开裂扭矩近似等于破坏扭矩(极限扭矩),钢筋混凝土纯扭构件的开裂扭矩可近似按素混凝土纯扭构件的极限扭矩计算。

Experimental studies have shown that the stress of the torsional longitudinal bars and torsional stirrups of reinforced concrete pure torsional members is very small before the cracks appear, and the influence of steel bars on the cracking torque is very small. Therefore, the effect of reinforcement can be ignored when studying the cracking torque. The torque is mainly borne by concrete, and the plain concrete pure torsional members are prone to failure at the first crack, and the cracking torque is approximately equal to the failure torque (ultimate torque). The cracking torque of reinforced concrete pure torsional members can be approximately calculated as the ultimate torque of plain concrete pure torsional members.

矩形截面纯扭构件在扭矩 T 作用下在截面上产生扭剪应力,在与构件轴线成 45°和 135°的方向,相应产生主拉应力 σ_{tp} 与主压应力 σ_{cp}。如果混凝土为理想弹性材料(应力和应变成正比),在弹性阶段,构件截面上的剪应力分布如图 8-4(a)所示,最大扭剪应力 τ_{max} 及最大主应力都发生在长边中点。试验表明,若按弹性应力分布计算素混凝土构件的受扭承载力,会低估其开裂扭矩。因此,通常按理想塑性材料估算素混凝土受扭构件的开裂扭矩。

Under the action of torque T torsional shear stress is generated on the section of the rectangular section pure torsional member. At 45° and 135° with the axis of the member, the corresponding principal tensile stress σ_{tp} and principal compressive stress σ_{cp} are generated. If concrete is an ideal elastic material(the stress proportional to the strain), the shear stress distribution on the section of the member during the elastic stage is shown in Fig. 8-4(a), and maximum torsional shear stress τ_{max} and maximum principal stress both occur at the midpoint of the long side. Experiments have shown that if the torsional capacity of plain concrete members is calculated based on the elastic stress distribution, the cracking torque will be underestimated. Therefore, the cracking torque of plain concrete torsional members is usually estimated based on ideal plastic materials.

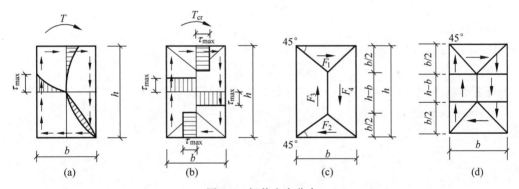

图 8-4 扭剪应力分布
(a) 理想弹性材料;(b) 理想塑性材料;(c) 开裂扭矩计算;(d) 开裂扭矩计算分块

Figure 8-4 Torsional shear stress distribution
(a) Ideal elastic material; (b) Ideal plastic material; (c) Cracking torque calculation;
(d) Cracking torque calculation block

对于按理想塑性材料考虑的素混凝土受扭构件,当截面长边中点的最大扭剪应力值达到 τ_{max} 或者相应的主拉应力 σ_{tp} 达到混凝土抗拉强度 f_t 时,只意味着构件进入塑性状态,并未发生破坏,扭矩还可以继续增加,直到截面边缘混凝土的拉应变达到混凝土极限拉应变,截面上各点的剪应力全部达到最大值,构件才丧失承载力而破坏。此时,截面承受的扭矩称为开裂扭矩 T_{cr}。截面上剪应力分布如图 8-4(b)所示。

For plain concrete torsional members considering as ideal plastic materials, when the maximum torsional shear stress value at the midpoint of the long side of the section reaches τ_{max} or corresponding principal tensile stress σ_{tp} reaches the tensile strength of concrete f_t, it only means that the member enters the plastic state without any damage occurring, and the torque can continue to increase until the tensile strain of concrete at the edge of the section reaches the ultimate tensile strain of concrete, and the shear stress at all points on the section reaches its maximum value, then the member loses its bearing capacity and is damaged. At this time, the torque borne by the section is called the cracking torque T_{cr}. The shear stress distribution on the section is shown in Fig. 8-4(b).

根据塑性力学理论，可将截面上的扭剪应力分布划分成四个部分，如图 8-4(c)所示。为了便于计算，将图 8-4(c)改为图 8-4(d)，计算各部分扭剪应力的合力对截面扭转中心的力矩，其总和为开裂扭矩 T_{cr}，即

According to the theory of plastic mechanics, the distribution of torsional shear stress on the section can be divided into four parts, as shown in Fig. 8-4(c). For the convenience of calculation, Fig. 8-4(c) is changed to Fig. 8-4(d), calculate the moment of the resultant force of each part of the torsional shear stress on the torsion center of the section, and the sum is the cracking torque T_{cr}, that is

$$T_{cr} = \tau_{max}\left\{2\times\frac{b}{2}(h-b)\times\frac{b}{4}+4\times\frac{1}{2}\left(\frac{b}{2}\right)^2\times\frac{2}{3}\times\frac{b}{2}+2\times\frac{1}{2}\times b\times\frac{b}{2}\left[\frac{2}{3}\times\frac{b}{2}+\frac{1}{2}(h-b)\right]\right\}=\frac{b^2}{6}(3h-b)\tau_{max} \quad (8\text{-}1)$$

构件开裂时：

When the member cracks:

$$\sigma_{tp} = \tau_{max} = f_t \quad (8\text{-}2)$$

对矩形截面，令

For rectangular sections, let

$$W_t = \frac{b^2}{6}(3h-b) \quad (8\text{-}3)$$

则

$$T_{cr} = f_t W_t \quad (8\text{-}4)$$

式中，σ_{tp} 为主拉应力，N/mm^2；τ_{max} 为最大扭剪应力，N/mm^2；f_t 为混凝土轴心抗拉强度设计值，N/mm^2；b、h 分别为矩形截面的短边、长边尺寸，mm；T_{cr} 为开裂扭矩，N·mm；W_t 为受扭构件的截面受扭塑性抵抗矩，对矩形截面，取 $\frac{b^2}{6}(3h-b)$，mm^3。

Where, σ_{tp} is the principal tensile stress, N/mm^2; τ_{max} is the maximum torsional shear stress, N/mm^2; f_t is the axial tensile strength design value of concrete, N/mm^2; b、h are the short side and long side size of rectangular section respectively, mm; T_{cr} is the cracking torque, N·mm; W_t is the section torsional plastic resistance moment of torsional member, for rectangular section, taking $\frac{b^2}{6}(3h-b)$, mm^3.

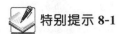 **特别提示 8-1**

为使用方便，开裂扭矩可以近似采用理想塑性材料的计算结果，但混凝土抗拉强度要乘以小于 1 的系数予以修正。试验表明，对高强混凝土，其修正系数约为 0.7；对低强混凝土，其修正系数接近 0.8；对素混凝土纯扭构件，修正系数为 0.87～0.97；对钢筋混凝土纯扭构件，修正系数为 0.86～1.06。《混凝土结构设计标准》(GB/T 50010—2010)偏于安全地取混凝土抗拉强度修正系数为 0.7，以综合反映混凝土塑性发挥的程度和拉压复合应力下混凝土强度降低的影响，所以矩形截面钢筋混凝土纯扭构件的开裂扭矩 T_{cr} 为

For convenience of use, the cracking torque can be approximated using the calculation results of ideal plastic materials, but the concrete tensile strength should be corrected by multiplying the coefficient less than 1. The test shows that the correction coefficient of high strength concrete is about 0.7; for low strength concrete, the correction coefficient is close to 0.8; for plain concrete pure torsional members, the correction coefficient is 0.87 to 0.97; for reinforced concrete pure torsional members, the correction coefficient is 0.86～1.06. In the "Standard for design of concrete structures" (GB/T 50010—2010) it is safe to take correction coefficient of concrete tensile strength as 0.7, so as to comprehensively reflect the development degree of concrete plasticity and the impact of concrete strength reduction under tensile and compressive composite stress, so the cracking torque T_{cr} of reinforced concrete pure torsional members with rectangular section is

$$T_{cr} = 0.7 f_t W_t \tag{8-5}$$

8.1.4 按变角度空间桁架模型的扭曲截面受扭承载力
Twisted section torsional capacity based on variable-angle space truss model

1. 变角度空间桁架模型
Variable-angle space truss model

试验表明，素混凝土受扭构件一旦出现斜裂缝就立即破坏。如果配置适量的受扭纵筋和受扭箍筋，不但其受扭承载力有显著提高，而且构件破坏时具有较好的延性。钢筋混凝土受扭构件扭曲截面受扭承载力主要有两种计算方法：以变角度空间桁架模型和以斜弯理论（扭曲破坏面极限平衡理论）为基础，《混凝土结构设计标准》(GB/T 50010—2010)采用的是前者。

Experiments have shown that the plain concrete torsional members fail immediately once diagonal cracks appear. If an appropriate amount of torsional longitudinal bars and torsional stirrups are configured, not only has the torsional bearing capacity significantly improved, but also the member has good ductility when it is damaged. There are two main calculation methods for the twisted section torsional capacity of reinforced concrete torsional members: based on the variable-angle space truss model and based on the diagonal flexure theory (limit equilibrium theory of twisted failure surface). The former is adopted in "Standard for design of concrete structures"(GB/T 50010—2010).

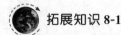 **拓展知识 8-1**

变角度空间桁架模型（如图 8-5 所示）的基本假定如下：

Basic assumptions of variable-angle space truss model (as shown in Fig. 8-5) are as follows：

（1）混凝土只承受压力，具有螺旋形裂缝的混凝土外壳组成桁架的斜压杆，其倾角为 α；

（2）纵筋与箍筋只承受拉力，分别为桁架的弦杆和腹杆；

（3）忽略核心混凝土的受扭作用和钢筋的销栓作用。

(1) The concrete are only subjected to compression, and the concrete in the outer shell with spiral cracks forms the diagonal compression struts of the truss, and its inclination angle is α；

(2) The longitudinal bars and stirrups only bear tensile force, which are the chord and web members of the truss respectively；

(3) Ignoring the torsional action of core concrete and the dowel action of reinforcement.

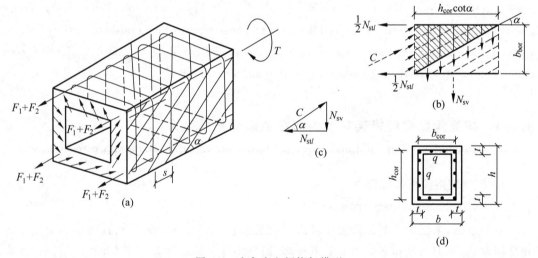

图 8-5　变角度空间桁架模型

Figure 8-5　Variable-angle space truss model

变角度空间桁架模型的基本思路：试验研究表明，当裂缝充分发展，钢筋应力接近屈服强度时，矩形截面受扭构件的截面核心混凝土部分退出工作，实心截面的钢筋混凝土受扭构件可用一个空心的箱形截面构件来代替，它由具有螺旋形裂缝的混凝土箱壁（外壳）、受扭纵筋以及受扭箍筋三者共同组成空间桁架以抵抗扭矩。其中受扭纵筋为空间桁架的弦杆，箍筋为受拉腹杆，被裂缝分割的斜向混凝土为斜压腹杆，它与构件纵轴线的夹角为 α。

The basic idea of the variable-angle space truss model: the experimental studies have shown that when the cracks fully develop and the steel stress is close to the yield strength, part of the core concrete of the section of the rectangular section torsional members exits the work, and the reinforced concrete torsional member with solid section can be replaced by the hollow box section member, which consists of a concrete box wall (outer shell) with spiral cracks, torsional longitudinal bars and torsional stirrups to form a space truss to resist the torque. The torsional longitudinal reinforcement is the chord of the space truss, the stirrup is the tensile web member, and the diagonal concrete divided by cracks is the diagonal compression web member, and its angle with the longitudinal axis of the member is α.

 拓展知识 8-2

2. 受扭构件纵向钢筋与箍筋的配筋强度比 ζ

Reinforcement strength ratio of the longitudinal bars to the stirrups of torsional members ζ

空间桁架模型中混凝土受压斜腹杆与构件纵轴线的夹角 α 并不是一个定值,它与受扭纵筋和受扭箍筋的配置数量有关,在 30°～60°变化,这也是变角度空间桁架模型的由来。

In the space truss model, angle α between concrete compression diagonal web member and longitudinal axis of the member is not a fixed value, it is related to the configuration quantity of torsional longitudinal bars and torsional stirrups, which changes from 30° to 60°. This is also the origin of the variable-angle space truss model.

1) 受扭构件纵向钢筋与箍筋的配筋强度比 ζ 的由来

1) Origin of reinforcement strength ratio of the longitudinal bars to the stirrups of torsional members ζ

从图 8-5(a)所示空间桁架模型的箱形截面中任取一个侧壁(如右侧壁)来分析,如图 8-5(b)所示。取斜裂缝以上部分为隔离体(阴影部分),作用在隔离体上的力有三个:受扭纵筋的拉力 N_{stl}、受扭箍筋的拉力 N_{sv}、混凝土斜压杆压应力的合力 C。N_{stl}、N_{sv} 和 C 构成平衡力系,如图 8-5(c)所示。可见,受扭纵筋和受扭箍筋对变角度空间桁架模型是不可缺少的,缺少任一个都不能形成变角度空间桁架以抵抗扭矩。也就是说,受扭纵筋和受扭箍筋必须同时配置才能起抗扭作用,仅配置其中一种是不能起抵抗扭矩作用的,并且只有这两种钢筋配置数量合适时才能充分发挥各自作用。如图 8-5(d)所示为作用于侧壁的剪力流 q 所引起的桁架内力。

如图 8-5(a)所示,变角度空间桁架模型由 2 榀竖向的变角度平面桁架与 2 榀水平的变角度平面桁架组成,若属于适筋受扭构件,构件破坏时受扭纵筋和受扭箍筋均能达到屈服强度,则受扭纵筋的拉力为

$$N_{stl} = f_y A_{stl} \frac{h_{cor}}{u_{cor}} \tag{8-6}$$

受扭箍筋的拉力为

$$N_{sv} = f_{yv} A_{st1} \frac{h_{cor}\cot\alpha}{s} \tag{8-7}$$

式中,f_y 为受扭纵筋的抗拉强度设计值,N/mm²;f_{yv} 为受扭箍筋的抗拉强度设计值,N/mm²;A_{stl} 为受扭计算中取对称布置的全部纵向普通钢筋截面面积,mm²;A_{st1} 为受扭计算中沿截面周边配置的箍筋单肢截面面积,mm²;s 为受扭箍筋的间距,mm;α 为混凝土

受压斜腹杆与构件纵轴线的夹角；b_{cor}、h_{cor}分别为箍筋内表面范围内截面核心部分的短边、长边尺寸，mm；u_{cor}为截面核心部分的周长，取$u_{cor}=2(b_{cor}+h_{cor})$，mm。

由图 8-5(c)得

$$\cot\alpha = \frac{N_{stl}}{N_{sv}} \tag{8-8}$$

由式(8-6)、式(8-7)得

$$\cot\alpha = \sqrt{\frac{f_y A_{stl} s}{f_{yv} A_{st1} u_{cor}}} \tag{8-9}$$

令

$$\zeta = \frac{f_y A_{stl} s}{f_{yv} A_{st1} u_{cor}} \tag{8-10}$$

则

$$\cot\alpha = \sqrt{\zeta} \tag{8-11}$$

式中，ζ为受扭构件纵向钢筋与箍筋的配筋强度比。

ζ是沿截面核心周长单位长度内受扭纵筋强度与沿构件长度方向单位长度内的单肢受扭箍筋强度之间的比值。

2) 受扭纵筋和受扭箍筋的配置数量必须合适

2) The number of torsional longitudinal reinforcement and torsional stirrup must be appropriate

如果受扭纵筋与受扭箍筋的配筋强度比ζ不同，变角度空间桁架模型中混凝土受压斜腹杆的倾角α也不同。当ζ大时α小，ζ小时α大。当受扭纵筋配置数量过多时，ζ很大，计算所得的α很小，但受扭破坏时实际的α却不会这么小，因为在受扭纵筋达到抗拉屈服强度前，受压的混凝土斜腹杆已经被压坏，构件发生了部分超筋破坏(受扭纵筋超筋)。同样，当受扭箍筋配置数量过多时，ζ很小，计算所得的α很大，但受扭破坏时实际的α并不会那么大，因为在受扭箍筋达到抗拉屈服强度前，受压混凝土斜腹杆已经被压坏，构件发生了部分超筋破坏(受扭箍筋超筋)。因此受扭纵筋和受扭箍筋的配置数量必须合适，否则，破坏时配置过多的那种钢筋的抗拉强度就不能充分发挥出来。

If the reinforcement strength ratio of the torsional longitudinal bars to the torsional stirrups ζ is different, the inclination angle α of the concrete diagonal compression web member in the variable-angle space truss model is also different. When ζ is large, α is small, and when ζ is small, α is large. When the configuration quantity of torsional longitudinal bars is too large, ζ is very large, the calculated α is very small, but the actual α is not so small when subjected to torsional failure, because before the torsional longitudinal bars reach the tensile yield strength, the compression concrete diagonal web member has been crushed, and partially overreinforced failure occurred in the member (torsional longitudinal bars are overreinforced). Similarly, when the configuration quantity of torsional stirrups is too large, ζ is very small, the calculated α is very large, but the actual α is not so large when subjected to torsional failure, because before the torsional stirrups reach the tensile yield strength, the compression concrete diagonal web member has been crushed, and partially overreinforced failure occurred in the member(torsional longitudinal bars are overreinforced). Therefore, the number of torsional longitudinal bars and torsional stirrups must be appropriate, otherwise, during damage the tensile strength of the reinforcement with excessive configuration can not be fully utilized.

3) 受扭构件纵向钢筋与箍筋的配筋强度比 ζ

3) Reinforcement strength ratio of the longitudinal bars to the stirrups of torsional members ζ

当 ζ 不等于 1 时，在受扭纵筋(或箍筋)屈服后产生内力重分布，混凝土受压斜腹杆的倾角 α 也会改变。试验研究表明，如果受压斜腹杆的倾角 α 在 30°和 60°之间，则 $\zeta=3\sim 0.333$。试验表明，构件破坏时，若受扭纵筋和箍筋用量适当，$0.5\leqslant\zeta\leqslant 2.0$，则两种钢筋的应力基本上都能达到屈服强度。为了进一步限制构件在使用荷载作用下的裂缝宽度，通常取受压斜腹杆倾角 α 的限制范围为

When ζ is not equal to 1, redistribution of internal force occurs after yielding of torsional longitudinal bars (or stirrups), and the inclination angle α of the concrete compression diagonal web member will also change. Experimental studies have shown that if the inclination angle α of compression diagonal web member is between 30° and 60°, then $\zeta=3\sim 0.333$. The experiment shows that when the member is destroyed, if the amount of torsional longitudinal bars and stirrups is appropriate, $0.5\leqslant\zeta\leqslant 2.0$, the stress of both types of reinforcement can basically reach the yield strength. In order to further limit the crack width of members under service loads, the limit range for the inclination angle α of compression diagonal web members is usually taken as follows

$$\frac{3}{5}\leqslant \tan\alpha \leqslant \frac{5}{3} \tag{8-12}$$

或

or

$$0.36\leqslant \zeta \leqslant 2.778 \tag{8-13}$$

式中，ζ 为受扭构件纵向钢筋与箍筋的配筋强度比。

Where, ζ is the reinforcement strength ratio of the longitudinal bars to the stirrups of torsional members.

《混凝土结构设计标准》(GB/T 50010—2010)规定 ζ 的取值为 $0.6\leqslant\zeta\leqslant 1.7$，$\zeta$ 不应小于 0.6，当 $\zeta>1.7$ 时，取 $\zeta=1.7$，工程设计时通常取 $\zeta=1.0\sim 1.3$。试验表明，当 ζ 在 1.2 左右时，受扭纵筋与受扭箍筋配合最佳，基本上能同时达到屈服强度，因此设计时取 $\zeta=1.2$ 比较合理。

"Standard for design of concrete structures" (GB/T 50010—2010) stipulates that the value of ζ is $0.6\leqslant\zeta\leqslant 1.7$, ζ should not be less than 0.6, when $\zeta>1.7$, take $\zeta=1.7$, usually take $\zeta=1.0\sim 1.3$ in engineering design. The experiment shows that when ζ is about 1.2 or so, the torsional longitudinal bars and the torsional stirrups fit best, and can basically reach the yield strength at the same time. Therefore, it is reasonable to take $\zeta=1.2$ in design.

3. 按变角度空间桁架模型的矩形截面纯扭构件受扭承载力

Torsional capacity of rectangular section pure torsional members based on variable-angle space truss model

根据变角度空间桁架模型得到的矩形截面纯扭构件的受扭承载力计算公式为

Obtained from the variable-angle space truss model, the formula for calculating the torsional capacity of pure torsional members with rectangular section is

$$T_u = 2A_{cor}\sqrt{\frac{f_y A_{stl} f_{yv} A_{stl}}{u_{cor} \cdot s}} = 2\sqrt{\zeta}\frac{f_{yv} A_{stl}}{s}A_{cor} \qquad (8\text{-}14)$$

式中，ζ 为受扭构件纵向钢筋与箍筋的配筋强度比；A_{cor} 为截面核心部分的面积，$A_{cor} = b_{cor}h_{cor}$，mm^2。

Where: ζ is the reinforcement strength ratio of the torsional longitudinal bars to the stirrups of torsional members; A_{cor} is the area of the core part of the section, $A_{cor} = b_{cor}h_{cor}$, mm^2.

当受扭构件为非对称配筋时，按照较少一侧配筋的对称配筋截面计算。

When the torsional member is equipped with asymmetrical reinforcement, it is calculated based on the symmetrical reinforcement section with fewer reinforcement on one side.

试验表明，混凝土强度等级、截面形状与尺寸、受扭纵筋的数量、屈服强度、配筋形式、箍筋用量、箍筋间距以及受扭纵筋与受扭箍筋的配置比例直接影响受扭构件的破坏形态，决定构件承受扭矩的能力。

Experiments have shown that the concrete strength grade, cross-sectional shapes and dimensions, the number of torsional longitudinal bars, yield strength, the form of reinforcement, amount of stirrups, stirrup spacing, and the configuration ratio of the torsional longitudinal bars to the torsional stirrups directly affect the destruction form of torsional members and determine the ability of members to withstand torque.

8.1.5 按《混凝土结构设计标准》的纯扭构件受扭承载力

Torsional capacity of pure torsional members according to "Code for design of concrete structures"

1. 按变角度空间桁架模型的受扭承载力计算结果与实测值差异的主要原因

The main reasons for the difference between the calculation results and the measured values of the torsional capacity based on the variable-angle space truss model

根据国内对矩形截面受扭构件试验结果的分析，构件的受扭承载力与变角度空间桁架模型的计算结果之间存在较大差异，在配筋率较低时，计算值一般小于实测值；在配筋率较高时，计算值一般大于实测值。主要有以下三个原因：

（1）变角度空间桁架模型忽略核心混凝土的作用，实际上构件开裂后,受扭钢筋对斜裂缝开展有一定的约束作用,使斜裂缝间混凝土的骨料之间存在咬合作用,对抗扭有一定的贡献。同时,斜裂缝只在构件表面一定深度出现,未贯穿整个截面,构件尚未被割裂成可动机构,混凝土仍有一定的抗扭能力。

(2) 按变角度空间桁架模型计算的受扭承载力与混凝土强度无关,未反映构件受扭承载力随混凝土强度提高而增大的规律,大大低估了受扭承载力。试验研究表明,截面尺寸和配筋完全相同的受扭构件,混凝土强度等级对极限扭矩是有影响的,混凝土强度等级高的受扭承载力也较大。

(3) 与斜裂缝相交的钢筋不可能全部达到屈服强度,这和变角度空间桁架模型中认为钢筋均达到屈服强度的假定存在差别。

2. 按《混凝土结构设计标准》(GB/T 50010—2010)的矩形截面纯扭构件受扭承载力
   Torsional capacity of pure torsional members with rectangular section according to "Code for design of concrete structures(GB 50010—2010)"

《混凝土结构设计标准》(GB/T 50010—2010)根据对试验资料的统计分析,基于变角度空间桁架模型,并且考虑可靠度的要求,分别给出了矩形截面、箱形截面、T 形和 I 形截面纯扭构件的受扭承载力经验计算公式。

Based on the statistical analysis of experimental data, and the variable-angle space truss model, and considering the requirements of reliability, "Standard for design of concrete structures" (GB/T 50010—2010) respectively gives empirical formulas for calculating the torsional capacity of rectangular section, box section, T-shaped section and I-shaped section pure torsional members.

纯扭构件的受扭承载力 T_u 由混凝土的受扭承载力 T_c 与受扭纵筋、受扭箍筋的受扭承载力 T_s 组成,即

The torsional capacity T_u of pure torsional members is composed of the torsional capacity T_c of concrete and the torsional capacity T_s of torsional longitudinal bars and torsional stirrups, that is

$$T_u = T_c + T_s \tag{8-15}$$

其中 T_c 可参照式(8-5)写成

Where T_c is written with reference to Eq. (8-5) as

$$T_c = \alpha_1 f_t W_t \tag{8-16}$$

参考根据变角度空间桁架模型得到的纯扭构件的受扭承载力计算公式(8-14),T_s 可写成

With reference to the calculation formula Eq. (8-14) of the torsional capacity of pure torsional members obtained from the variable-angle space truss model, T_s can be written as

$$T_s = \alpha_2 \sqrt{\zeta} \frac{f_{yv} A_{st1}}{s} A_{cor} \tag{8-17}$$

则

So

$$T_u = T_c + T_s = \alpha_1 f_t W_t + \alpha_2 \sqrt{\zeta} \frac{f_{yv} A_{st1}}{s} A_{cor} \quad (8\text{-}18)$$

$$\frac{T_u}{f_t W_t} = \alpha_1 + \alpha_2 \sqrt{\zeta} \frac{f_{yv} A_{st1}}{f_t W_t s} A_{cor} \quad (8\text{-}19)$$

图 8-6 中黑点为配有不同数量受扭钢筋的普通混凝土纯扭构件受扭承载力试验结果，根据对试验结果的统计回归，得式(8-19)中系数 $\alpha_1 = 0.35, \alpha_2 = 1.2$。

The black dots in Fig. 8-6 are the experimental results of the torsional capacity of ordinary concrete pure torsional members with different quantities of torsional bars. According to the statistical regression of the experimental results, the coefficient in Eq. (8-19) is $\alpha_1 = 0.35, \alpha_2 = 1.2$.

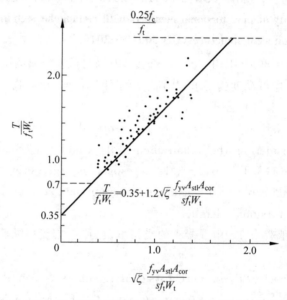

图 8-6 配有不同数量受扭钢筋的普通混凝土纯扭构件受扭承载力试验结果

Figure 8-6 Experimental results of torsional capacity of ordinary concrete pure torsional members with different quantities of torsional reinforcement

式(8-19)是在对试验结果统计回归的基础上，考虑了可靠指标 β 值的要求，由试验点偏下限得出的，如图 8-6 所示。建立公式时，包括了少量部分超筋构件的试验点。

Eq. (8-19) is based on the statistical regression of the experimental results, taking into account the requirements for the reliability index β value, and is obtained from the lower limit of the test points, as shown in Fig. 8-6. When establishing the formula, a small number of test points of partially overreinforced members were included.

由式(8-18)和系数 $\alpha_1 、 \alpha_2$ 的值，矩形截面纯扭构件的受扭承载力 T_u 按式(8-20)计算：

From Eq. (8-18) and the value of coefficient $\alpha_1 、 \alpha_2$, the torsional capacity T_u of pure torsional members with rectangular section is obtained from Eq. (8-20):

$$T_u = T_c + T_s = 0.35 f_t W_t + 1.2 \sqrt{\zeta} \frac{f_{yv} A_{st1}}{s} A_{cor} \quad (8\text{-}20)$$

式中,f_t 为混凝土轴心抗拉强度设计值,N/mm²; W_t 为受扭构件的截面受扭塑性抵抗矩,mm³; ζ 为受扭构件纵筋与箍筋的配筋强度比,要求 $0.6 \leqslant \zeta \leqslant 1.7$; f_{yv} 为受扭箍筋的抗拉强度设计值,取值不应大于 360N/mm², N/mm²; A_{st1} 为受扭计算中取沿截面周边配置的箍筋单肢截面面积,mm²; A_{cor} 为截面核心部分的面积,$A_{cor}=b_{cor}h_{cor}$, mm², 其中 b_{cor} 和 h_{cor} 分别为按箍筋内表面计算的截面核心部分的短边、长边尺寸,mm; s 为受扭箍筋间距,mm。

Where, f_t is the axial tensile strength design value of concrete, N/mm²; W_t is the section torsional plastic resistance moment of torsional member, mm³; ζ is the reinforcement strength ratio of the longitudinal bars to the stirrups of torsional members, requiring $0.6 \leqslant \zeta \leqslant 1.7$; f_{yv} is the design value of tensile strength of the torsional stirrup that should not be greater than 360N/mm², N/mm²; A_{st1} is the cross-sectional area of single-leg of stirrup configured along the perimeter of the section in the torsional calculation, mm²; A_{cor} is the area of the core part of the section, $A_{cor}=b_{cor}h_{cor}$, mm², where b_{cor} and h_{cor} are the dimensions of the short side and the long side of the section core calculated according to the inner surface of the stirrup respectively, mm; s is the torsional stirrup spacing, mm.

式(8-20)中,等式右边的第一项表示开裂后混凝土的受扭承载力 T_c(指斜裂缝间的混凝土,不包括已退出工作的截面核心混凝土),第二项表示受扭纵筋和受扭箍筋的受扭承载力 T_s。

In Eq. (8-20), the first item on the right side of the equation represents the torsional capacity T_c of concrete after cracking (referring to the concrete between diagonal cracks, excluding the core concrete of the section that has been retired from the work), and the second item indicates the torsional capacity T_s of the torsional longitudinal bars and the torsional stirrups.

 拓展知识 8-3

【例题 8-1】 某钢筋混凝土矩形截面纯扭构件,环境类别为一类,结构安全等级为二级,截面尺寸为 $b \times h = 250\text{mm} \times 500\text{mm}$,承受的扭矩设计值 $T=35\text{kN} \cdot \text{m}$。混凝土强度等级为 C30,纵筋和箍筋均采用 HRB400 钢筋,计算纯扭构件的配筋。

【Example 8-1】 Reinforced concrete pure torsional member with rectangular section, environmental category is Class I, structural safety level is Level 2, cross-sectional dimensions are $b \times h = 250\text{mm} \times 500\text{mm}$, and the design value of the torque is $T=35\text{kN} \cdot \text{m}$. The concrete strength grade is C30, HRB400 reinforcement is used for longitudinal bars and stirrups, and calculate the reinforcement of the pure torsional member.

解:(1) 确定计算所用基本参数

C30 混凝土,HRB400 钢筋,查附表 1-3、附表 1-4、附表 1-6 可知,$f_c = 14.3\text{N/mm}^2$, $f_t = 1.43\text{N/mm}^2$, $f_y = f_{yv} = 360\text{N/mm}^2$。

环境类别为一类，查附表 3-2 混凝土保护层厚度 $c=20\text{mm}$，选用双肢箍筋，直径 $d_v=8\text{mm}$，则

$$b_{\text{cor}}=b-2c-2d_v=(250-2\times 20-2\times 8)\text{mm}=194\text{mm}$$

$$h_{\text{cor}}=h-2c-2d_v=(500-2\times 20-2\times 8)\text{mm}=444\text{mm}$$

$$A_{\text{cor}}=194\times 444\text{mm}^2=86136\text{mm}^2,\ u_{\text{cor}}=(2\times 194+2\times 444)\text{mm}=1276\text{mm}$$

（2）验算截面限制条件和构造配筋条件

截面受扭塑性抵抗矩

$$W_t=\frac{b^2}{6}(3h-b)=\frac{250^2}{6}\times(3\times 500-250)\text{mm}^3=13.02\times 10^6\text{mm}^3$$

$a_s=40\text{mm}$, $h_w/b=h_0/b=(500-40)/250=1.84<4$

$$\frac{T}{0.8W_t}=\frac{35\times 10^6}{0.8\times 13.02\times 10^6}\text{N}/\text{mm}^2$$
$$=3.36\text{N}/\text{mm}^2<0.25\beta_c f_c=0.25\times 1.0\times 14.3\text{N}/\text{mm}^2=3.575\text{N}/\text{mm}^2$$

$$\frac{T}{W_t}=\frac{35\times 10^6}{13.02\times 10^6}\text{N}/\text{mm}^2=2.69\text{N}/\text{mm}^2>0.7f_t=0.7\times 1.43\text{N}/\text{mm}^2=1.001\text{N}/\text{mm}^2$$

应按计算配筋。

（3）计算箍筋

取受扭构件纵筋与箍筋的配筋强度比 $\zeta=1.2$，由式(8-20)，取 $T=T_u$，即

$$\frac{A_{st1}}{s}=\frac{T-0.35f_tW_t}{1.2\sqrt{\zeta}f_{yv}A_{\text{cor}}}=\frac{35\times 10^6-0.35\times 1.43\times 13.02\times 10^6}{1.2\sqrt{1.2}\times 360\times 86136}\text{mm}^2/\text{mm}$$
$$=0.699\text{mm}^2/\text{mm}$$

验算箍筋的配筋率

$$\rho_{sv}=\frac{nA_{st1}}{bs}=\frac{2\times 0.699}{250}=0.559\%>\rho_{sv,\min}=0.28\frac{f_t}{f_{yv}}=0.28\times\frac{1.43}{360}=0.11\%$$

所以，箍筋的配筋率满足要求。

（4）计算纵筋

由式(8-10)得

$$A_{stl}=\frac{\zeta f_{yv}A_{st1}u_{\text{cor}}}{f_ys}=\frac{1.2\times 360\times 0.699\times 1276}{360}\text{mm}^2=1070\text{mm}^2$$

$$\rho_{tl}=\frac{A_{stl}}{bh}=\frac{1070}{250\times 500}=0.856\%>\rho_{tl,\min}=0.6\sqrt{\frac{T}{Vb}}\cdot\frac{f_t}{f_y}$$
$$=0.6\times\sqrt{2}\times\frac{f_t}{f_y}=\frac{0.85\times 1.43}{360}=0.337\%$$

上式中 $\frac{T}{Vb}$ 的取值见 8.3.2 受扭钢筋的构造要求。

所以，纵筋的配筋率满足要求。

（5）选配钢筋

受扭箍筋选双肢Φ8，单肢受扭箍筋截面面积 $A_{st1}=50.3\text{mm}^2$，箍筋间距 $s=50.3/0.699\text{mm}=72\text{mm}$，取 $s=70\text{mm}$；考虑受扭纵筋沿截面周边均匀布置，选 6Φ16，$A_{stl}=1206\text{mm}^2$。

8.2 复合受扭构件截面承载力计算
Calculation of section bearing capacity of composite torsional members

实际工程中纯扭构件很少,基本上是弯矩、剪力和扭矩同时作用(如梁),或者是弯矩、剪力、轴力和扭矩同时作用(如柱、墙)的复合受扭构件,构件处于复合受力状态。试验表明,复合受扭构件的受扭承载力与受弯、受剪承载力是相互影响的,即构件的受扭承载力随同时作用的弯矩、剪力的大小而变化,构件的受弯、受剪承载力也随同时作用的扭矩大小而变化,这种相互影响称为构件承载力之间的相关性。这种相互影响非常复杂,要完全考虑这种相关性并用统一的方程表达非常困难。《混凝土结构设计标准》(GB/T 50010—2010)对复合受扭构件的承载力计算采用部分相关、部分叠加的计算方法,即对混凝土的抗力部分考虑相关性,对钢筋的抗力部分采用叠加的方法。

There are few pure torsional members in practical engineering, basically they are composite torsional members under the simultaneous action of bending moment, shear force and torque (such as beam), or the simultaneous action of bending moment, shear force, axial force (such as column and wall). The member is in a composite stress state. Experiments have shown that the torsional bearing capacity is influenced by their bending and shear bearing capacity of composite torsional members, that is, the torsional capacity of the member changes with the magnitude of bending moment and shear force applied at the same time. The flexural and shear bearing capacity of the member also changes with the magnitude of torque applied at the same time. This mutual influence is called the correlation between the bearing capacity of members. This mutual influence is very complex. It is very difficult to fully consider this correlation and express it with a unified equation. "Standard for design of concrete structures" (GB/T 50010—2010) adopts the calculation method of partial correlation and partial superposition for the bearing capacity calculation of composite torsional members, that is, the correlation is considered for the resistance of concrete, and the superposition method is adopted for the resistance of steel bars.

8.2.1 剪扭构件承载力计算
Calculation of bearing capacity of shear torsional members

试验表明:在剪力和扭矩共同作用下,混凝土的抗剪能力和抗扭能力分别降低,随着扭矩的增大,构件的受剪承载力逐渐降低;同时随着剪力的增大,构件的受扭承载力逐渐降低,这种现象称为剪力和扭矩的相关性。

Experiments have shown that under the combined action of shear force and torque, the shear capacity and torsion resistance of concrete decrease respectively. With the increase of torque, the shear bearing capacity of the member gradually decreases; at the same time, with the increase of shear force, the torsional capacity of the member decreases gradually, this phenomenon is called the correlation between shear force and torque.

1. 矩形截面无腹筋构件剪扭承载力相关性
Correlation of shear torsional capacity of rectangular section members without web reinforcement

无腹筋构件在不同扭矩与剪力比值下的承载力试验结果如图8-7(a)所示。图8-7中无量纲坐标系的横坐标为V_c/V_{c0}，纵坐标为T_c/T_{c0}，其中V_{c0}和T_{c0}分别为无腹筋构件在单纯受剪力或扭矩作用时的混凝土受剪承载力和受扭承载力，V_c和T_c分别为同时受剪力和扭矩作用时混凝土的受剪承载力和受扭承载力。

The bearing capacity experimental results of members without web reinforcement under different torque and shear force ratio are shown in Fig. 8-7(a). The horizontal ordinate of the nondimensional coordinate system in Fig. 8-7 is V_c/V_{c0}, and the vertical ordinate is T_c/T_{c0}, where V_{c0} and T_{c0} are respectively the shear bearing capacity and torsional bearing capacity of concrete members without web reinforcement under the action of pure shear or pure torque, and V_c and T_c are respectively the shear capacity and torsional capacity of concrete under the action of both shear force and torque.

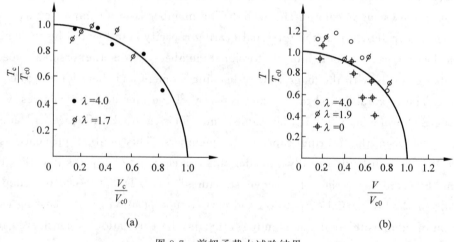

图 8-7 剪扭承载力试验结果
(a) 无腹筋构件；(b) 有腹筋构件

Figure 8-7 Shear torsional capacity experimental results
(a) Members without web reinforcement; (b) Members with web reinforcement

坐标系中的V_{c0}和T_{c0}分别取受弯构件受剪承载力公式中的混凝土作用项和纯扭构件受扭承载力公式中的混凝土作用项，即

V_{c0} and T_{c0} in the coordinate system take respectively the concrete action term in the shear bearing capacity formula of bending members and the concrete action term in the torsional capacity formula of pure torsional members, namely

$$V_{c0} = 0.7 f_t b h_0 \tag{8-21}$$

$$T_{c0} = 0.35 f_t W_t \tag{8-22}$$

式中，V_{c0}为剪扭构件在单纯受剪力作用(纯剪)时的混凝土受剪承载力，N；T_{c0}为剪扭构

件在单纯受扭矩作用(纯扭)时的混凝土受扭承载力,N·mm。

Where, V_{c0} is the shear bearing capacity of concrete in shear torsional member under pure shear force action (pure shear), N; T_{c0} is the torsional capacity of concrete in shear torsional member under pure torsion action (pure torsion), N·mm.

无腹筋构件的受剪和受扭承载力相关关系大致按 1/4 圆弧规律变化,如图 8-8 所示 AD 曲线,即随着同时作用的扭矩增大,构件的受剪承载力逐渐降低,当扭矩达到构件的受扭承载力时(即 $T_c = T_{c0}$, $T_c/T_{c0} = 1$),其受剪承载力下降为零($V_c = 0$),反之亦然。

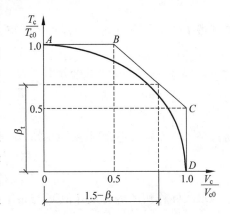

The correlation between shear and torsional capacity of members without web reinforcement changes roughly according to the rule of 1/4 arc, the AD curve as shown in Fig. 8-8, that is, with the increase of torque acting at the same time, the shear bearing capacity of members gradually decreases. When the torque reaches the torsional capacity of members ($T_c = T_{c0}$, $T_c/T_{c0} = 1$), the shear bearing capacity decreases to zero ($V_c = 0$), and vice versa.

图 8-8 混凝土剪扭承载力相关关系

Figure 8-8 Correlation between shear and torsional bearing capacity of concrete

2. 混凝土受扭承载力降低系数 β_t
Torsional capacity reduction coefficient of concrete β_t

配有箍筋的矩形截面有腹筋剪扭构件的试验结果如图 8-7(b)所示,混凝土部分所提供的受扭承载力 T_c 和受剪承载力 V_c 之间的剪扭承载力相关曲线通常也可近似以 1/4 圆弧曲线表示(如图 8-8 所示)。为简化计算,《混凝土结构设计标准》(GB/T 50010—2010)建议用图 8-8 所示的三段折线关系(直线段 AB、BC、CD)近似代替 1/4 圆弧关系。此三段折线表明:

CD 段:

$$\frac{T_c}{T_{c0}} \leqslant 0.5 \text{ 时}, \quad \frac{V_c}{V_{c0}} = 1.0 \tag{8-23}$$

BC 段:

$$\frac{T_c}{T_{c0}}, \frac{V_c}{V_{c0}} > 0.5 \text{ 时}, \quad \frac{T_c}{T_{c0}} + \frac{V_c}{V_{c0}} = 1.5 \tag{8-24}$$

式中,V_c 为同时受剪力和扭矩作用(剪扭共同作用)时混凝土的受剪承载力,N; T_c 为同时受剪力和扭矩作用(剪扭共同作用)时混凝土的受扭承载力,N·mm。

(1) 直线段 AB 表示当 $V_c/V_{c0} \leqslant 0.5$ 时,取 $T_c/T_{c0} = 1.0$; 或者当 $V_c \leqslant 0.5 V_{c0} = 0.35 f_t b h_0$ 时,取 $T_c = T_{c0} = 0.35 f_t b h_0$,即此时可忽略剪力的影响,混凝土的受扭承载力不予降低,仅按纯扭构件的受扭承载力公式进行计算。

(2) 直线段 CD 表示当 $T_c/T_{c0} \leqslant 0.5$ 时,取 $V_c/V_{c0} = 1.0$; 或者当 $T_c \leqslant 0.5 T_{c0} = 0.175 f_t W_t$ 时,取 $V_c = V_{c0} = 0.7 f_t b h_0$,即此时可忽略扭矩的影响,混凝土的受剪承载力不

予降低,仅按受弯构件的斜截面受剪承载力公式进行计算。

(3) 直线段 BC 表示当 $0.5 < T_c/T_{c0} \leq 1.0$ 或 $0.5 < V_c/V_{c0} \leq 1.0$ 时,要考虑剪扭相关性,混凝土的受剪及受扭承载力均予以降低,但以线性相关(直线段 BC)代替 1/4 圆弧相关。将斜线段 BC 上任意点到横坐标轴的距离用 β_t 表示,即

$$T_c/T_{c0} = \beta_t \tag{8-25}$$

斜线段 BC 上任意点到纵坐标轴的距离为

$$V_c/V_{c0} = 1.5 - \beta_t \tag{8-26}$$

式(8-25)、式(8-26)也可分别写为

$$T_c = \beta_t T_{c0} \tag{8-27}$$

$$V_c = (1.5 - \beta_t) V_{c0} \tag{8-28}$$

将式(8-26)等号两边分别除式(8-25)等号两边,即

$$\frac{V_c/V_{c0}}{T_c/T_{c0}} = \frac{1.5 - \beta_t}{\beta_t} \tag{8-29}$$

近似取实际作用的剪力设计值 V 与扭矩设计值 T 之比 $V/T = V_c/T_c$,由式(8-29)得

$$\beta_t = \frac{1.5}{1 + \dfrac{V_c/V_{c0}}{T_c/T_{c0}}} = \frac{1.5}{1 + \dfrac{V_c/T_c}{V_{c0}/T_{c0}}} = \frac{1.5}{1 + \dfrac{V}{T} \cdot \dfrac{T_{c0}}{V_{c0}}} \tag{8-30}$$

将式(8-21)、式(8-22)代入式(8-30),得

$$\beta_t = \frac{1.5}{1 + \dfrac{V}{T} \cdot \dfrac{0.35 f_t W_t}{0.7 f_t b h_0}} \tag{8-31}$$

简化后得

$$\beta_t = \frac{1.5}{1 + 0.5 \dfrac{V}{T} \cdot \dfrac{W_t}{b h_0}} \tag{8-32}$$

式中,β_t 为一般剪扭构件混凝土受扭承载力降低系数,当 β_t 小于 0.5 时,取 0.5;当 β_t 大于 1.0 时,取 1.0。

对集中荷载作用下的独立剪扭构件(包括作用有多种荷载且其集中荷载对支座截面所产生的剪力值占总剪力值的 75% 以上的情况),取 $V_{c0} = [1.75/(\lambda+1)] f_t b h_0$,由式(8-22)、式(8-30)得混凝土受扭承载力降低系数

$$\beta_t = \frac{1.5}{1 + 0.2(\lambda+1) \dfrac{V}{T} \cdot \dfrac{W_t}{b h_0}} \tag{8-33}$$

式中,β_t 为集中荷载作用下剪扭构件混凝土受扭承载力降低系数;λ 为计算截面的剪跨比,其取值范围为 $1.5 \leq \lambda \leq 3$。

由图 8-8 可见,对斜线段 BC,$0.5 \leq \beta_t \leq 1.0$;当 $\beta_t < 0.5$ 时,取 $\beta_t = 0.5$,可不考虑扭矩对混凝土受剪承载力的影响;当 $\beta_t > 1.0$ 时,取 $\beta_t = 1.0$,可不考虑剪力对混凝土受扭承载力的影响。

虽然,按式(8-32)或式(8-33)计算的混凝土受扭承载力降低系数 β_t 比按 1/4 圆弧(如图 8-8 所示)的计算值稍大,但采用此 β_t 值后,构件的剪扭承载力相关曲线与 1/4 圆弧比较

接近。

3. 矩形截面剪扭构件承载力计算

Calculation of bearing capacity of shear torsional members with rectangular section

对于矩形截面剪扭构件,《混凝土结构设计标准》(GB/T 50010—2010)采用的受剪和受扭承载力设计表达式是以有腹筋构件的剪扭承载力相关关系为 1/4 圆弧作为校正线,采用混凝土部分相关,钢筋部分不相关的近似拟合公式。

For shear torsional members with rectangular section, the design expressions of shear and torsional capacity adopted in the "Standard for design of concrete structures" (GB/T 50010—2010) are based on using the correlation between the shear capacity and torsional capacity of members with web reinforcement as a 1/4 arc as the correction line. The approximate fitting formula of partial correlation of concrete and partial non-correlation of reinforcement is adopted.

当需要考虑剪力和扭矩的相关性时,对构件的受剪承载力公式和受扭承载力公式分别按下述规定予以修正:按式(8-34)对受剪承载力公式中的混凝土作用项乘以$(1.5-\beta_t)$,按式(8-35)对受扭承载力公式中的混凝土作用项乘以β_t。

When it is necessary to consider the correlation between shear force and torque, the formula of shear capacity and torsional capacity of the member should be modified according to the following provisions: according to Eq. (8-34) multiply the concrete action term in the shear capacity formula by $(1.5-\beta_t)$, and according to Eq. (8-35) multiply the concrete action term in the torsional capacity formula by β_t.

1) 矩形截面一般剪扭构件的受剪和受扭承载力设计表达式

1) Design expressions of shear and torsional capacity of general shear torsional members with rectangular section

受剪承载力

Shear capacity

$$V \leqslant V_u = 0.7(1.5-\beta_t)f_t b h_0 + f_{yv}\frac{A_{sv}}{s}h_0 \tag{8-34}$$

受扭承载力

Torsional capacity

$$T \leqslant T_u = 0.35\beta_t f_t W_t + 1.2\sqrt{\zeta}\frac{f_{yv}A_{stl}A_{cor}}{s} \tag{8-35}$$

2) 集中荷载作用下独立剪扭构件的受剪和受扭承载力设计表达式

2) Design expressions of shear and torsional capacity of independent shear torsional members under concentrated load

集中荷载作用下独立剪扭构件的受扭承载力仍按式(8-35)计算,但受剪承载力应按式(8-36)计算:

The torsional capacity of independent shear torsional members under concentrated load is still obtained from Eq. (8-35), but the shear bearing capacity is obtained from Eq. (8-36)

$$V \leqslant (1.5-\beta_t)\frac{1.75}{\lambda+1}f_t b h_0 + f_{yv}\frac{A_{sv}}{s}h_0 \qquad (8\text{-}36)$$

式中,β_t 为集中荷载作用下剪扭构件混凝土受扭承载力降低系数,按式(8-33)计算;当 β_t 小于 0.5 时,取 0.5;当 β_t 大于 1.0 时,取 1.0。

Where, β_t is the torsional capacity reduction coefficient of shear torsional members under concentrated load, calculated according to Eq. (8-33); when β_t is less than 0.5, take 0.5; when β_t is greater than 1.0, take 1.0.

8.2.2 弯扭构件承载力计算
Calculation of bearing capacity of flexural torsional members

与剪扭构件相似,在弯矩和扭矩共同作用下的弯扭构件的受力比较复杂,弯扭承载力也存在相关性,设计过程将非常复杂。为了简化计算,在试验基础上,《混凝土结构设计标准》(GB/T 50010—2010)建议采用叠加法计算弯扭构件的承载力,即分别按纯弯矩 M 和纯扭矩 T 计算所需的纵筋和箍筋,然后将相应的钢筋截面面积进行叠加,即弯扭构件的纵筋用量为受弯所需的纵筋截面面积(A_s、A'_s)和受扭所需的纵筋截面面积 A_{stl} 之和,而箍筋用量则由抗扭所需的受扭箍筋决定。

Similar to the shear torsional members, the flexural torsional members under the combined action of bending moment and torque is more complex, and the flexural torsional capacity is also relevant, so the design process will be very complex. For simplified calculation, on the basis of tests, "Standard for design of concrete structures" (GB/T 50010—2010) recommends that the bearing capacity of flexural torsional members be calculated by the superposition method, that is, the required longitudinal bars and stirrups are calculated according to the pure bending moment M and pure torque T respectively, and then the corresponding cross-sectional area of reinforcement is superposed, that is, the amount of longitudinal bars for flexural torsional members is the sum of the cross-sectional area of longitudinal bars required for bending (A_s, A'_s) and the cross-sectional area of longitudinal bars required for torsion A_{stl}, while the amount of stirrups is determined by the torsional stirrups required for torsion resistance.

8.2.3 弯剪扭构件承载力计算
Calculation of bearing capacity of bending-shear-torsion members

1. 弯剪扭构件的破坏形态
Destruction forms of bending-shear-torsion members

钢筋混凝土构件在弯矩、剪力和扭矩共同作用下,其受力状态及破坏形态十分复杂,构件的破坏形态和承载力不仅与扭弯比 T/M 和扭剪比 $T/(Vb)$ 有关,还与构件截面形状、截面尺寸、配筋形式及数量和材料强度等因素有关。试验表明,受扭构件随弯矩、剪力和扭矩的比值及配筋不同,有弯型破坏、扭型破坏和剪扭型破坏三种破坏形态,如图 8-9 所示。

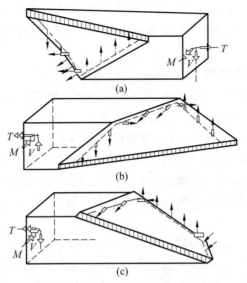

图 8-9 弯剪扭共同作用下构件破坏形态
(a) 弯型破坏；(b) 扭型破坏；(c) 剪扭型破坏

Figure 8-9　Destruction forms of members under combined action of bending, shear and torsion
(a) Bending failure；(b) Torsional failure；(c) Shear torsional failure

1) 弯型破坏

1) Bending failure

试验表明,弯剪扭共同作用下,在配筋适量的情况下,当弯矩较大、扭矩较小(即扭弯比较小)时,裂缝首先在弯曲受拉底面出现,然后发展至两侧面。三个面上的螺旋形裂缝形成一个扭曲破坏面,而第四面即弯曲受压顶面无裂缝。截面下部弯拉区的受拉纵筋首先开始屈服,构件破坏时与螺旋形裂缝相交的纵筋及箍筋均受拉且达到屈服强度。构件顶部受压,扭矩产生的拉应力减少了截面上部弯压区的钢筋压应力,如图 8-9(a)所示,其破坏形态通常称为弯型破坏。

2) 扭型破坏

2) Torsional failure

构件在弯剪扭共同作用下,当扭矩较大、弯矩较小(即扭弯比较大),而构件顶部纵筋少于底部纵筋时,虽然由于弯矩作用使顶部纵筋受压,但因为弯矩较小,则顶部纵筋的压应力也较小。由于顶部纵筋少于底部纵筋,所以扭矩产生的拉应力就有可能抵消弯矩产生的压应力,并使顶部纵筋先达到屈服强度,最后构件底部受压而破坏,如图 8-9(b)所示,其破坏形态通常称为扭型破坏。

3) 剪扭型破坏

3) Shear torsional failure

若剪力和扭矩起控制作用,裂缝首先在侧面出现(在这个侧面上,剪力和扭矩产生的主应力方向是相同的),然后向顶面及底面扩展,这三个面上的螺旋形裂缝构成扭曲破坏面,破坏时与螺旋形裂缝相交的受扭纵筋和受扭箍筋受拉且达到屈服强度,而受压区则靠近另一侧面,在这个侧面上,剪力与扭矩产生的主应力方向是相反的。

没有扭矩作用的受弯构件的斜截面会发生剪压破坏。对于弯剪扭共同作用下的构件，试验表明，当剪力较大、扭矩较小（即扭剪比较小）时，可能发生类似于剪压破坏的剪切破坏，如图 8-9(c)所示。

2. 弯剪扭构件承载力计算
Calculation of bearing capacity of bending-shear-torsion members

弯、剪、扭共同作用下复合受力构件的相关关系比较复杂，按变角度空间桁架模型或斜弯理论进行承载力计算十分烦琐，目前研究得不够深入。《混凝土结构设计标准》(GB/T 50010—2010)根据大量试验研究和变角度空间桁架模型分析，以剪扭和弯扭构件承载力计算方法为基础，建立了弯剪扭构件承载力的计算方法。弯剪扭构件承载力计算可采用叠加法，即将分别按受弯和受扭计算的纵筋截面面积相叠加，分别按受剪和受扭计算的箍筋截面面积相叠加。

Under the combined action of bending, shear and torsion, the correlation of composite load-bearing members is relatively complex. It is very complex to calculate the bearing capacity based on the variable-angle space truss model or diagonal bending theory, and the current research is not deep enough. "Standard for design of concrete structures" (GB/T 50010—2010) based on a very large number of experimental research and analysis of variable-angle space truss model establishes the calculation method for the bearing capacity of bending-shear-torsion member based on the bearing capacity calculation methods of shear torsional and flexural torsional members. The bearing capacity of bending shear torsion members can be calculated by superposition method, that is, the cross-sectional area of longitudinal bars calculated by bending and torsion is superposed, and the cross-sectional area of stirrups calculated by shear and torsion is superposed.

《混凝土结构设计标准》(GB/T 50010—2010)6.4.13 条规定：矩形、T 形、I 形和箱形截面弯剪扭构件，其纵向钢筋截面面积应分别按受弯构件的正截面受弯承载力和剪扭构件的受扭承载力计算确定，并应配置在相应的位置。箍筋截面面积应分别按剪扭构件的受剪承载力和受扭承载力计算确定，并应配置在相应的位置。

Article 6.4.13 of "Standard for design of concrete structures" (GB/T 50010—2010) stipulates that for rectangular, T-shaped, I-shaped and box section bending-shear-torsion members, the cross-sectional area of longitudinal bars should be calculated and determined according to the normal section flexural capacity of flexural members and the torsional capacity of shear torsional members respectively, and should be configured in corresponding positions. The cross-sectional area of stirrups should be determined according to the calculation of shear bearing capacity and torsional capacity of shear torsional members respectively, and should be arranged at corresponding positions.

矩形截面弯剪扭构件的承载力计算可按以下步骤进行：

The bearing capacity of bending-shear-torsion members with rectangular section can be calculated as follows:

1) 绘制有关图形，初选截面尺寸和材料强度等级

绘制构件的设计弯矩图（M 图）、设计剪力图（V 图）和设计扭矩图（T 图），初步选定截

面尺寸和材料强度等级。验算截面尺寸限制条件，若不满足要求应加大截面尺寸或提高混凝土强度等级。

1) Draw the relevant diagrams, primarily select the cross-sectional dimensions and material strength grade

Draw the design moment diagram (M diagram), design shear diagram (V diagram) and design torque diagram (T diagram) of the member, and tentatively select the cross-sectional dimensions and material strength grade. Check the constraint conditions of section size, if the requirements are not met, the section size should be increased or the concrete strength grade should be improved.

2) 验算是否应按计算配置剪扭钢筋

2) Check whether shear-torsion reinforcement should be configured according to the calculation

当满足式(8-37)的要求时，可不进行剪扭承载力计算，但为了防止构件开裂后产生突然的脆性破坏，必须按构造要求配置剪扭所需的纵筋和箍筋，而且受弯所需的纵筋还是应按计算配置。

When the requirements of Eq. (8-37) are met, the shear torsional capacity calculation may not be carried out. However, in order to prevent sudden brittle failure after the cracking of the member, the longitudinal bars and stirrups required for shear and torsion must be configured according to the structural requirements, and the longitudinal bars required for bending should still be configured according to the calculation.

$$\frac{V}{bh_0} + \frac{T}{W_t} \leqslant 0.7f_t \tag{8-37}$$

3) 验算配筋计算是否可以忽略剪力 V 或扭矩 T 的作用

3) Check whether the effect of shear force V or torque T can be ignored in the calculation of reinforcement

《混凝土结构设计标准》(GB/T 50010—2010)规定，在弯矩、剪力和扭矩共同作用下的矩形、T形、I形和箱形截面的弯剪扭构件，当符合以下条件时，可按下列规定进行承载力计算：

"Standard for design of concrete structures" (GB/T 50010—2010) stipulates that the bearing capacity of bending-shear-torsion members of rectangular, T-shaped, I-shaped and box sections under the combined action of bending moment, shear force and torque can be calculated according to the following provisions when meeting the following conditions：

当 $V \leqslant 0.35f_t bh_0$ 或 $V \leqslant 0.875f_t bh_0/(\lambda+1)$ 时，可以忽略剪力的作用，仅按受弯构件的正截面受弯承载力和纯扭构件的扭曲截面受扭承载力分别进行计算，叠加后配置。

When $V \leqslant 0.35f_t bh_0$ or $V \leqslant 0.875f_t bh_0/(\lambda+1)$, the effect of shear force can be ignored, and the normal section flexural capacity of the flexural member and the torsional capacity of the torsional section of the pure torsional member are calculated respectively, and the configuration can be made after superposition.

当 $T \leqslant 0.175f_t W_t$ 或对于箱形截面构件 $T \leqslant 0.175\alpha_h f_t W_t$ (α_h 为箱形截面壁厚影响系

数)时,可以忽略扭矩的作用,不进行受扭承载力计算,仅按受弯构件的正截面受弯承载力和斜截面受剪承载力分别进行计算,叠加后配置。

When $T \leqslant 0.175 f_t W_t$ or for box-section members $T \leqslant 0.175 \alpha_h f_t W_t$ (α_h is the influence coefficient of the wall thickness of the box section), the effect of torque can be ignored, and the torsional capacity is not calculated, only the flexural capacity of the normal section and the shear capacity of the inclined section of the flexural members are calculated respectively and the configuration can be made after superposition.

4)确定箍筋用量

4) Determine the amount of stirrups

(1) 选定受扭构件纵筋与箍筋的配筋强度比 ζ,要求 $0.6 \leqslant \zeta \leqslant 1.7$,一般可取 1.2 左右。

(1) Select the reinforcement strength ratio of the longitudinal bars to the stirrups of torsional members ζ, which requires $0.6 \leqslant \zeta \leqslant 1.7$, generally about 1.2.

(2) 按式(8-32)或式(8-33)计算混凝土受扭承载力降低系数 β_t。

(2) According to Eq. (8-32) or Eq. (8-33), calculate the torsional capacity reduction coefficient β_t of concrete.

(3) 计算受剪所需的箍筋单肢用量 A_{sv1}/s_v(如图 8-10(a)所示):将 ζ、β_t 及其他参数代入剪扭构件的受剪承载力计算公式(8-34)或式(8-36),若采用双肢箍筋,箍筋肢数 $n=2$,A_{sv1}/s_v 即为需要量 A_{sv}/s_v 的一半;若采用四肢箍筋,$n=4$,A_{sv1}/s_v 即为需要量 A_{sv}/s_v 的 $1/4$。其中 A_{sv} 为受剪承载力所需的箍筋截面面积,$A_{sv}=nA_{sv1}$。

(3) Calculate the amount of stirrups per limb A_{sv1}/s_v required for shear(as shown in Fig. 8-10(a)): ζ、β_t and other parameters are substituted into the shear bearing capacity calculation formula Eq. (8-34) or Eq. (8-36) of shear torsional members. If two-leg stirrup is adopted with the number of stirrup legs $n=2$, A_{sv1}/s_v is one-half the required A_{sv}/s_v; if four-leg stirrup is adopted with $n=4$, A_{sv1}/s_v is $1/4$ of the required A_{sv}/s_v, and A_{sv} is the cross-sectional area of stirrups required for shear bearing capacity, $A_{sv}=nA_{sv1}$.

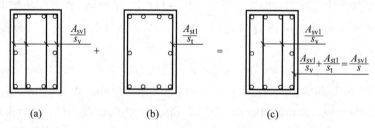

图 8-10 弯剪扭构件的箍筋配置

(a) 受剪所需的受剪箍筋;(b) 受扭所需的受扭箍筋;(c) 剪扭箍筋叠加后

Figure 8-10 Stirrup configuration of bending-shear-torsion members

(a) Shear stirrups required for shear; (b) Torsional stirrups required for torsion;

(c) After superposition of shear-torsion stirrups

(4) 计算受扭所需的箍筋单肢用量 A_{st1}/s_t(如图 8-10(b)所示):将 ζ、β_t 及其他参数代入剪扭构件的受扭承载力计算公式(8-35)。

(4) Calculate the amount of stirrups per limb A_{st1}/s_t required for torsion(as shown in

Fig. 8-10(b)：ζ、β_t and other parameters are substituted into Eq. (8-35) for calculating the torsional capacity of shear torsional members.

(5) 将(3)、(4)计算所得的箍筋用量叠加，得箍筋单肢总用量 $A_{sv1}/s = A_{sv1}/s_v + A_{st1}/s_t$，如图 8-10(c)所示。

(5) The amount of stirrups calculated in (3) and (4) is superimposed to get the total amount of stirrups per limb $A_{sv1}/s = A_{sv1}/s_v + A_{st1}/s_t$, as shown in Fig. 8-10(c).

(6) 按箍筋单肢总用量 A_{sv1}/s 选用箍筋直径和间距，并符合构造要求，验算最小配箍率。

(6) The diameter and spacing of stirrups should be selected according to the total amount of stirrups per limb A_{sv1}/s, and should meet the structural requirements, check the minimum stirrup ratio.

 特别提示 8-2

5) 计算纵筋用量

5) Calculate the amount of longitudinal bars

(1) 计算受弯纵筋

(1) Calculate the bending longitudinal bars

按单筋或双筋受弯构件正截面受弯承载力公式单独计算在弯矩作用下所需的受弯纵筋截面面积 A_s、A_s'，受弯纵筋应布置在截面的受拉区(按 A_s 配置)、受压区(按 A_s' 配置)，如图 8-11(a)所示，并验算受弯纵筋的最小配筋率。

The cross-sectional area A_s、A_s' of the bending longitudinal bars required under the action of the bending moment is calculated separately according to the flexural capacity formula of the normal section of the singly reinforced or doubly reinforced flexural member, and the bending longitudinal bars should be arranged in the tension zone of the section (configured as A_s) and compression zone (configured as A_s'), as shown in Fig. 8-11(a), and the minimum reinforcement ratio of the bending longitudinal bars should be checked.

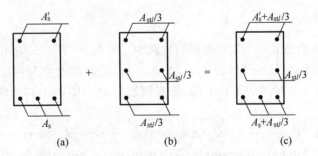

图 8-11 弯剪扭构件的纵向钢筋配置

(a) 受弯所需的受弯纵筋；(b) 受扭所需的受扭纵筋；(c) 弯扭纵筋叠加后

Figure 8-11 Longitudinal bars arrangement of bending-shear-torsion members

(a) Bending longitudinal bars required for bending; (b) Torsional longitudinal bars required for torsion;

(c) After superposition of bending and torsional longitudinal bars

（2）计算受扭纵筋

(2) Calculate the torsional longitudinal bars

应根据受扭所需的箍筋单肢用量 A_{stl}/s_t 和选定的受扭构件纵筋与箍筋配筋强度比 ζ，按式(8-10)计算受扭纵筋截面面积 A_{stl}。受扭纵筋应沿截面四周对称布置，如果受扭纵筋 A_{stl} 在截面内分上、中、下三层配置，则每一层受扭纵筋的截面面积为 $A_{stl}/3$，如图 8-11(b)所示，并验算受扭纵筋的最小配筋率。

The cross-sectional area A_{stl} of the torsional longitudinal bars should be calculated according to Eq. (8-10) and according to the amount of single-leg stirrups per limb A_{stl}/s_t and the reinforcement strength ratio of the longitudinal bars to the stirrups of torsional members ζ. The torsional longitudinal bars should be symmetrically placed along the section. If the torsional longitudinal bars A_{stl} is configured in three layers in the section: upper, middle and lower layers, the cross-sectional area of each layer of torsional longitudinal bars is $A_{stl}/3$, as shown in Fig. 8-11(b), and the minimum reinforcement ratio of the torsional longitudinal bars should be checked.

（3）计算截面受拉区和受压区的纵筋总量

(3) Calculate the total amount of longitudinal bars in the tension zone and compression zone of the section

按照叠加原则可知，截面底部受拉区纵筋总截面面积为 $A_s+A_{stl}/3$，截面顶部受压区纵筋总截面面积为 $A_s'+A_{stl}/3$，中间层纵筋截面面积为 $A_{stl}/3$，如图 8-11(c)所示。所配纵筋应满足纵筋的各项构造要求。

According to the principle of superposition, the total cross-sectional area of longitudinal bars in the tension zone at the bottom of the section is $A_s+A_{stl}/3$, the total cross-sectional area of longitudinal bars in the compression zone at the top of the section is $A_s'+A_{stl}/3$, and the cross-sectional area of longitudinal bars in the middle layer is $A_{stl}/3$, as shown in Fig. 8-11(c). The longitudinal bars should meet the structural requirements of the longitudinal bars.

拓展知识 8-4

【例题 8-2】 某钢筋混凝土矩形截面构件截面尺寸为 $b \times h = 250\text{mm} \times 600\text{mm}$，在均布荷载作用下，截面承受的弯矩设计值 $M = 95\text{kN} \cdot \text{m}$，剪力设计值 $V = 105\text{kN}$，扭矩设计值 $T = 10\text{kN} \cdot \text{m}$，混凝土强度等级采用 C30，所有钢筋均为 HRB400，进行该弯剪扭构件截面配筋计算。

【Example 8-2】 Cross-sectional dimensions of reinforced concrete rectangular section member are $b \times h = 250\text{mm} \times 600\text{mm}$. Under the uniform load, the design value of bending moment borne by the section is $M = 95\text{kN} \cdot \text{m}$, the design value of shear force is $V = 105\text{kN}$, and the design value of torque is $T = 10\text{kN} \cdot \text{m}$, the concrete strength grade is C30, and all the steel bars are HRB400. Calculate the section reinforcement of the bending-shear-torsion member.

解：环境类别为一类，查附表 3-2，混凝土保护层厚度 $c = 20\text{mm}$，选用双肢箍筋，直径

$d_v = 8$mm,则

$$b_{cor} = b - 2c - 2d_v = (250 - 2 \times 20 - 2 \times 8)\text{mm} = 194\text{mm}$$

$$h_{cor} = h - 2c - 2d_v = (600 - 2 \times 20 - 2 \times 8)\text{mm} = 544\text{mm}$$

$$A_{cor} = 194 \times 544 \text{mm}^2 = 105536\text{mm}^2$$

$$u_{cor} = 2 \times 194 + 2 \times 544 \text{mm} = 1476\text{mm}$$

(1) 验算截面尺寸

$$h_0 = h - a_s = (600 - 40)\text{mm} = 560\text{mm}$$

$$W_t = \frac{b^2}{6}(3h - b) = \frac{250^2}{6}(3 \times 600 - 250)\text{mm}^3 = 1.6 \times 10^7 \text{mm}^3$$

$$\frac{V}{bh_0} + \frac{T}{0.8W_t} = \left(\frac{105 \times 10^3}{250 \times 560} + \frac{10 \times 10^6}{0.8 \times 1.6 \times 10^7}\right)\text{N/mm}^2 = 1.53\text{N/mm}^2$$

$$< 0.25\beta_c f_c = 0.25 \times 1.0 \times 14.3\text{N/mm}^2 = 3.575\text{N/mm}^2$$

截面尺寸满足要求。

(2) 验算是否可以忽略剪力和扭矩的作用

$$0.35 f_t bh_0 = 0.35 \times 1.43 \times 250 \times 560 = 70.07 \times 10^3 \text{N} = 70.07\text{kN} < V = 105\text{kN}$$

$$0.175 f_t W_t = 0.175 \times 1.43 \times 1.6 \times 10^7 \text{N} \cdot \text{mm} = 4 \times 10^6 \text{N} \cdot \text{mm} < T$$

$$= 10 \times 10^6 \text{N} \cdot \text{mm}$$

所以,剪力和扭矩均不可忽略。

(3) 验算是否需要进行剪扭承载力计算

$$\frac{V}{bh_0} + \frac{T}{W_t} = \left(\frac{105 \times 10^3}{250 \times 560} + \frac{10 \times 10^6}{1.6 \times 10^7}\right)\text{N/mm}^2 = 1.375\text{N/mm}^2$$

$$> 0.7 f_t = 0.7 \times 1.43\text{N/mm}^2 = 1.001\text{N/mm}^2$$

所以,需进行剪扭承载力计算。

(4) 计算受弯纵筋

$$\alpha_s = \frac{M}{\alpha_1 f_c bh_0^2} = \frac{95 \times 10^6}{1.0 \times 14.3 \times 250 \times 560^2} = 0.085$$

$$\xi = 1 - \sqrt{1 - 2\alpha_s} = 1 - \sqrt{1 - 2 \times 0.085} = 0.089 < \xi_b = 0.518$$

$$A_s = \frac{\alpha_1 f_c b\xi h_0}{f_y} = \frac{1.0 \times 14.3 \times 250 \times 0.089 \times 560}{360}\text{mm}^2 = 495\text{mm}^2$$

(5) 计算箍筋

$$\beta_t = \frac{1.5}{1 + 0.5 \frac{V}{T} \cdot \frac{W_t}{bh_0}} = \frac{1.5}{1 + 0.5 \times \frac{105 \times 10^3}{10 \times 10^3} \times \frac{1.6 \times 10^7}{250 \times 560}} = 0.938 < 1.0$$

取 $\beta_t = 0.938$。

由

$$V \leq V_u = 0.7(1.5 - \beta_t) f_t bh_0 + f_{sv} \frac{A_{sv}}{s} h_0$$

得

$$105\times10^3=0.7\times(1.5-0.938)\times1.43\times250\times560+360\times\frac{A_{sv}}{s}\times560$$

$$\frac{A_{sv}}{s}=0.13\text{mm}^2/\text{mm}$$

由 $T\leqslant T_u=0.35\beta_t f_t W_t+1.2\sqrt{\zeta}f_{yv}\dfrac{A_{stl}}{s}A_{cor}$,取 $\zeta=1.2$ 代入,得

$$10\times10^6=0.35\times0.938\times1.43\times1.6\times10^7+1.2\sqrt{1.2}\times360\times\frac{A_{stl}}{s}\times105536$$

$$\frac{A_{stl}}{s}=0.05\text{mm}^2/\text{mm}$$

（6）计算受扭纵筋

由式(8-10)得

$$A_{stl}=\frac{\zeta f_{yv}A_{stl}u_{cor}}{sf_y}=\frac{1.2\times360\times0.05\times1476}{360}\text{mm}^2=89\text{mm}^2$$

（7）配置钢筋

计算箍筋：

$$\frac{A_{sv}}{s}=\frac{A_{stl}}{s}+\frac{A_{stl}}{s}=(0.13+0.05)\text{mm}^2/\text{mm}=0.18\text{mm}^2/\text{mm}$$

选 Φ 8 双肢箍筋, $A_{sv1}=50.3\text{mm}^2$,箍筋间距 $s=50.3/0.18\text{mm}=279\text{mm}$。

取箍筋为双肢Φ8@200,验算配箍率：

$$\rho_{sv}=\frac{nA_{sv1}}{bs}=\frac{2\times50.3}{250\times200}=0.201\%>\rho_{min}=0.28\frac{f_t}{f_{yv}}=0.28\times\frac{1.43}{360}=0.11\%$$

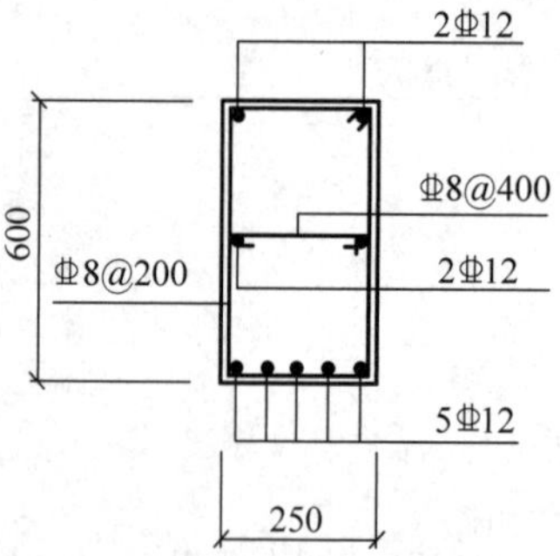

计算纵筋：根据截面尺寸,拟将受扭纵筋分 3 层布置,则每层受扭纵筋的截面面积为

$$\frac{1}{3}A_{stl}=\frac{1}{3}\times89\text{mm}^2=30\text{mm}^2$$

截面上部和中部纵筋均采用 2 Φ 12（$A_s=226\text{mm}^2$）；截面下部,受扭纵筋与受弯纵筋叠加为

$$\frac{1}{3}A_{stl}+A_s=(30+495)\text{mm}^2=525\text{mm}^2$$

选 5 Φ 12（$A_s=565\text{mm}^2$）,截面配筋如图 8-12 所示。

图 8-12　例题 8-2 的图

Figure 8-12　Figure of Example 8-2

【例题 8-3】　某承受均布荷载的矩形截面框架边梁,截面尺寸为 $b\times h=300\text{mm}\times600\text{mm}$,支座截面的负弯矩设计值 $M=190\text{kN}\cdot\text{m}$,剪力设计值 $V=130\text{kN}$,扭矩设计值 $T=50\text{kN}\cdot\text{m}$,混凝土强度等级为 C30,纵向钢筋、箍筋采用 HRB400,求梁截面的纵向钢筋和箍筋。

【Example 8-3】 A rectangular section frame edge beam is subjected to uniform load, the cross-sectional dimensions are $b\times h=400\text{mm}\times600\text{mm}$, the design value of negative moment at the support section is $M=190\text{kN}\cdot\text{m}$, the design value of shear force is $V=130\text{kN}$, the design value of torque is $T=50\text{kN}\cdot\text{m}$, the concrete strength grade is C30, the longitudinal bars and stirrups are HRB400. Calculate the longitudinal bars and stirrups of the beam section.

解：(1) 验算截面尺寸

根据混凝土强度等级和钢筋强度级别，查附表 1-3、附表 1-4、附表 1-6 得 $f_c = 14.3\text{N/mm}^2$，$f_t = 1.43\text{N/mm}^2$，$f_y = 360\text{N/mm}^2$，$f_{yv} = 360\text{N/mm}^2$，取 $a_s = 40\text{mm}$，则 $h_0 = h - a_s = (600 - 40)\text{mm} = 560\text{mm}$。

截面受扭塑性抵抗矩

$$W_t = \frac{b^2}{6}(3h - b) = \frac{300^2}{6} \times (3 \times 600 - 300)\text{mm}^3 = 22.5 \times 10^6 \text{mm}^3$$

$$\frac{V}{bh_0} + \frac{T}{0.8W_t} = \left(\frac{130 \times 10^3}{300 \times 560} + \frac{50 \times 10^6}{0.8 \times 22.5 \times 10^6}\right)\text{N/mm}^2 = 3.55\text{N/mm}^2$$

$$< 0.25\beta_c f_c = 0.25 \times 1 \times 14.3\text{N/mm}^2 = 3.58\text{N/mm}^2$$

截面尺寸符合要求。

(2) 计算受弯纵向钢筋 $A_{s,M}$

$$\alpha_s = \frac{M}{\alpha_1 f_c b h_0^2} = \frac{190 \times 10^6}{1 \times 14.3 \times 300 \times 560^2} = 0.14 < \alpha_{s,\max} = 0.384$$

$$\gamma_s = 0.5(1 + \sqrt{1 - 2\alpha_s}) = 0.5 \times (1 + \sqrt{1 - 2 \times 0.14}) = 0.924$$

$$A_{s,M} = \frac{M}{f_y \gamma_s h_0} = \frac{190 \times 10^6}{360 \times 0.924 \times 560}\text{mm}^2$$

$$= 1020\text{mm}^2 > 0.45\frac{f_t}{f_y}bh = 0.45 \times \frac{1.43}{360} \times 300 \times 600\text{mm}^2 = 322\text{mm}^2$$

$$> 0.2\%bh = 0.2\% \times 300 \times 600\text{mm}^2 = 360\text{mm}^2$$

所以，配筋率满足要求。

(3) 按剪扭构件计算配筋

① 验算是否可按构造配置受扭和受剪钢筋

$$\frac{V}{bh_0} + \frac{T}{W_t} = \left(\frac{130 \times 10^3}{300 \times 560} + \frac{50 \times 10^6}{22.5 \times 10^6}\right)\text{N/mm}^2 = 2.99\text{N/mm}^2$$

$$> 0.7f_t = 0.7 \times 1.43\text{N/mm}^2 = 1.00\text{N/mm}^2$$

所以，不能按构造配筋，应按计算配筋。

② 计算剪扭构件的混凝土受扭承载力降低系数

$$\beta_t = \frac{1.5}{1 + 0.5\frac{VW_t}{Tbh_0}} = \frac{1.5}{1 + 0.5 \times \frac{130 \times 10^3 \times 22.5 \times 10^6}{50 \times 10^6 \times 300 \times 560}} = 1.28 > 1.0$$

所以，取 $\beta_t = 1.0$。

③ 计算受剪所需的箍筋 A_{sv1}/s

采用双肢箍筋，$\Phi 8$，箍筋肢数 $n = 2$，$A_{sv} = nA_{sv1}$。

由式(8-34)得

$$\frac{A_{sv1}}{s} = \frac{V - 0.7(1.5-\beta_t)f_t b h_0}{n f_{yv} h_0}$$

$$= \frac{130 \times 10^3 - 0.7(1.5-1) \times 1.43 \times 300 \times 560}{2 \times 360 \times 560} \text{mm}^2/\text{mm}$$

$$= 0.114 \text{mm}^2/\text{mm}$$

④ 计算受扭所需的箍筋 A_{st1}/s 和受扭纵筋 A_{stl}

设受扭纵向钢筋与箍筋的配筋强度比 $\zeta = 1.2$。混凝土保护层厚度 $c = 20$mm，截面核心混凝土尺寸为

$$b_{cor} = b - 2c - 2d_{箍筋} = (300 - 2 \times 20 - 2 \times 8)\text{mm} = 244\text{mm}$$

$$h_{cor} = h - 2c - 2d_{箍筋} = (600 - 2 \times 20 - 2 \times 8)\text{mm} = 544\text{mm}$$

由式(8-35)得受扭箍筋：

$$\frac{A_{st1}}{s} = \frac{T - 0.35\beta_t f_t W_t}{1.2\sqrt{\zeta} f_{yv} A_{cor}} = \frac{50 \times 10^6 - 0.35 \times 1 \times 1.43 \times 22.5 \times 10^6}{1.2 \times \sqrt{1.2} \times 360 \times 244 \times 544} \text{mm}^2/\text{mm}$$

$$= 0.617 \text{mm}^2/\text{mm}$$

由式(8-10)得受扭纵筋：

$$A_{stl} = \zeta \frac{f_{yv}}{f_y} u_{cor} \frac{A_{st1}}{s} = 1.2 \times \frac{360}{360} \times 2 \times (244 + 544) \times 0.617 \text{mm}^2 = 1167 \text{mm}^2$$

验算受扭纵筋最小配筋率：

$$\frac{T}{Vb} = \frac{50 \times 10^6}{130 \times 10^3 \times 300} = 1.28 < 2$$

$$\rho_{stl,\min} = 0.6\sqrt{\frac{T}{Vb}} \cdot \frac{f_t}{f_y} = 0.6 \times \sqrt{1.28} \times \frac{1.43}{360} = 0.269\%$$

$$A_{stl} = 1167 \text{mm}^2 > \rho_{stl,\min} bh = 0.269\% \times 300 \times 600 \text{mm}^2 = 484 \text{mm}^2$$

所以，受扭纵筋配筋率满足要求。

(4) 验算最小配箍率及截面配筋

$$\rho_{sv} = \frac{2\left(\dfrac{A_{sv1} + A_{st1}}{s}\right)}{b} = \frac{2 \times (0.114 + 0.617)}{300} = 0.0049$$

$$= 0.49\% > \rho_{sv,\min} = 0.28 \frac{f_t}{f_{yv}} = 0.28 \times \frac{1.43}{360} = 0.11\%$$

所以，配箍率满足要求。

采用双肢箍筋Φ8，$A_{sv1} = 50.3 \text{mm}^2$，则

$$s = \frac{A_{sv1}}{A_{st1}/s + A_{sv1}/s} = \frac{50.3}{0.617 + 0.114}\text{mm} = 69\text{mm}$$

采用 $s = 60$mm。

选择受扭纵向钢筋：沿周边均匀布置，必须有角筋，间距不大于200mm，考虑梁顶面和底面各为2Φ14，梁侧边各2Φ14，共8Φ14，$A_{stl} = 1231 \text{mm}^2$，大于计算值1167mm²，误差略超过5%。

选择梁顶承受负弯矩及受扭的纵向钢筋：抗扭需要的纵筋为2Φ14，受扭纵筋截面面

积 $A_{stl} = 308\text{mm}^2$,受弯纵筋面积 $A_{s,M} = 1020\text{mm}^2$,合计 1328mm^2,采用 $3\underline{\Phi}16 + 3\underline{\Phi}18$,纵筋截面面积 1366mm^2。梁截面配筋如图 8-13 所示。

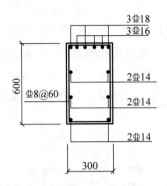

图 8-13 例题 8-3 的图
Figure 8-13　Figure of Example 8-3

8.3 受扭构件的构造要求
Structural requirements of torsional members

为避免钢筋配置过多,或截面尺寸太小或混凝土强度等级过低而发生超筋破坏,《混凝土结构设计标准》(GB/T 50010—2010)规定受扭构件的截面应满足以下限制条件。

8.3.1 截面尺寸控制条件
Section size control conditions

为防止弯剪扭构件在弯矩、剪力、扭矩共同作用或各自作用下在钢筋屈服前发生构件腹部混凝土局部斜向压坏,必须控制构件的截面尺寸不能过小。

《混凝土结构设计标准》(GB/T 50010—2010)6.4.1 条规定,在弯矩、剪力和扭矩共同作用下,h_w/b 不大于 6 的矩形、T 形、I 形截面和 h_w/t_w 不大于 6 的箱形截面构件,其截面应符合下列条件。

1. 当 h_w/b(或 h_w/t_w)$\leqslant 4$ 时
   When h_w/b (or h_w/t_w) $\leqslant 4$

截面应符合以下公式:
The cross section should conform to the following formula:

$$\frac{V}{bh_0} + \frac{T}{0.8W_t} \leqslant 0.25\beta_c f_c \tag{8-38}$$

2. 当 h_w/b(或 h_w/t_w)$= 6$ 时
   When h_w/b (or h_w/t_w) $= 6$

截面应符合以下公式:
The cross section should conform to the following formula:

$$\frac{V}{bh_0} + \frac{T}{0.8W_t} \leqslant 0.2\beta_c f_c \tag{8-39}$$

式中，V 为剪力设计值，N；T 为扭矩设计值，N·mm；b 为矩形截面的宽度，如图 8-14(a) 所示，T 形或 I 形截面的腹板宽度，如图 8-14(b) 所示，箱形截面的两侧壁总厚度 $2t_w$，如图 8-14(c) 所示，mm；W_t 为受扭构件的截面受扭塑性抵抗矩，mm^3；h_0 为截面有效高度，mm；h_w 为截面腹板高度，矩形截面取截面有效高度，T 形截面取截面有效高度减去翼缘高度，I 形截面和箱形截面取腹板净高，mm；t_w 为箱形截面壁厚，其值不应小于 $b_h/7$，mm，此处，b_h 为箱形截面的宽度，mm。

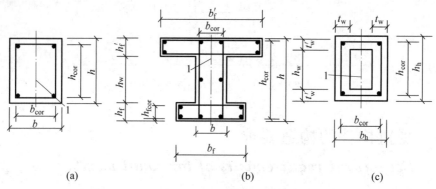

图 8-14 受扭构件的计算截面
(a) 矩形截面；(b) I 形截面；(c) 箱形截面
Figure 8-14 Calculated section of torsional members
(a) Rectangular section；(b) I-shaped section；(c) Box section

当 $4 < h_w/b$（或 h_w/t_w）< 6 时，按线性插值法确定；当 h_w/b（或 h_w/t_w）$\geqslant 6$ 时，受扭构件的截面尺寸及扭曲截面承载力计算应符合专门规定。

特别提示 8-3

8.3.2 受扭钢筋的构造要求
Structural requirements of torsional reinforcement

为防止弯剪扭构件发生少筋破坏，保证构件破坏时具有一定的延性，《混凝土结构设计标准》(GB/T 50010—2010) 规定应按构造要求配置受扭纵向钢筋和受扭箍筋。受扭构件的配筋应满足最小配筋率的要求，受扭纵筋和受扭箍筋的最小配筋量可根据钢筋混凝土构件所能承受的扭矩 T 不低于相同截面素混凝土构件开裂扭矩 T_{cr} 的原则确定。为防止受扭构件发生少筋破坏，受扭钢筋的配置必须大于其最小配筋率，箍筋还应满足最大间距和最小直径的构造要求。

1. 受扭纵筋的最小配筋率和构造要求
   Minimum reinforcement ratio and structural requirements of torsional longitudinal bars

弯剪扭构件受扭纵向钢筋的配筋率 ρ_{tl} 应符合以下规定：

$$\rho_{tl} = \frac{A_{stl}}{bh} \geqslant \rho_{tl,\min} = 0.6\sqrt{\frac{T}{Vb}} \cdot \frac{f_t}{f_y} \tag{8-40}$$

式中，A_{stl} 为沿截面周边布置的受扭纵向钢筋总截面面积，mm^2；b 为受剪的截面宽度，按式(8-39)的规定取用，对箱形截面，取 $b=b_h$，mm；h 为截面高度，mm；$\rho_{tl,min}$ 为受扭纵向钢筋的最小配筋率；T 为扭矩设计值，N·mm；V 为剪力设计值，N；f_t 为混凝土轴心抗拉强度设计值，N/mm^2；f_y 为受扭纵向钢筋抗拉强度设计值，N/mm^2。

当 $T/(Vb)>2$ 时，取 $T/(Vb)=2$；对箱形截面构件，式中的 b 应以 b_h 代替。

在弯剪扭构件中，配置在截面弯曲受拉边的纵向受力钢筋截面面积不应小于按受弯构件的纵向受拉钢筋最小配筋率计算的钢筋截面面积与按受扭纵向钢筋配筋率计算并分配到弯曲受拉边的钢筋截面面积之和。

纵向钢筋同时具有抗弯和抗扭的双重作用，所以纵向钢筋的配置应同时满足各自的配筋要求。沿截面周边布置的受扭纵向钢筋的间距不应大于 200mm 和梁截面短边长度 b。除应在梁截面四角设置受扭纵向钢筋外，其余受扭纵向钢筋宜沿截面周边均匀对称布置。

拓展知识 8-5

2. 受扭箍筋的最小配筋率和构造要求

Minimum reinforcement ratio and structural requirements of torsional stirrups

在弯剪扭构件中，箍筋的配筋率 ρ_{sv} 应满足下列要求：

$$\rho_{sv}=\frac{A_{sv}}{bs}\geqslant \rho_{sv,min}=0.28\frac{f_t}{f_{yv}} \tag{8-41}$$

截面宽度 b 按式(8-39)规定取用，但对箱形截面构件，式中的 b 应以 b_h 代替。

箍筋间距应符合《混凝土结构设计标准》(GB/T 50010—2010)表 9.2.9 的规定，受扭所需的箍筋应做成封闭式，且应沿截面周边布置。当采用多肢箍筋时，位于截面内部的箍筋不应计入受扭所需的箍筋面积。受扭所需箍筋的末端应做成 135°弯钩，弯钩端头平直段长度不应小于 $10d$（d 为箍筋直径）。在超静定结构中，考虑协调扭转而配置的箍筋，其间距不宜大于 $0.75b$。

拓展知识 8-6

当梁侧面需配置受扭纵向钢筋时，梁平法标注中此项注写值以大写字母"N"开头，接着注写配置在梁两个侧面的总配筋值，且对称配置。例如，N6⊈22 表示梁的两个侧面共配置 6⊈22 的受扭纵向钢筋，每侧各配置 3⊈22。

受扭纵向钢筋应满足梁侧面纵向构造钢筋的间距要求，且不再重复配置纵向构造钢筋。当作为梁侧面构造钢筋时，其搭接与锚固长度可取为 $15d$；当作为梁侧面受扭纵向钢筋时，其搭接长度为 l_l（非抗震）或 l_{lE}（抗震），锚固长度为 l_a（非抗震）或 l_{aE}（抗震），其锚固方式同框架梁下部钢筋。

3. 按构造要求配置受扭纵向钢筋和受扭箍筋的条件

Conditions for configuring torsional longitudinal bars and torsional stirrups according to structural requirements

在弯矩、剪力和扭矩作用下的构件，当符合下列条件时，可不进行构件剪扭承载力的计算，但为了防止构件开裂后突然发生脆性破坏，必须按构造要求配置受扭纵向钢筋和受扭箍筋：

$$\frac{V}{bh_0} + \frac{T}{W_t} \leqslant 0.7 f_t \tag{8-42a}$$

或

$$\frac{V}{bh_0} + \frac{T}{W_t} \leqslant 0.7 f_t + 0.07 \frac{N}{bh_0} \tag{8-42b}$$

式中，N 为与剪力、扭矩设计值 V、T 相应的轴向压力设计值，当 $N > 0.3 f_c A$ 时，取 $N = 0.3 f_c A$，N；f_c 为混凝土轴心抗压强度设计值，N/mm^2；A 为构件截面面积，mm^2。

本 章 小 结

（1）受扭构件是受扭矩作用的构件，工程中常见弯矩、剪力和扭矩同时作用的复合受扭构件，纯扭构件在建筑结构中很少。扭转分为平衡扭转和协调扭转，平衡扭转需满足静力平衡条件，协调扭转的扭矩则由变形协调条件确定。钢筋混凝土受扭构件由混凝土、受扭钢筋来抵抗构件截面的扭矩，受扭钢筋包括受扭纵筋和受扭箍筋，两者缺一不可，且配置数量须相互匹配，否则不能充分发挥两者的抗扭作用。

（2）根据受扭钢筋配置数量，钢筋混凝土纯扭构件的破坏形态主要有超筋破坏、部分超筋破坏、适筋破坏和少筋破坏四种。配筋适当的矩形截面钢筋混凝土纯扭构件，裂缝始于截面长边中点附近且与纵轴线约成 45°角，此后形成螺旋形裂缝。随着扭矩的增大，纵筋、箍筋达到屈服，混凝土被压碎而破坏，这与适筋受弯构件正截面受弯破坏类似，属于延性破坏。

（3）受扭构件的适筋破坏是计算受扭承载力的依据。通过验算最小箍筋配筋率和最小纵筋配筋率来防止少筋破坏，通过限制截面尺寸防止完全超筋破坏，通过控制受扭纵向钢筋与受扭箍筋的配筋强度比防止部分超筋破坏。实际工程中，受扭构件应避免设计成超筋和少筋构件。

（4）变角度空间桁架模型是重要的受扭承载力计算模型。《混凝土结构设计标准》（GB/T 50010—2010）根据变角度空间桁架模型并对大量试验数据进行统计回归，得出了受扭承载力计算的经验公式。矩形截面纯扭构件的计算包括开裂扭矩计算、承载力计算，还应满足裂缝宽度限值及构造要求。

（5）在弯剪扭构件中，混凝土的抗剪能力随扭矩的增大而降低，而混凝土的抗扭能力随剪力的增大而降低，《混凝土结构设计标准》（GB/T 50010—2010）通过采用剪扭构件混凝土受扭承载力降低系数 β_t，来考虑剪扭构件混凝土受剪和受扭之间的相关性。

（6）弯扭、剪扭和弯剪扭构件承载力计算的理论基础与纯扭构件是相同的，弯剪扭构件的配筋可用叠加法进行计算，即纵向钢筋截面面积由受弯承载力和受扭承载力所需纵向钢筋截面面积进行叠加得到，箍筋截面面积由受剪承载力和受扭承载力所需箍筋截面面积进行叠加得到。受弯所需纵筋截面面积按受弯构件正截面受弯承载力计算公式进行计算，受剪和受扭承载力计算公式中混凝土承担的部分则采用混凝土受扭承载力降低系数 β_t 来考虑剪扭相关性。

第8章 受扭构件的扭曲截面承载力

Twisted section bearing capacity of torsional members

第8章拓展知识和特别提示

视频：第8章小结讲解

习 题

8-1 填空题

1. 矩形截面素混凝土构件在扭矩作用下破坏，首先在其_____中点最薄弱处产生一条斜裂缝，然后向两边延伸，形成一个_____开裂、_____受压的空间扭曲斜裂面，其破坏性质属于_____。

2. 素混凝土梁矩形截面纯扭构件的受扭塑性抵抗矩为_____，其开裂扭矩表达式为_____。

3. 钢筋混凝土纯扭构件根据其所配纵筋和箍筋的多少，有_____破坏、_____破坏、_____破坏和_____破坏四种破坏形态。

4. 少筋纯扭构件的破坏扭矩近似等于_____扭矩。

5. 在钢筋混凝土受扭构件中一般采用_____钢筋和_____组成的空间骨架来提高构件的受扭承载力。

6. 剪扭相关性体现在由于扭矩的存在，截面抗剪承载力_____；由于剪力的存在，截面抗扭承载力_____。

7. 对同时承受扭矩和剪力作用的钢筋混凝土构件，计算中引入系数 β_t，称为_____。

8. 钢筋混凝土受扭构件计算中应满足 $0.6 \leqslant \zeta = \dfrac{f_y A_{stl} s}{f_{yv} A_{st1} u_{cor}} \leqslant 1.7$，要求 $\zeta \geqslant 0.6$ 的目的是_____，要求 $\zeta \leqslant 1.7$ 的目的是_____。

9. 为了使受扭构件能有效承受扭矩，所配置的受扭箍筋必须采用_____的形式，其末端应做成不小于135°的_____。

8-2 选择题

1. 矩形截面纯扭构件中，最大剪应力发生在（　　）。
   A. 短边中点　　　B. 长边中点　　　C. 角部

2. 受扭构件中受扭裂缝的特点是（　　）。
   A. 与构件轴线大致成45°的断断续续的螺旋形裂缝
   B. 与构件轴线大致成45°的连续的螺旋形裂缝
   C. 与构件轴线大致成45°的斜裂缝

3. 计算素混凝土纯扭构件开裂扭矩的公式是（　　）。
   A. 根据弹性理论推导的
   B. 根据截面各点剪应力都达到 f_t 推导的

C. 经验公式

D. 在弹性理论基础上考虑塑性影响

4. 矩形截面纯扭构件的截面尺寸为 $b \times h = 250\text{mm} \times 500\text{mm}$，则受扭纵筋的最少根数应为（　）。

A. 4　　　　　　　　　　　　　　B. 6

C. 8　　　　　　　　　　　　　　D. 只要钢筋截面面积够即可

5. 纯扭构件中配置受扭箍筋时，要求为（　）箍筋。

A. 双肢　　　B. 单肢　　　C. 复合　　　D. 无确定要求

6. 受扭构件的配筋方式为（　）。

A. 仅配置受扭箍筋

B. 仅配置受扭纵筋

C. 配置受扭纵筋和箍筋

7. 受扭纵筋一般沿构件截面（　）。

A. 上面布置　　　B. 下面布置　　　C. 周边均匀布置　　　D. 上下布置

8. 受扭纵筋与箍筋的配筋强度比 ζ 在 $0.6 \sim 1.7$ 时，（　）。

A. 均布纵筋、箍筋部分屈服　　　　B. 均布纵筋、箍筋均屈服

C. 仅箍筋屈服　　　　　　　　　　D. 不对称纵筋、箍筋均屈服

9. 设计剪扭构件时，应（　）。

A. 增大截面尺寸　　　　　　　　　B. 增加受扭纵筋

C. 增加受扭箍筋　　　　　　　　　D. 增大受扭纵筋和箍筋的配筋强度比 ζ

10. 《混凝土结构设计标准》(GB/T 50010—2010)对于剪扭构件承载力计算采用的计算模式是（　）。

A. 混凝土和钢筋均考虑相关关系

B. 混凝土和钢筋均不考虑相关关系

C. 混凝土不考虑相关关系，钢筋考虑相关关系

D. 混凝土考虑相关关系，钢筋不考虑相关关系

11. 剪扭构件混凝土承载力降低系数 $\beta_t = 1$ 时，（　）。

A. 混凝土受扭承载力为纯扭时的一半

B. 混凝土受剪承载力不变

C. 混凝土受剪及受扭承载力均不变

D. 混凝土受剪承载力为纯剪时的一半

12. 当符合条件 $T \leqslant 0.175 f_t W_t$ 时，钢筋混凝土弯剪扭构件按下列哪种作用计算承载力？（　）

A. 弯矩作用　　　　　　　　　　　B. 弯矩和剪力作用

C. 弯矩和扭矩作用　　　　　　　　D. 扭矩作用

8-3 名词解释

1. 受扭构件	2. 纯扭构件	3. 弯剪扭构件	4. 平衡扭转
5. 协调扭转	6. 全超筋破坏	7. 部分超筋破坏	8. 开裂扭矩
9. 变角度空间桁架模型	10. 纵筋与箍筋的配筋强度比	11. 剪扭构件	12. 弯扭构件

8-4 思考题

1. 受扭构件中的受扭斜裂缝与受弯构件中的受剪斜裂缝有何不同？
2. 素混凝土和适筋混凝土纯扭构件的破坏特点是什么？
3. 受扭纵向钢筋为什么沿截面周边均匀布置，四角是否必须布置？
4. 从受扭构件的配筋合理性看，采用螺旋式配筋比较合理，但实际上为什么采用封闭式箍筋加纵筋抗扭？
5. 说明变角度空间桁架模型理论的假定。
6. 如何计算剪扭构件的混凝土受扭承载力降低系数 β_t？其取值范围是什么？
7. 简述剪扭构件扭曲截面承载力的计算步骤。
8. 《混凝土结构设计标准》(GB/T 50010—2010)中规定的矩形截面纯扭构件受扭承载力计算公式与按变角度空间桁架模型推导的受扭承载力公式有什么不同？
9. 《混凝土结构设计标准》(GB/T 50010—2010)在受扭构件扭曲截面承载力计算中是如何考虑弯矩、剪力、扭矩共同作用的？纵向钢筋与箍筋的配筋量是如何计算的？
10. 弯剪扭构件中受弯、受剪和受扭钢筋各应配置在截面的什么位置？哪些钢筋可以合并设置？
11. 弯剪扭构件在什么条件下可不进行受扭钢筋的计算，而只按构造要求配筋？
12. 受扭构件的截面尺寸和截面配筋构造有哪些主要要求？

8-5 计算题

1. 某矩形截面纯扭构件的截面尺寸为 $b \times h = 300\text{mm} \times 500\text{mm}$，配有 4⌽16 纵向钢筋，箍筋⌽8@100，混凝土强度等级为 C30。取 $a_s = 40\text{mm}$，求扭曲截面的受扭承载力。

2. 某矩形截面纯扭构件的截面尺寸为 $b \times h = 250\text{mm} \times 500\text{mm}$，扭矩设计值 $T = 12\text{kN} \cdot \text{m}$，混凝土强度等级为 C25，箍筋 HRB400，纵向钢筋 HRB400。取 $a_s = 40\text{mm}$，进行构件截面配筋设计。

3. 某矩形截面弯扭构件的截面尺寸为 $b \times h = 250\text{mm} \times 500\text{mm}$，弯矩设计值 $M = 70\text{kN} \cdot \text{m}$，扭矩设计值 $T = 12\text{kN} \cdot \text{m}$，混凝土强度等级为 C25，箍筋 HRB400，纵向钢筋 HRB400。取 $a_s = 40\text{mm}$，进行构件截面配筋设计。

4. 某矩形截面弯剪扭构件的截面尺寸为 $b \times h = 250\text{mm} \times 500\text{mm}$，截面上弯矩组合设计值 $M = 70\text{kN} \cdot \text{m}$，剪力组合设计值 $V = 100\text{kN}$（剪力由均布荷载产生），扭矩组合设计值 $T = 12\text{kN} \cdot \text{m}$，箍筋内表皮至构件表面距离为 30mm，采用强度等级为 C25 的混凝土和 HRB400 钢筋。取 $a_s = 40\text{mm}$，进行构件截面配筋设计。

第 9 章

裂缝、变形及耐久性

Crack, deformation and durability

教学目标：

1. 了解钢筋混凝土构件产生裂缝的原因及其影响因素；
2. 了解钢筋混凝土受弯构件截面弯曲刚度的计算；
3. 熟悉受弯构件最大裂缝宽度的计算方法，熟悉减少裂缝宽度的措施；
4. 熟悉构件挠度的验算方法，熟悉减少构件挠度的措施；
5. 了解耐久性的概念，熟悉影响混凝土结构耐久性的因素，理解混凝土碳化和钢筋锈蚀的原理；
6. 知道耐久性设计的主要内容和技术措施；
7. 了解混凝土结构耐久性的有关规定。

导读：

混凝土的抗拉强度很低，许多混凝土结构在建设和使用过程中出现了不同程度、不同形式的裂缝，裂缝问题是长期困扰土木工程界的技术难题。虽然结构设计是建立在承载力基础上的，但大多数工程的使用标准却是由裂缝控制的，结构的破坏和倒塌也都是从裂缝的扩展开始的。有些裂缝虽然没有达到使建筑物倒塌的危险程度，但由于人们的心理作用以及建筑装修及美观方面的原因，常常影响到建筑物的正常使用。裂缝可以引起渗漏、长期强度和耐久性的降低，如混凝土保护层剥落、混凝土碳化、钢筋锈蚀等，将大大缩短结构的使用寿命。工程中应将裂缝的有害程度控制在允许范围内。混凝土结构需要控制裂缝的宽度，提高混凝土的耐久性。

挠度形成的原因主要有：荷载或温度造成的材料变形，建造过程的疏误，基础沉降，其他因素如材料徐变、干缩等。过大的挠度会引起使用者的不安，影响某些设备的正常使用，并可能引起过宽的裂缝。

混凝土结构的耐久性不是直接由力学因素引起的。在自然环境和使用条件下，混凝土材料的老化导致结构性能劣化和承载力下降。耐久性是材料抵抗自身和自然环境双重因素长期破坏作用的能力，即保证其经久耐用的能力。耐久性越好，材料的使用寿命越长。混凝土工程大多是永久性的，工程量大，耗资多，由于技术和施工水平限制以及主观认识等方面的原因，人们对混凝土结构的设计、选材、制作和养护等往往只重视强度而忽视了其耐久性。若耐久性差将会给社会造成极为沉重的负担。因此，从节约资金、有效利用资源及保护环境等方面综合考虑,必须深入研究混凝土的耐久性问题。

引例:

第9章引例

核心词汇:

正常使用极限状态	serviceability limit state	海水腐蚀	seawater corrosion
裂缝宽度	crack width	干燥收缩	drying shrinkage
裂缝控制	crack control	抗裂度	crack resistance capacity
变形控制	deformation control	变形验算	deformation check
荷载效应	load effect	准永久组合	quasi-permanent combination
跨中挠度	mid-span deflection	计算跨度	effective span
长期刚度	long-term stiffness	短期刚度	short-term stiffness
挠度限值	limit value of deflection	挠度验算	deflection calculation
最小刚度原则	principle of minimum stiffness	钢筋锈蚀	steel corrosion
耐久性	durability	碱骨料反应	alkali-aggregate reaction
混凝土碳化	concrete carbonization	化学侵蚀	chemical corrosion
冻融破坏	freeze-thaw damage	硫酸盐腐蚀	sulfate corrosion
酸腐蚀	acid corrosion	耐久性设计	durability design

9.1 概述

Summary

钢筋混凝土结构和构件除应按承载能力极限状态进行设计外,还应进行正常使用极限状态的验算,以满足结构的正常使用功能和耐久性要求。对一般常见的工程结构,正常使用极限状态的设计主要是验算结构构件的抗裂度、裂缝宽度和变形等,以满足结构适用性和耐久性的需要。当结构构件达到或超过正常使用极限状态时,结构将不能正常使用,但危害程度不如承载力不足引起结构破坏造成的损失大,对其可靠度的要求可适当降低。

In addition to the design of reinforced concrete structures and members according to the ultimate limit state, the checking calculation of serviceability limit state should also be carried out to meet the requirements of normal service function and durability of the structure. For common engineering structures, the design of serviceability limit state is mainly to check the crack resistance, crack width and deformation of structural members to meet the needs of structural applicability and durability. When the structural members reach or exceed the serviceability limit state, the structure will not be used normally, but the damage is not as great as the loss caused by structural damage due to insufficient bearing capacity. The requirements for its reliability can be appropriately reduced.

1. 正常使用极限状态验算的特点

Characteristics of serviceability limit state checking

(1) 考虑到结构超过正常使用极限状态对人民生命财产的危害远比超过承载能力极限状态的要小，《混凝土结构设计标准》(GB/T 50010—2010)规定裂缝宽度及变形验算均采用荷载标准值和材料强度的标准值，即不考虑荷载分项系数和材料分项系数，也不考虑结构重要性系数。

(1) Considering that the damage to life and property caused by the structure exceeding the serviceability limit state is far less than that caused by the structure exceeding the ultimate limit state, "Standard for design of concrete structures" (GB/T 50010—2010) stipulates that the crack width and deformation check should adopt the standard value of load and the standard value of material strength, that is, the load partial coefficient and material partial coefficient should not be considered, and the structural importance coefficient should not be considered.

(2) 由于可变荷载作用时间的长短对裂缝宽度和变形的大小有影响，验算裂缝宽度和变形时应采用荷载短期效应组合值并考虑荷载长期效应的影响。

(2) Since the duration of variable load has an effect on the crack width and deformation, when checking the crack width and deformation the combined value of short-term load effect should be adopted and the influence of long-term load effect should be considered.

2. 正常使用极限状态设计表达式

Design expression of serviceability limit state

对于正常使用极限状态，混凝土结构构件应分别按荷载的准永久组合、标准组合并考虑长期作用的影响，按下列极限状态设计表达式进行验算：

For the serviceability limit state, the concrete structural members should be checked according to the quasi-permanent combination, standard combination of loads respectively and the influence of long-term action, and should be checked and calculated according to the following limit state design expression:

$$S \leqslant C \tag{9-1}$$

式中，S 为正常使用极限状态荷载组合效应设计值；C 为结构构件达到正常使用要求的规定限值，如变形、裂缝宽度、应力等的限值。

Where, S is the load combination effect design value of serviceability limit state; C is the specified limit value for structural members to meet the requirements of normal service, such as the limit values of deformation, crack width, stress and so on.

3. 荷载效应的标准组合、准永久组合

Standard combination and quasi-permanent combination of load effect

在正常使用极限状态下，可变荷载作用时间的长短对裂缝和变形的大小是有影响的，可变荷载的最大值并非长期作用在结构上，裂缝和变形也是随着时间而增长的。对于正常使用极限状态，对结构构件应分别按荷载效应的标准组合、准永久组合并考虑长期作用的影响，验算其是否超过正常使用规定的限值。

Under the serviceability limit state, the duration of variable load action has an impact on the size of cracks and deformation. The maximum value of variable load does not act on the structure for a long time, and the cracks and deformations also increase with time. For the serviceability limit state, the structural members should be checked and calculated according to the standard combination and quasi-permanent combination of load effects and the influence of long-term effects should be considered to verify whether they exceed the limit values specified for normal service.

标准组合是采用在设计基准期内根据正常使用条件可能出现最大可变荷载时的荷载标准值进行组合而确定的，在一般情况下均采用这种组合。

The standard combination is determined by using the load standard value when the maximum variable load may occur under normal service conditions in the design reference period, this combination is generally used.

作用在结构上的荷载有多种，例如作用在楼面梁上的荷载有结构自重（永久荷载）和楼面可变荷载。可变荷载中有一部分荷载值随时间变化不大，这部分荷载称为准永久荷载，如住宅中的家具等。在书库等建筑物的楼面活荷载中，准永久荷载占的比例将达到80%。将永久荷载产生的弯矩与可变荷载中的准永久荷载产生的弯矩叠加起来，就称为弯矩的准永久组合。准永久组合是采用设计基准期内持久作用的准永久值进行组合而确定的，它是考虑可变荷载的长期作用起主要影响并具有自己独立性的一种组合形式。

There are many kinds of loads acting on the structure, for example, the loads acting on the floor beams include the structural self weight (permanent load) and floor variable loads. Some of the variable load values are not time-dependent, and this portion of the load is called quasi-permanent load, such as furniture in residential buildings, etc. Among the floor live load of buildings such as bookstores, the proportion of quasi-permanent load will reach 80%. The superposition of bending moment generated by permanent load and that generated by quasi-permanent load in variable load is called quasi-permanent combination of bending moment. Quasi-permanent combination is determined by using the quasi-permanent value of the permanent action in the design reference period. It is a combination form that considers that the long-term action of variable load plays a major role and has its own independence.

9.2 钢筋混凝土构件的裂缝宽度计算
Calculation of crack width of reinforced concrete members

9.2.1 混凝土构件裂缝概述
Overview of cracks in concrete members

1. 混凝土构件裂缝按成因分类
Classification of cracks in concrete members according to their causes

钢筋混凝土结构的裂缝根据成因可分为荷载裂缝和非荷载裂缝。荷载裂缝是由使用荷载产生的,如图 9-1 所示;非荷载裂缝也称变形裂缝,一般是由温度变化,混凝土收缩、膨胀,不合理的施工和构造,如养护不周、拆模时间不当、构造形式不妥引起应力集中,不均匀沉降等原因产生。宽度小于 0.02mm 的裂缝对使用(防水、防腐、承重)都无危险性,可假定具有小于 0.02mm 裂缝的结构为无裂缝结构。结构抗裂只是把裂缝控制在一定范围内。下面介绍几种裂缝的成因。

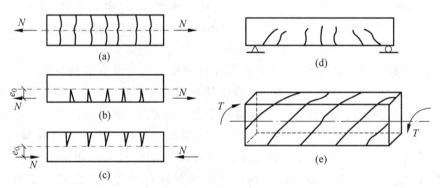

图 9-1 荷载引起的钢筋混凝土受力构件的裂缝
(a)轴心受拉构件;(b)偏心受拉构件;(c)偏心受压构件;(d)受弯构件;(e)受扭构件
Figure 9-3 Cracks in reinforced concrete stressed members caused by loads
(a) Axial tension member;(b) Eccentric tension member;(c) Eccentric compression member;
(d) Flexural member;(e) Torsional member

1)塑性收缩裂缝
1) Plastic shrinkage crack

塑性裂缝多在新浇筑的混凝土构件暴露于空气中的上表面出现。塑性收缩是指混凝土在凝结之前,表面因失水较快而产生的收缩。塑性收缩裂缝一般在干热或大风天气出现,裂缝多呈中间宽、两端细且长短不一,互不连贯状态,较短的裂缝一般长 20~30cm,较长的裂缝可达 2~3m,宽 1~5mm。

塑性裂缝产生的主要原因是:混凝土在终凝前几乎没有强度或强度很小,或者混凝土刚刚终凝而强度很小时,受高温或较大风力的影响,混凝土表面失水过快,造成毛细管中产生较大的负压而使混凝土体积急剧收缩,而此时混凝土的强度又无法抵抗其本身收缩,因此产生开裂。影响混凝土塑性收缩开裂的主要因素有水灰比、混凝土的凝结时间、环境温度、风速、相对湿度等,掺工程纤维可解决混凝土塑性裂缝问题。

2) 沉降收缩裂缝

2) Settlement shrinkage crack

沉降裂缝的产生是由于地基土质不匀、松软、回填土不实或浸水而造成不均匀沉降所致,或者因为模板刚度不足,模板支撑间距过大或支撑底部松动等导致,特别是在冬季,模板支撑在冻土上,冻土化冻后产生不均匀沉降,使混凝土结构产生裂缝。此类裂缝多为深进或贯穿性裂缝,裂缝呈梭形,其走向与沉降情况有关,一般沿与地面垂直或成 30°～45°方向发展。较大的沉降裂缝往往有一定的错位,裂缝宽度往往与沉降量成正比关系。裂缝宽度为 0.3～0.4mm,受温度变化的影响较小。地基变形稳定之后,沉降裂缝也基本趋于稳定。

3) 温度裂缝

3) Temperature crack

温度裂缝多发生在大体积混凝土表面或温差变化较大地区的混凝土结构中。混凝土浇筑后,在硬化过程中,水泥水化产生大量的水化热。当水泥用量从 350～550kg/m³ 时,每立方米混凝土将释放出 17500～27500kJ 的热量,从而使混凝土内部温度升至 70℃左右甚至更高。由于混凝土的体积较大,大量的水化热聚集在混凝土内部不易散发,导致内部温度急剧上升,而混凝土表面散热较快,这样就形成内外的较大温差,造成内部与外部热胀冷缩的程度不同,使混凝土表面产生一定的拉应力。当拉应力超过混凝土的轴心抗拉强度时,混凝土表面就会产生裂缝,这种裂缝多发生在混凝土施工中后期。

混凝土构件成型后,若未及时覆盖,则表面水分散失快,体积收缩大,而混凝土内部温度变化小,收缩也小,因而表面收缩变形受到内部混凝土的约束,出现拉应力,引起混凝土表面的收缩。大体积混凝土的温升高,应按相应规范施工,控制混凝土内外温差。

4) 干缩裂缝

4) Drying shrinkage crack

混凝土中的水泥在硬化过程中失水使混凝土发生干缩,产生干缩裂缝,常用加膨胀剂的方法来减少或防止干缩裂缝的发生,还可用细而密的布筋减少干缩裂缝。

2. 裂缝的形状

Shape of cracks

裂缝按其形状分为表面的、贯穿的、纵向的、横向的、上宽下窄、下宽上窄、枣核形、对角线式、斜向的、外宽内窄的和纵深的(深度达构件厚度的一半)等。裂缝形状和结构受力状态有直接关系,一般裂缝的方向同主拉应力方向垂直。

 拓展知识 9-1, 特别提示 9-1

3. 影响裂缝宽度的因素

Factors affecting crack width

《混凝土结构设计标准》(GB/T 50010—2010)定义的裂缝开展宽度是指受拉钢筋重心对应水平处构件外侧表面混凝土的裂缝宽度。影响裂缝宽度的因素很多,主要因素有:受拉钢筋的应力、受拉钢筋直径、受拉钢筋配筋率、混凝土保护层厚度、混凝土抗拉强度、受拉钢筋的黏着特征和荷载特征等。黏结-滑移理论认为钢筋的黏结性能、应变、直径和配筋率是主要影响因素,而无滑移理论认为混凝土保护层厚度是主要因素。

多数研究者认为混凝土抗拉强度对裂缝宽度影响不大,可略去不计;裂缝截面的受拉钢筋应力是影响裂缝宽度的最重要因素;裂缝宽度随钢筋直径而变化,在钢筋直径相同、钢筋应力大致相等的情况下,裂缝宽度随配筋率的增加而减小,当配筋率接近某一数值时,裂缝宽度基本不变;在使用荷载作用下,裂缝的间距不随荷载作用时间而变化,但裂缝宽度随时间以逐渐减小的比率在增加;重复荷载作用下不断发展的裂缝宽度是初始使用荷载下裂缝宽度的 $1 \sim 1.5$ 倍。

4. 裂缝控制等级

Crack control level

《混凝土结构设计标准》(GB/T 50010—2010)将钢筋混凝土结构构件的裂缝控制等级划分为 3 级。一级:严格要求不出现裂缝的构件,构件受拉边缘混凝土不产生拉应力;二级:一般要求不出现裂缝的构件,构件受拉边缘拉应力小于混凝土轴心抗拉强度标准值 f_{tk};三级:允许出现裂缝的构件,最大裂缝宽度小于裂缝宽度允许值。一级和二级抗裂要求的构件一般要采用预应力混凝土;普通的钢筋混凝土构件抗裂要求为三级,一般都是带裂缝工作的。

9.2.2 构件裂缝宽度的验算
Checking calculation of crack width of members

根据国内外调查资料可知,钢筋混凝土构件裂缝中由荷载引起的占 20% 左右,由变形引起的占 80% 左右。前者主要通过裂缝宽度验算进行控制;而后者主要通过采取合理的设计、施工、构造、材料等措施进行控制,大部分裂缝是可以克服和避免的,不必进行裂缝计算。

According to the survey data at home and abroad, about 20% of the cracks in reinforced concrete members are caused by load, and about 80% of the cracks are caused by deformation. The former is mainly controlled by checking the crack width; while the latter is mainly controlled by adopting reasonable design, construction, structure, materials and other measures. Most of the cracks can be overcome and avoided, and it is not necessary to calculate the cracks.

1. 裂缝宽度的计算理论

Calculation theory of crack width

由于裂缝的开展是混凝土收缩和钢筋伸长导致混凝土与钢筋之间不断产生相对滑移造成的,因此裂缝宽度就等于裂缝间钢筋的伸长减去混凝土的伸长,所以,裂缝间距小,裂缝宽度就小,即裂缝密而细,这是工程中所希望的。试验表明,裂缝宽度的离散性比裂缝间距更大些,平均裂缝宽度的确定必须以平均裂缝间距为基础。

构件裂缝宽度的验算属于正常使用极限状态的验算。国内外关于钢筋混凝土受弯构件裂缝宽度的计算公式很多,它们都是在一定理论基础上根据大量实测或试验统计资料确定的公式,各自符合本国的国情,有一定的适用场合。由于取不同的影响因素作为主要参数,不同的分析方法其计算公式也不同,大致分为两大类。第一类是以"黏结-滑移"理论为基础的半经验半理论公式,裂缝的间距取决于钢筋与混凝土间黏结应力的分布,裂缝的开展是由于钢筋与混凝土间的变形不再维持协调,出现相对滑动而产生;第二类是以统计分析方法为基础的经验公式,《混凝土结构设计标准》(GB/T 50010—2010)推荐采用的裂缝宽度计算公式就属于这种。

 特别提示 9-2

2. 最大裂缝宽度计算公式
Calculation formula of maximum crack width

实际工程中,由于混凝土质量不均匀,裂缝的间距有疏有密,每条裂缝开展的宽度也不同,因此,验算裂缝宽度应以最大裂缝宽度为准。研究表明,短期荷载作用下的裂缝宽度可根据平均裂缝宽度乘以裂缝宽度扩大系数得到,长期荷载作用下的最大裂缝宽度可由短期荷载作用下的最大裂缝宽度乘以裂缝扩大系数得到。根据试验结果,将相关的各种系数归并后,《混凝土结构设计标准》(GB/T 50010—2010)7.1.2 条给出了矩形、T 形、倒 T 形和 I 形截面的钢筋混凝土受拉、受弯和偏心受压构件,按荷载标准组合或准永久组合并考虑长期作用影响的最大裂缝宽度计算公式:

In practical engineering, due to uneven concrete quality, the spacing between cracks is sparse or dense, and the width of each crack is also different. Therefore, the calculation of crack width should be based on the maximum crack width. The research shows that the crack width under short-term load can be obtained by multiplying the average crack width by the crack width expansion coefficient, and the maximum crack width under long-term load can be obtained by multiplying the maximum crack width under short-term load by the crack expansion coefficient. According to the experimental results, after merging various relevant coefficients, Article 7.1.2 of "Standard for design of concrete structures" (GB/T 50010—2010) gives the formula for calculating the maximum crack width of reinforced concrete members in tension, bending and eccentric compression with rectangular, T-shaped, inverted T-shaped and I-shaped sections according to the load standard combination or quasi-permanent combination and considering the influence of long-term effect:

$$\omega_{\max} = \alpha_{cr}\psi \frac{\sigma_s}{E_s}\left(1.9c_s + 0.08\frac{d_{eq}}{\rho_{te}}\right) \quad (9-2)$$

式中,α_{cr} 为构件受力特征系数,轴心受拉构件取 2.7,偏心受拉构件取 2.4,受弯构件和偏心受压构件取 1.9;ψ 为裂缝间纵向受拉钢筋应变不均匀系数,反映裂缝间混凝土参与受拉的程度,ψ 越小,表明混凝土承受拉力的程度越大。ψ 的表达式为

Where, α_{cr} is the force characteristic coefficient of members, 2.7 for axial tension members, 2.4 for eccentric tension members, 1.9 for flexural members and eccentric compression members; ψ is the uneven coefficient of longitudinal tensile reinforcement strain between cracks, reflects the degree of concrete participation in tension between cracks, the smaller the coefficient ψ is, the greater the degree to which the concrete bears tensile force. The expression of ψ is

$$\psi = 1.1 - \frac{0.65 f_{tk}}{\rho_{te} \sigma_s} \quad (0.2 \leqslant \psi \leqslant 1.0) \tag{9-3}$$

式中,当 ψ<0.2 时,取 ψ=0.2;当 ψ>1.0 时,取 ψ=1.0;对直接承受重复荷载的构件,取 ψ=1.0;σ_s 为按荷载准永久组合计算的钢筋混凝土构件纵向受拉普通钢筋应力,N/mm^2;E_s 为钢筋的弹性模量,N/mm^2;c_s 为最外层纵向受拉钢筋外边缘至受拉区底边的距离,mm,当 c_s<20mm 时,取 c_s=20mm,当 c_s>65mm 时,取 c_s=65mm;f_{tk} 为混凝土轴心抗拉强度标准值,N/mm^2;d_{eq} 为受拉区纵向钢筋的等效直径,当采用不同直径的钢筋时,采用等效直径,mm。

Where, when ψ<0.2, take ψ=0.2; when ψ>1.0, take ψ=1.0; for members directly bear repeated load, ψ=1.0 is taken. σ_s is the longitudinal tensile stress of ordinary steel bars of reinforced concrete members calculated based on quasi-permanent load combination, N/mm^2; E_s is the elastic modulus of reinforcement, N/mm^2; c_s is the distance from the outer edge of the outermost longitudinal tensile reinforcement to the bottom edge of the tension zone, mm, when c_s<20mm, take c_s=20mm; when c_s>65mm, take c_s=65mm; f_{tk} is the axial tensile strength standard value of concrete, N/mm^2; d_{eq} is the equivalent diameter of longitudinal bars in the tension zone, when using different diameters of steel bars, the equivalent diameter should be used, mm.

$$d_{eq} = \frac{\sum n_i d_i^2}{\sum n_i \nu_i d_i} \tag{9-4}$$

式中,n_i 为受拉区第 i 种纵向钢筋的根数;d_i 为受拉区第 i 种纵向钢筋的公称直径,mm;ν_i 为受拉区第 i 种纵向钢筋的相对黏结特性系数,光圆钢筋取 0.7,带肋钢筋取 1.0;ρ_{te} 为按有效受拉混凝土截面面积计算的纵向受拉钢筋配筋率,当计算值小于 0.01 时,取 0.01,计算公式为

Where, n_i is the number of the ith type of longitudinal bars in the tension zone; d_i is the nominal diameter of the ith type of longitudinal bars in the tension zone, mm; ν_i is the relative bonding characteristic coefficient of the ith type of longitudinal bars in the tension zone, 0.7 for round bar, and 1.0 for ribbed bar; ρ_{te} is the reinforcement ratio of longitudinal tensile reinforcement calculated based on the effective tensile concrete cross-sectional area, when the calculated value is less than 0.01, take 0.01, the calculation formula is

$$\rho_{te} = \frac{A_s}{A_{te}} \tag{9-5}$$

式中,A_{te} 为有效受拉混凝土截面面积,对轴心受拉构件,取构件截面面积;对受弯、偏心受拉和偏心受压构件,取 $A_{te}=0.5bh+(b_f-b)h_f$,mm^2,其 b_f、h_f 分别为受拉翼缘的宽度、高度,如图 9-2 所示,mm。

Where, A_{te} is the effective tensile concrete cross-sectional area, for axial tension members, the cross-sectional area of the member is taken; for bending, eccentric tension and eccentric compression members, $A_{te}=0.5bh+(b_f-b)h_f$ is taken, mm^2, b_f、h_f is the width and height of the tension flange respectively, as shown in Fig. 9-2(b), mm.

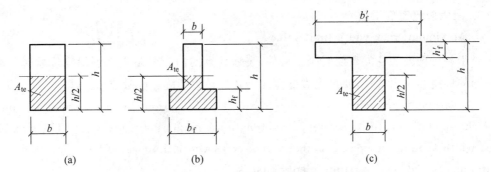

图 9-2 梁的截面形式和有效受拉混凝土截面面积
(a) 矩形截面；(b) 倒 T 形截面；(c) T 形截面

Figure 9-2 Section form of beam and effective tensile concrete cross-sectional area
(a) Rectangular section; (b) Inverted T-shaped section; (c) T-shaped section

对不同荷载作用情况下的钢筋混凝土构件，在荷载标准组合或准永久组合下，钢筋混凝土构件受拉区纵向普通钢筋的应力可按下列公式计算。

For reinforced concrete members under different loads, under standard combination or quasi-permanent combination of the load, the stress of longitudinal ordinary steel bars in the tension zone of reinforced concrete members can be calculated according to the following formula.

受弯构件：
Flexural member：

$$\sigma_{sq}=\frac{M_q}{0.87A_s h_0} \tag{9-6a}$$

轴心受拉构件：
Axial tension member：

$$\sigma_{sq}=\frac{N_q}{A_s} \tag{9-6b}$$

式中，A_s 为受拉区纵向普通钢筋截面面积，mm^2，对轴心受拉构件，取全部纵向普通钢筋截面面积；对偏心受拉构件，取受拉较大边的纵向普通钢筋截面面积；对受弯、偏心受压构件，取受拉区纵向普通钢筋截面面积。N_q、M_q 分别为按荷载准永久组合计算的轴向力值（单位：N）和弯矩值，$N \cdot mm$。

Where, A_s is the cross-sectional area of longitudinal ordinary bars in the tension, mm^2, for axial tension members, taking the cross-sectional area of all longitudinal ordinary reinforcement; for eccentric tension members, taking the cross-sectional area of longitudinal ordinary reinforcement at the edge of larger tension; for flexural and eccentric compression members, taking the cross-sectional area of longitudinal ordinary reinforcement zone. N_q、M_q are the axial force value (N) and bending moment value, $N \cdot mm$ calculated according to the load quasi-permanent combination respectively.

特别提示 9-3

3. 最大裂缝宽度限值
Maximum crack width limit value

对于各种截面的各类钢筋混凝土构件,裂缝计算的内容就是最大裂缝宽度验算。应根据不同的环境和使用要求,使最大裂缝宽度不超过最大裂缝宽度限值 $\omega_{\lim}$,即

For various types reinforced concrete members of various cross-sections, the content of crack calculation is the checking calculation of the maximum crack width. The maximum crack width should not exceed the maximum crack width limit value according to different environments and service requirements, that is

$$\omega_{\max} \leqslant \omega_{\lim} \tag{9-7}$$

式中, $\omega_{\max}$ 为按荷载效应的标准组合并考虑荷载长期作用影响计算的最大裂缝宽度,mm; $\omega_{\lim}$ 为结构构件最大裂缝宽度限值,见表 9-1,mm。

Where, $\omega_{\max}$ is the maximum crack width calculated according to the standard combination of load effect and considering the long-term effect of load, mm; $\omega_{\lim}$ is the maximum crack width limit value of structural members, as shown in Table 9-1, mm.

确定最大裂缝宽度限值,主要考虑外观要求和耐久性要求,并以后者为主。从外观上看,裂缝过宽将给人不安全感,同时也影响对结构质量的评价。根据国内外的调查和试验结果,耐久性所要求的裂缝宽度限值应着重考虑环境条件及结构构件的工作条件。

The maximum crack width limit value should be determined mainly considering the appearance requirements and durability requirements, and the latter is the main consideration. From the appearance, too wide cracks will give people a sense of insecurity, but also affect the evaluation of structural quality. According to the investigation and experimental results at home and abroad, the crack width limit value required by durability should focus on the environmental conditions and the working conditions of structural members.

表 9-1 结构构件的裂缝控制等级及最大裂缝宽度限值 $\omega_{\lim}$

Table 9-1 Crack control level and maximum crack width limit value of structural members $\omega_{\lim}$

环境类别	钢筋混凝土结构		预应力混凝土结构	
	裂缝控制等级	$\omega_{\lim}$/mm	裂缝控制等级	$\omega_{\lim}$/mm
一	三级	0.30(0.40)	三级	0.20
二 a		0.20		0.10
二 b			二级	—
三 a、三 b			一级	—

注:①表中最大裂缝宽度限值,用于验算荷载作用引起的最大裂缝宽度;②表中规定适用于采用热轧钢筋的钢筋混凝土构件和采用预应力钢丝、钢绞线及螺纹钢筋的预应力混凝土构件;③对处于年平均相对湿度小于 60% 地区一类环境下的受弯构件,其最大裂缝宽度可采用括号内数值。

任务 9-1：判断你所在教室中钢筋混凝土板所处的环境类别、板的裂缝控制等级，并从表 9-1 中查出其最大裂缝宽度限值。

Task 9-1: Determine the environment category of the reinforced concrete slab and the crack control level of the slab in your classroom, and find out the maximum crack width limit value from Table 9-1.

4. 裂缝宽度的控制措施
Control measures of crack width

（1）裂缝宽度的验算是在满足构件承载力的前提下进行的，因为截面尺寸、配筋率等均已确定，验算中可能会出现满足挠度要求，却不满足裂缝宽度要求的情况，这通常在配筋率较低而选用的钢筋直径较大的情况下出现。因此，当计算的最大裂缝宽度超过裂缝宽度限值不大时，可选择直径较小的钢筋，宜采用带肋钢筋，必要时可适当增大配筋量或提高混凝土强度等级。

英文翻译 9-3

（2）对于受拉及受弯构件，当承载力要求较高时，往往会出现不能同时满足裂缝宽度或变形限制要求的情况，这时增大截面尺寸或增加用钢量显然是不经济也是不合理的，有效的措施是施加预应力。若最大裂缝宽度超过裂缝宽度限值较大时，也应施加预应力。

（3）在施工过程中遇到钢筋代换时，除必须满足承载力要求外，还应保证构件的裂缝宽度满足限值要求，必要时应进行裂缝宽度验算。

（4）表 9-1 中所列的裂缝宽度限值指的是作用（荷载）的短期效应组合并考虑长期效应组合影响引起的裂缝宽度，不包括施工中混凝土收缩过大、养护不当、掺入氯盐过多等原因引起的非受力裂缝。在设计和施工中应采取切实可行的构造措施和施工措施，防止非受力裂缝的发生和发展，如设计中采用合理的混凝土保护层厚度、合理地选用和布置钢筋；施工中保证混凝土的密实性，选用合理的水灰比、水泥、外加剂，严格养护制度等，都能有效地把非受力裂缝控制在允许的范围内，这比通过计算控制裂缝宽度效果明显得多。

【例题 9-1】 某教学楼钢筋混凝土矩形截面简支梁的结构安全等级为二级，环境类别为一类，截面尺寸为 $b \times h = 300\text{mm} \times 600\text{mm}$，混凝土保护层厚度 $c = 25\text{mm}$，计算跨度 $l_0 = 5\text{m}$，混凝土强度等级为 C30，纵筋为 HRB400 钢筋，实配钢筋 4⌀25，箍筋⌀8，均布永久荷载标准值（包括梁自重）$g_k = 20\text{kN/m}$，均布可变荷载标准值 $q_k = 30\text{kN/m}$。验算梁的最大裂缝宽度是否满足要求。

视频 9-1

【Example 9-1】 The structural safety level of reinforced concrete simply supported rectangular section beam in the teaching building is Level 2, the environmental category is Class I. The cross-sectional dimensions are $b \times h = 300\text{mm} \times 600\text{mm}$, thickness of concrete cover is $c = 25\text{mm}$, effective span is $l_0 = 5\text{m}$, concrete strength grade is C30, longitudinal reinforcement is HRB400, actual reinforcement is 4⌀25, stirrup ⌀8, the uniformly distributed permanent load standard value (including self-weight of the beam) is $g_k = 20\text{kN/m}$, and the uniformly distributed variable load standard value is $q_k = 30\text{kN/m}$. Check whether the maximum crack width of the beam meets the requirements.

解：（1）确定计算参数

混凝土强度等级为C30，查附表1-2得混凝土轴心抗拉强度标准值 $f_{tk}=2.01\text{N/mm}^2$，混凝土弹性模量 $E_c=3.0\times10^4\text{N/mm}^2$，箍筋ϕ8，纵向受力钢筋保护层厚度 $c_s=c+d=(25+8)\text{mm}=33\text{mm}$。

HRB400钢筋，钢筋弹性模量 $E_s=2\times10^5\text{N/mm}^2$，纵筋 4⌽25，$A_s=1964\text{mm}^2$，相对黏结特征系数 $\nu_i=1.0$。

箍筋ϕ8，截面有效高度 $h_0=(600-25-8-25/2)\text{mm}=554.5\text{mm}$。梁为受弯构件，构件受力特征系数 $\alpha_{cr}=1.9$，准永久值系数 $\psi_q=0.5$。

（2）计算纵向受拉钢筋的有效配筋率 ρ_{te} 和应力 σ_{sq}

按荷载效应准永久组合作用计算的跨中弯矩值：

$$M_q=\frac{1}{8}(g_k+\psi_q q_k)l_0^2=\frac{1}{8}\times(20+0.5\times30)\times5^2\text{kN·m}=109\text{kN·m}$$

$$\rho_{te}=\frac{A_s}{A_{te}}=\frac{A_s}{0.5bh}=\frac{1964}{0.5\times300\times600}=0.022>0.01$$

$$\sigma_{sq}=\frac{M_q}{0.87A_s h_0}=\frac{109\times10^6}{0.87\times1964\times554.5}\text{N/mm}^2=115\text{N/mm}^2$$

（3）计算纵向钢筋应变的不均匀系数 ψ

$$\psi=1.1-\frac{0.65 f_{tk}}{\rho_{te}\sigma_{sq}}=1.1-\frac{0.65\times2.01}{0.022\times115}=1.05$$

不均匀系数 ψ 大于1，取1。

（4）计算最大裂缝宽度 ω_{max}

$$\omega_{max}=\alpha_{cr}\psi\frac{\sigma_{sq}}{E_s}\left(1.9c_s+0.08\frac{d_{eq}}{\rho_{te}}\right)$$

$$=1.9\times1.00\times\frac{115}{2\times10^5}\left(1.9\times33+0.08\times\frac{25}{0.022}\right)\text{mm}=0.17\text{mm}$$

（5）验算最大裂缝宽度

$$\omega_{max}=0.17\text{mm}<\omega_{lim}=0.3\text{mm}(\omega_{lim}\text{ 可查表9-1})$$

所以，裂缝宽度满足要求。

 拓展知识9-2，拓展知识9-3，拓展知识9-4， 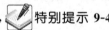 特别提示9-4

9.3 受弯构件的挠度计算
Deflection calculation of flexural members

9.3.1 受弯构件的挠度和刚度
Deflection and stiffness of flexural members

1. 挠度和截面刚度的概念

Concept of deflection and section stiffness

挠度一般指梁、板等受弯构件在荷载作用下的最大竖向变形。挠度形成的因素有自重或外力，温度造成材料的变形，建造过程的疏误，基础沉降，其他因素如材料徐变、干缩等。挠度与荷载大小、构件截面尺寸以及构件材料的物理性能等有关。

结构或构件受力后在截面上产生内力，并使截面产生变形，截面上的材料抵抗变形的能力就是截面刚度。对于承受弯矩的截面，抵抗截面转动的能力就是截面弯曲刚度。截面的转动是以截面曲率来量度的，因此截面弯曲刚度就是使截面产生单位曲率需要施加的弯矩值。当弯矩一定时，截面弯曲刚度越大，其截面曲率就越小，挠度也越小。

2. 简支梁的跨中挠度

Mid-span deflection of simply supported beams

由力学分析可知，弹性材料简支梁在荷载作用下，跨中最大挠度为

$$f = S\frac{Ml_0^4}{EI} \tag{9-8}$$

式中，f 为梁跨中最大挠度，mm；S 为与荷载形式有关的荷载效应系数，均布荷载时 $S=5/48$，跨中集中荷载时 $S=1/12$；M 为梁跨中最大弯矩，N·mm；l_0 为梁的计算跨度，mm；EI 为匀质材料梁的截面弯曲刚度，N·mm^2，其中 E 为材料的弹性模量，N/mm^2，I 为截面惯性矩，mm^4。

特别提示 9-5

由于钢筋混凝土不是匀质弹性材料，其弹性模量随着荷载的增大而减小，在受拉区混凝土开裂后，开裂截面的惯性矩也将发生变化，钢筋混凝土受弯构件的截面抗弯刚度不是一个常数，而是随着弯矩的增大而逐渐减小，其挠度随弯矩增大变化的规律也和匀质弹性材料梁不同。钢筋混凝土梁的弯矩 M 与挠度 f 的关系是不断变化的，为曲线关系，如图 9-3 中实线所示，不能用 EI 来表示。

为区别于匀质弹性材料梁的抗弯刚度，将钢筋混凝土受弯构件在荷载效应标准组合作用下的截面弯曲刚度用 B_s 表示，称短期刚度；考虑荷载长期作用影响后的刚度称长期刚度，用 B 表示。求得截面的长期刚度 B 后，钢筋混

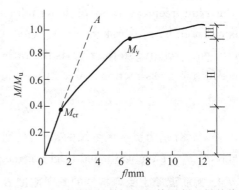

图 9-3 匀质弹性材料梁和钢筋混凝土梁的挠度曲线

Figure 9-3 Deflection curve of homogeneous elastic material beam and reinforced concrete beam

凝土受弯构件的挠度就可按弹性材料梁的挠度计算公式进行计算。钢筋混凝土简支梁跨中挠度的表达式为

$$f = S\frac{Ml_0^4}{B} \tag{9-9}$$

3. 钢筋混凝土受弯构件的短期截面弯曲刚度 B_s

Short-term section flexural stiffness of reinforced concrete flexural members B_s

截面弯曲刚度不仅随弯矩增大而减小,而且还随荷载作用时间的增长而减小。《混凝土结构设计标准》(GB/T 50010—2010)7.2.3 条根据理论推导和试验资料分析,给出不考虑长期荷载长期作用的影响,按裂缝控制等级要求的荷载组合作用下,钢筋混凝土受弯构件短期刚度的计算公式为

The section flexural stiffness decreases not only with the increase of the bending moment, but also with the increase of the loading time. According to the theoretical derivation and test data analysis, Article 7.2.3 of "Standard for design of concrete structures" (GB/T 50010—2010) gives the the formula for calculating the short-term stiffness of reinforced concrete flexural members under the load combination according to the crack control level requirements without considering the influence of long-term effect of long-term load

$$B_s = \frac{E_s A_s h_0^2}{1.15\psi + 0.2 + \dfrac{6\alpha_E \rho}{1 + 3.5 r'_f}} \tag{9-10}$$

式中,α_E 为钢筋弹性模量与混凝土弹性模量的比值,即 E_s/E_c;ρ 为纵向受拉钢筋配筋率,对钢筋混凝土受弯构件,取为 $A_s/(bh_0)$;γ'_f 为受压翼缘加强系数(相对于肋部),是受压翼缘截面面积与腹板有效截面面积的比值,其表达式为

Where, α_E is the ratio of elastic modulus of reinforcement to that of concrete, i.e. E_s/E_c; ρ is the reinforcement ratio of longitudinal tensile reinforcement, for reinforced concrete flexural members, is taken as $A_s/(bh_0)$; γ'_f is the compressive flange strengthening coefficient (relative to the rib), is the ratio of the compressive flange cross-sectional area to the effective cross-sectional area of web, and the expression is

$$\gamma'_f = \frac{(b'_f - b)h'_f}{bh_0} \tag{9-11}$$

式中,b'_f、h'_f 分别为受压区翼缘的宽度、高度,mm。

Where, b'_f, h'_f are the width and height of the flange in the compression zone respectively, mm.

4. 钢筋混凝土受弯构件的长期刚度 B

Long-term stiffness of reinforced concrete flexural members B

在荷载长期作用下,构件截面弯曲刚度将会降低,使构件的挠度增大。长期荷载作用下受弯构件挠度不断增长的原因是在荷载的长期作用下,受压区混凝土发生徐变(最重要的因素),使受压应变随时间增长,受拉钢筋与受拉混凝土之间的黏结滑移徐变,受拉混凝土的应力松弛以及裂缝的向上发展,导致受拉混凝土不断退出工作,使受拉钢筋平均应变随时间增大。

以上因素使构件截面弯曲刚度降低,挠度增大。在实际工程中,总有部分荷载长期作用在构件上,因此,《混凝土结构设计标准》(GB/T 50010—2010)7.2.2 条规定,矩形、T 形、倒 T 形和 I 形截面受弯构件考虑荷载长期作用影响的刚度 B 可按下列规定

计算。

1) 采用荷载标准组合时

1) When using load standard combination

$$B = \frac{M_k}{M_q(\theta-1)+M_k}B_s \tag{9-12}$$

式中,M_k 为按荷载的标准组合计算的弯矩,取计算区段内的最大弯矩值,N·mm;M_q 为按荷载准永久组合计算的弯矩,取计算区段内的最大弯矩值,N·mm;B_s 为按标准组合计算的预应力混凝土受弯构件的短期刚度,N·mm²;θ 为考虑荷载长期作用对挠度增大的影响系数,为长期荷载下的挠度与短期荷载下的挠度之比。

2) 采用荷载准永久组合时

2) When using load quasi-permanent combination

$$B = \frac{B_s}{\theta} \tag{9-13}$$

式中,B_s 为按荷载准永久组合计算的钢筋混凝土受弯构件的短期刚度,N·mm²;挠度增大系数 θ 可按《混凝土结构设计标准》(GB/T 50010—2010)7.2.5 条的规定取用:当受压钢筋配筋率 $\rho'=0$ 时,取 $\theta=2.0$;当 $\rho'=\rho$ 时,取 $\theta=1.6$;当 ρ' 为中间数值时,θ 按线性内插法取用,此处 $\rho'=A_s'/(bh_0)$,$\rho=A_s/(bh_0)$。对翼缘位于受拉区的倒 T 形截面,θ 应增加 20%。

式(9-13)实际上是考虑荷载中长期作用的那部分使构件的刚度降低,对短期刚度进行修正。

9.3.2 受弯构件的挠度验算
Deflection calculation of flexural members

1. 提出构件挠度限值要求的原因

Reasons for proposing deflection limit value requirements for members

随着高强混凝土和钢筋的采用,构件截面尺寸相应减小,变形问题更为突出。一般建筑对混凝土构件的挠度有一定的限值要求,主要是出于以下四方面的考虑。

(1) 保证建筑的使用功能要求。例如,楼盖梁、板的挠度过大,将使仪器设备难以保持水平;吊车梁的挠度过大会妨碍吊车的正常运行;屋面构件和挑檐的挠度过大会造成积水和渗漏等。

(2) 防止对结构构件产生不良影响。例如,梁端的旋转将使支承面积减小;当梁支承在砖墙上时,可能使墙体沿梁顶、梁底出现内外水平缝,严重时将产生局部承压破坏或墙体失稳破坏;构件挠度过大,在可变荷载下可能出现因动力效应引起的共振等。

(3) 防止对非结构构件产生不良影响。例如,防止结构构件变形过大而使门窗等活动部件不能正常开关,防止非结构构件如隔墙及天花板的开裂、压碎、膨出或其他形式的破坏等。

(4) 保证人们的感觉在可接受的程度之内。例如,防止梁、板挠度过大、明显下垂引起的不安全感;防止可变荷载引起的振动及噪声使人产生不良感觉等。

2. 最小刚度原则
Principle of minimum stiffness

钢筋混凝土挠度计算的关键是要解决抗弯刚度 B 的合理取值问题。由于钢筋混凝土梁的弯矩一般沿梁轴线是变化的,梁是带裂缝工作的,各截面裂缝的开展程度和配筋通常都不同,因此其刚度不仅是随时间变化的,而且沿梁长度也是变化的。显然,按照梁长变化的刚度来计算挠度是十分烦琐的。为了简化计算,《混凝土结构设计标准》(GB/T 50010—2010)7.2.1 条规定,对于等截面构件,可假定各同号弯矩区段内的刚度相等,并取用该区段内最大弯矩处的刚度(即该区段内的最小刚度),然后按力学方法计算挠度。这就是最小刚度原则,如图 9-4 所示。

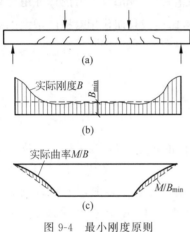

图 9-4　最小刚度原则

Figure 9-4　Principle of minimum stiffness

3. 受弯构件挠度的验算步骤
Checking steps of deflection of flexural members

持久状况下挠度验算应按正常使用极限状态的要求,采用荷载的短期效应组合、长期效应组合或者短期效应组合并考虑长期效应组合的影响。验算受弯构件的挠度是为了保证其正常使用,计算的最大竖向挠度不应超过《混凝土结构设计标准》(GB/T 50010—2010)规定的限值。

The deflection calculation under the persistent condition should adopt the short-term effect combination, long-term effect combination or short-term effect combination of load and consider the influence of long-term effect combination according to the requirements of serviceability limit state. The deflection of flexural members is checked to ensure their normal service. The maximum vertical deflection calculated should not exceed the limit value specified in "Standard for design of concrete structures"(GB/T 50010—2010).

受弯构件挠度验算按下列步骤进行：①计算荷载短期效应组合值 M_k 和荷载长期效应组合值 M_q；②计算短期刚度 B_s；③计算长期刚度 B；④用 B 代替力学位移公式中的 EI,计算构件的最大挠度 f_{max},并按公式 $f_{max} \leqslant f_{lim}$ 进行验算,公式如下：

The deflection calculation of flexural members should be carried out according to the following steps：①Calculate the combination value of load short-term effect M_k and the combination value of load long-term effect M_q；②Calculate the short-term stiffness B_s；③ Calculate the long-term stiffness B；④ Substitute B for EI in the mechanical displacement formula, calculate the maximum deflection f_{max} of the member and check the calculation according to the formula $f_{max} \leqslant f_{lim}$. The formula is as follows：

$$f_{max} = S \frac{M_q l_0^2}{B} \leqslant f_{lim} \tag{9-14}$$

式中,f_{max} 为受弯构件最大挠度,mm；S 为与荷载形式有关的荷载效应系数,如均布荷载作用时 $S=5/48$,跨中集中荷载作用时 $S=1/12$；M_q 为按荷载的准永久组合计算的弯矩,取计算区段内的最大弯矩值，N·mm；l_0 为梁的计算跨度，mm；B 为梁的长期刚度，N·mm²；$f_{\lim}$ 为受弯构件的挠度限值，见表 9-2，mm。

Where, $f_{\max}$ is the maximum deflection of the flexural member, mm; S is the load effect coefficient related to load form, such as $S=5/48$ under uniform load, and $S=1/12$ under concentrated load at midspan; M_q is the bending moment calculated according to the quasi-permanent load combination, taking the maximum bending moment value in the calculation section, N·mm; l_0 is the effective span of the beam, mm; B is the long-term stiffness of the beam, N·mm²; $f_{\lim}$ is the deflection limit value of the flexural member, see Table 9-2, mm.

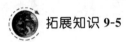
拓展知识 9-5

表 9-2　受弯构件的挠度限值 $f_{\lim}$
Table 9-2　Deflection limit value of flexural members $f_{\lim}$

构件类型		挠度限值/mm
吊车梁	手动吊车	$l_0/500$
	电动吊车	$l_0/600$
屋盖、楼盖及楼梯构件	$l_0<7\mathrm{m}$	$l_0/200(l_0/250)$
	$7\mathrm{m}\leqslant l_0\leqslant 9\mathrm{m}$	$l_0/250(l_0/300)$
	$l_0>9\mathrm{m}$	$l_0/300(l_0/400)$

注：①表中 l_0 为构件的计算跨度；②表中括号内数值适用于使用上对挠度有较高要求的构件；③计算悬臂构件的挠度限值时，其计算跨度按实际悬臂长度的 2 倍取用。

4. 减小构件挠度的措施
Measures to reduce the deflection of members

若验算结果 $f_{\max}>f_{\lim}$，不满足《混凝土结构设计标准》（GB/T 50010—2010）要求时，要减小挠度值，就需增大抗弯刚度。由短期刚度计算式(9-10)可知增大截面高度是提高截面弯曲刚度、减小构件挠度的最有效措施。对某些构件还可以充分利用纵向受压钢筋对长期刚度的有利影响，在受压区配置一定数量的受压钢筋，采用双筋截面；采用预应力混凝土构件也是提高受弯构件刚度的有效措施；还可以采取增大受拉钢筋配筋率、选择合理的截面形式(T 形、I 形)以及提高混凝土强度等级等措施。实际工程中，往往采用控制跨高比（跨度和高度之比）的方法来满足挠度控制条件的要求。

If the checking result $f_{\max}>f_{\lim}$ does not meet the requirements of "Standard for design of concrete structures" (GB/T 50010—2010), the flexural stiffness needs to be increased to reduce deflections. From the short-term stiffness calculation Eq. (9-10), it can be seen that increasing the height of the section is the most effective measure to improve the flexural stiffness of the section and reduce the deflection of the member. For some members, the beneficial effect of longitudinal compression reinforcement on long-term stiffness can also be fully utilized. A certain number of compression reinforcement can be arranged in the compression zone, and the doubly reinforced section is adopted; the use of prestressed concrete members is also an effective measure to improve the stiffness of flexural members. Measures such as increasing the reinforcement ratio of tensile reinforcement, selecting reasonable section forms (T-shaped and I-shaped) and improving the concrete strength grade and other measures can also be taken. In practical engineering, the method of controlling span-depth ratio (the ratio of span to depth) is often used to meet the requirements of deflection control conditions.

【例题 9-2】 验算例题 9-1 中梁的跨中最大挠度是否满足要求。

【Example 9-2】 Check whether the maximum deflection at mid-span of the beam in Example 9-1 meets the requirements.

解：(1) 确定计算参数

混凝土强度等级为 C30，混凝土轴心抗拉强度标准值 $f_{tk}=2.01\text{N/mm}^2$，混凝土弹性模量 $E_c=3.0\times10^4\text{N/mm}^2$，箍筋 $\phi 8$，纵向受力钢筋保护层厚度 $c_s=(25+8)\text{mm}=33\text{mm}$。

HRB400 钢筋，钢筋弹性模量 $E_s=2\times10^5\text{N/mm}^2$，纵筋 $4\underline{\Phi}25$，$A_s=1964\text{mm}^2$，相对黏结特征系数 $\nu_i=1.0$。

截面有效高度 $h_0=(600-25-8-25/2)\text{mm}=554.5\text{mm}$；梁为受弯构件，则 $\alpha_E=E_s/E_c=2\times10^5/3\times10^4=6.67$。

纵筋配筋率 $\rho=A_s/bh_0=1964/(300\times554.5)=0.012=1.2\%$；纵向钢筋应变的不均匀系数 $\psi=1$，$\rho_{te}=0.022>0.01$，纵向受拉钢筋的应力 $\sigma_{sq}=115\text{N/mm}^2$，按荷载准永久组合计算的弯矩 $M_q=109\text{kN}\cdot\text{m}$。

(2) 计算短期刚度

矩形截面 $\gamma'_f=0$，则短期刚度为

$$B_s=\frac{E_s A_s h_0^2}{1.15\psi+0.2+\dfrac{6\alpha_E\rho}{1+3.5\gamma'_f}}$$

$$=\frac{2\times10^5\times1964\times554.5^2}{1.15\times1+0.2+\dfrac{6\times6.67\times0.012}{1+3.5\times0}}\text{N}\cdot\text{mm}^2$$

$$=65.98\times10^{12}\text{N}\cdot\text{mm}^2$$

(3) 计算长期刚度

$$\theta=1.6+0.4\left(1-\frac{\rho'}{\rho}\right)=1.6+0.4\left(1-\frac{0}{0.012}\right)=2$$

$$B=\frac{B_s}{\theta}=\frac{65.98\times10^{12}}{2}\text{N}\cdot\text{mm}^2=32.99\times10^{12}\text{N}\cdot\text{mm}^2$$

(4) 计算梁的跨中最大挠度并验算

$$f_{max}=\frac{5}{48}\cdot\frac{M_q l_0^2}{B}=\frac{5}{48}\times\frac{109\times10^6\times5^2\times10^6}{32.99\times10^{12}}\text{mm}=8.6\text{mm}$$

楼盖梁的 $l_0=5\text{m}<7\text{m}$，查表 9-2 得挠度限值 $f_{lim}=l_0/200=5000/200\text{mm}=25\text{mm}$，则 $f_{max}=8.6\text{mm}<f_{lim}=25\text{mm}$，所以满足要求。

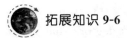
拓展知识 9-6

9.4 混凝土结构的耐久性
Durability of concrete structures

9.4.1 耐久性的一般概念
General concept of durability

混凝土结构应满足安全性、适用性和耐久性三方面的功能要求。承载力计算及裂缝宽度、变形验算分别是为了满足安全性和适用性要求的。混凝土结构的耐久性是指结构或构件在设计工作年限内,在正常维护条件下,不需要进行大修就可满足正常使用和安全功能要求的能力。结构在正常使用和正常维护条件下应具有足够的耐久性能,以保证结构能够正常使用到预定的设计工作年限,一般建筑结构的设计工作年限为 50 年,纪念性建筑和特别重要的建筑结构为 100 年及以上。例如,在设计规定的使用期内,钢筋不致因混凝土保护层过薄或裂缝过宽而发生锈蚀等,影响结构的使用年限。

英文翻译 9-7

混凝土耐久性可分为材料的耐久性和结构的耐久性两个层次。具有耐久性的混凝土当露置于使用环境时将保持其原来的形状、质量和适用性。混凝土的耐久性从广义上说应包括大气对混凝土的腐蚀(如干湿度、温度、冻融、碳化等)、水对混凝土的作用、碱骨料反应、环境水侵蚀和磨损等。抗渗性、抗冻性、抗侵蚀性均可反映混凝土耐久性。

混凝土结构的耐久性按耐久性极限状态控制,特别是随时间发展因材料劣化而引起材料性能衰退。耐久性极限状态表现为钢筋混凝土构件表面出现锈胀裂缝,预应力钢筋开始锈蚀,结构表面混凝土出现可见的耐久性损失(酥裂、粉化)等。

9.4.2 影响混凝土结构耐久性的因素
Factors affecting the durability of concrete structures

影响混凝土结构耐久性的因素很多,主要是内部因素和外部因素综合作用的结果。内部因素主要有混凝土的强度、密实性、水泥用量、水灰比、氯离子及碱含量、外加剂用量、混凝土保护层厚度等。外部因素主要是环境条件,如温度、湿度、CO_2 含量、侵蚀性介质等。另外,设计问题、施工质量、使用中维护不当也会影响耐久性。其中影响耐久性最重要的因素是混凝土碳化和钢筋锈蚀。对于影响混凝土结构耐久性的因素还需进一步研究,目前采用的是宏观的控制方法。

There are many factors that affect the durability of concrete structures, mainly due to the comprehensive effect of internal and external factors. Internal factors mainly include concrete strength, compactness, cement consumption, water cement ratio, chloride ion and alkali content, dosage of admixture and thickness of concrete cover, etc. External factors are mainly environmental conditions, such as temperature, humidity, CO_2 content, aggressive media, etc. In addition, design problems, construction quality and improper maintenance in use will also affect the durability. Among them, the most important factors affecting durability are concrete carbonization and steel corrosion. The factors that affect the durability of concrete structures need to be further studied. At present, the macro control method is adopted.

1. 影响混凝土耐久性的因素

Factors affecting the durability of concrete

1) 混凝土碳化

1) Carbonization of concrete

（1）碳化的概念

(1) Concept of carbonation

混凝土因水泥水化反应产生碱性水化物而呈碱性，但当外部酸性介质（如大气环境中的CO_2）不断向混凝土内部扩散时，会与混凝土中的碱性水化物（主要是氢氧化钙）发生中和反应生成碳酸钙，即混凝土中性化，称为混凝土的碳化。碳化对混凝土本身是无害的，但当碳化深度达到钢筋表面，会破坏钢筋表面的氧化膜（脱钝），为钢筋锈蚀创造条件。同时碳化会加剧混凝土的收缩，导致混凝土开裂，使钢筋容易锈蚀。

Concrete is alkaline due to the alkaline hydration produced by the cement hydration reaction. However, when the external acidic medium (such as CO_2 in the atmospheric environment) continuously diffuses into the concrete, it neutralizes the alkaline hydrate in the concrete (mainly calcium hydroxide), and generates calcium carbonate, that is, the neutralization of concrete, which is called concrete carbonation. Carbonation is harmless to the concrete itself, but when the carbonation depth reaches the surface of the reinforcement, the oxide film on the reinforcement surface will be damaged (depassivation), creating conditions for steel corrosion. At the same time, carbonation will aggravate the shrinkage of concrete leading to concrete cracking and making the reinforcement easy to rust.

（2）碳化深度的确定

(2) Determination of carbonation depth

碳化深度可用碳酸试液测定，当敲开混凝土滴入试液后，碳化的部分保持原色，未碳化部分混凝土呈浅红色。

The carbonization depth can be measured with the carbonic acid test solution. When the concrete is knocked open and the test solution is dropped, the carbonized part will remain the original color, and the uncorbonized part of the concrete will be light red.

（3）影响碳化的因素

(3) Factors affecting carbonization

影响碳化的因素主要有环境因素和材料本身因素。环境因素主要是空气中CO_2的浓度，一般室内的CO_2浓度比较高，所以室内混凝土的碳化比室外的快些。试验表明，混凝土周围相对湿度为50%~70%时，碳化速度较快。温度交替变化有利于CO_2的扩散，可加速混凝土的碳化。混凝土强度等级越高，内部结构越密实，孔隙率越低，孔径越小，碳化速度越慢。水灰比大会加速碳化。

The main factors affecting carbonization are environmental factors and material factors. The environmental factor is mainly the concentration of CO_2 dioxide in the air. Generally, the indoor concentration of CO_2 dioxide is relatively high, so the carbonation of indoor concrete is faster than that of outdoor concrete. The test shows that the carbonation rate is faster when the relative humidity around the concrete is 50%-70%. Alternating temperature changes are conducive to the diffusion of CO_2 which can accelerate the carbonation of concrete. The higher the concrete strength grade is, the denser the internal structure is, the lower the porosity is, the smaller the pore diameter is, the slower the carbonation rate is. Water-cement ratio will accelerate carbonation.

(4) 延缓碳化的主要措施

(4) Main measures to delay carbonation

合理设计混凝土配合比,规定水泥用量的下限值和水灰比的上限值,合理采用掺合料;提高混凝土的密实性和抗渗性;规定混凝土保护层最小厚度;采用覆盖面层(水泥砂浆或涂料等)。

Reasonably design the concrete mix proportion, specify the lower limit value of cement consumption and the upper limit value of water-cement ratio, and reasonably use admixtures; improve compactness and impermeability of concrete; specify the minimum thickness of concrete cover; adopt the covering surface layer (cement mortar or paint, etc).

2) 温度、湿度变化

2) Change of temperature and humidity

混凝土因温度变化产生热胀冷缩,因湿度变化产生干缩或湿胀,当这种胀缩受到约束影响时,混凝土就将开裂。

The concrete will expand with heat and shrink with cold due to temperature changes, and will shrink with dry or expand with wet due to humidity changes. When this expansion and shrinkage is affected by the constraint, the concrete will crack.

3) 化学侵蚀

3) Chemical corrosion

化学侵蚀是指混凝土结构处于有侵蚀性化学物质环境中时,导致混凝土酥松、剥落,丧失力学性能。

Chemical corrosion refers to that when the concrete structure is in an environment with corrosive chemicals, it will cause the concrete to become loosen, peeling and lose its mechanical properties.

4) 冻融破坏

4) Freeze-thaw damage

冻融破坏是指混凝土在低温下因内部的水结冰膨胀,破坏混凝土内部的结构,导致混凝土剥落破坏。随着冻融循环的反复,材料的破坏逐步加剧。

Freeze-thaw damage refer to the process in which concrete freezes and expands due to the interior water at low temperature, causing damage to the internal structure of the concrete and resulting in concrete spalling and failure. With the repeated freeze-thaw cycles, the damage of the material gradually increased.

5）碱骨料反应

5) Alkali aggregate reaction

混凝土中的碱金属与含有碱活性的骨料发生化学反应生成碱活性物质,这种物质吸水后体积急剧膨胀,造成混凝土破坏。

Alkali metals in the concrete react with aggregates containing alkali activity to produce alkali active substances, which expand rapidly after water absorption, causing concrete damage.

2. 影响钢筋耐久性的因素

Factors affecting the durability of reinforcement

1）钢筋锈蚀

1) Steel corrosion

钢筋表面氧化膜的破坏是使钢筋锈蚀的必要条件,如果含氧水分侵入,钢筋就会锈蚀。混凝土的碳化、侵蚀性的酸性介质,特别是氯离子都会引起钢筋锈蚀。钢筋锈蚀严重时,体积膨胀,沿钢筋长度方向出现纵向裂缝,使混凝土保护层胀裂、脱落,进一步加快钢筋的锈蚀速度,使钢筋截面削弱,受力面积减少,截面承载力降低,最终使结构构件破坏或失效。

The destruction of the oxide film on the surface of steel bars is a necessary condition for steel corrosion. If oxygen-containing water invades, the steel bars will rust. The carbonation of concrete, corrosive acidic media, especially chloride ions can cause corrosion of steel bars. When the reinforcement is seriously corroded, its volume expands and longitudinal cracks appear along the length direction of the reinforcement, which cause the concrete cover to expand and fall off, further accelerate the corrosion rate of the reinforcement, weaken the reinforcement section, reduce the load-bearing area, reduce the bearing capacity of the section, and finally cause damage or failure of structural members.

混凝土中钢筋的锈蚀机理是电化学腐蚀。钢筋中化学物质的不均匀分布、混凝土碱度的差异以及裂缝处氧气增浓等原因,使钢筋表面各部位之间产生电位差,从而形成了许多具有阳极和阴极的微电池。

The corrosion mechanism of reinforcement in concrete is electrochemical corrosion. Due to the uneven distribution of chemical substances in the reinforcement, the difference of concrete alkalinity and the increase of oxygen concentration at the crack, potential differences are generated between various parts of the reinforcement surface, thus forming many micro batteries with anodes and cathodes.

钢筋表面的氧化膜被破坏后,钢筋表面从空气中吸收溶有 CO_2、O_2 或 SO_2 的水分,在微电池中形成电解质水膜,在阴极和阳极间以电解方式产生电化学腐蚀反应,生成氢氧化亚铁,它在空气中又进一步被氧化成氢氧化铁,即铁锈。铁锈是疏松多孔的,体积比原来增加 2~4 倍,使混凝土保护层胀裂,进一步加快了锈蚀的发展。

After the oxide film on the surface of steel bars is destroyed, the steel bar surface absorbs water dissolved in CO_2, O_2 or SO_2 from the air, forms an electrolyte water film in the micro battery, and generates electrochemical corrosion reaction between the cathode and anode by electrolysis, and generates ferrous hydroxide, which is further oxidized into ferric hydroxide in the air, namely rust. The rust is loose and porous, and the volume is 2～4 times higher than that of the original, which makes the concrete cover expand and crack, and further accelerates the development of rust.

钢筋锈蚀是一个相当长的过程，先是在裂缝较宽的个别点上"坑蚀"逐渐形成"环蚀"，同时向两边扩展，形成锈蚀面，使钢筋截面削弱。锈蚀严重时，体积膨胀，导致沿钢筋长度方向的混凝土产生纵向裂缝，使混凝土保护层剥落，称为"暴筋"。通常可把大范围内出现沿钢筋长度方向的纵向裂缝作为判别混凝土结构构件寿命终结的标准。

Steel corrosion is a fairly long process. First, "pitting" gradually forms "annular corrosion" at individual points with wide cracks, and at the same time it expands to both sides to form a corrosion surface, weakening the reinforcement section. When the corrosion is serious, the volume expands, causing longitudinal cracks in the concrete along the length of the reinforcement, which cause the spalling of the concrete cover, which is called "exposed reinforcement". Generally, the longitudinal cracks along the length of the steel bar in a large range can be regarded as the criteria for judging the end of the service life of concrete structural members.

防止钢筋锈蚀的主要措施有：降低水灰比，增加水泥用量，提高混凝土的密实度；保证足够的混凝土保护层厚度；严格控制氯离子含量；采用覆盖层，防止 CO_2、O_2、Cl^- 的渗入等。

The main measures to prevent steel corrosion are: reduce the water-cement ratio, increase the cement consumption and improve the compactness of concrete; ensure sufficient thickness of concrete cover; strictly control the content of chloride ion; adopt the covering layer to prevent the infiltration of CO_2, O_2 and Cl^-, etc.

2) 应力腐蚀

2) Stress corrosion

钢筋在应力状态下会发生电位变化，使电化学作用引起的钢筋锈蚀速度加快。

The potential of reinforcement will change under stress state, which will accelerate the steel corrosion rate caused by electrochemical action.

9.4.3 混凝土结构的耐久性设计
Durability design of concrete structures

1. 耐久性设计的内容

   Content of durability design

由于影响混凝土结构材料性能劣化的因素比较复杂，不确定性很大，一般混凝土结构耐久性设计只能采用经验性的定性方法解决。参考《混凝土结构耐久性设计标准》(GB/T 50476—2019)的规定，根据调查研究及我国国情，《混凝土结构设计标准》(GB/T 50010—2010)规定了混凝土结构耐久性设计的基本内容：①确定结构所处的环境类别；②提出对混凝土材料的耐久性基本要求；③确定构件中钢筋的混凝土保护层厚度；④确定不同环境条件下的耐久性技术措施；⑤提出结构使用阶段的检测与维护要求。

对临时性的混凝土结构,可不考虑混凝土的耐久性要求。

2. 结构混凝土材料的耐久性基本要求

Basic requirements for durability of structural concrete materials

混凝土的密实性、氯离子含量和碱含量是影响耐久性的重要因素,《混凝土结构设计标准》(GB/T 50010—2010)对结构用混凝土提出相应的要求和限制,以保证其耐久性。设计工作年限为 50 年的混凝土结构,其混凝土材料的耐久性基本要求见表 9-3。

表 9-3 结构混凝土材料的耐久性基本要求

Table 9-3 Basic requirements for durability of structural concrete materials

环境类别	最大水胶比	最低强度等级	水溶性氯离子最大含量/%	最大碱含量/(kg/m³)
一	0.60	C25	0.30	不限制
二 a	0.55	C25	0.20	3.0
二 b	0.50(0.55)	C30(C25)	0.10	
三 a	0.45(0.50)	C35(C30)	0.10	
三 b	0.40	C40	0.06	

注:①氯离子含量系指其占胶凝材料的质量百分比,计算时辅助胶凝材料的量不应大于硅酸盐水泥的量;②预应力构件混凝土中的水溶性氯离子最大含量为 0.06%,其最低混凝土强度等级宜按表中的规定提高不少于两个等级;③素混凝土结构的混凝土水胶比及最低强度等级的要求可适当放松,但混凝土最低强度等级应符合本标准的有关规定;④有可靠工程经验时,二类环境中的最低混凝土强度等级为 C25;⑤处于严寒和寒冷地区二 b、三 a 类环境中的混凝土应使用引气剂,并可采用括号中的有关参数;⑥当使用非碱活性骨料时,对混凝土中的碱含量可不作限制。

处于一类环境中,设计工作年限为 100 年的混凝土结构应符合下列规定:

(1) 钢筋混凝土结构的最低强度等级为 C30,预应力混凝土结构的最低强度等级为 C40。

(2) 混凝土中的最大氯离子含量为 0.06%。

(3) 宜使用非碱活性骨料,当使用碱活性骨料时,混凝土中的最大碱含量为 3.0kg/m³。

(4) 最外层钢筋的混凝土保护层厚度应比设计工作年限为 50 年规定的保护层厚度增加 40%;当采取有效的表面防护措施时,混凝土保护层厚度可适当减少。

处于二类和三类环境中,设计工作年限为 100 年的混凝土结构,应采取专门的有效措施。处于四类和五类环境中的混凝土结构,其耐久性要求应符合有关标准的规定。

3. 混凝土结构及构件应采取的耐久性技术措施

Durability technical measures to be taken for concrete structures and members

(1) 预应力混凝土结构中的预应力筋应根据具体情况采取表面防护、孔道灌浆、加大混凝土保护层厚度等措施。外露的锚固端应采取封锚和混凝土表面处理等有效措施。

(2) 有抗渗要求的混凝土结构,混凝土的抗渗等级应符合有关标准的要求。

(3) 严寒及寒冷地区的潮湿环境中,结构混凝土应满足抗冻要求,混凝土抗冻等级应符合有关标准的要求。

(4) 处于二、三类环境中的悬臂构件宜采用悬挑梁-板的结构形式,或在其上表面增设防护层。

(5) 处于二类和三类环境中的结构构件,其表面的预埋件、吊钩、连接件等金属部位应采取可靠的防锈措施。后张预应力混凝土外露金属锚具的防护要求见《混凝土结构设计标准》(GB/T 50010—2010)10.3.13 条。

(6) 处于三类环境中的混凝土结构构件,可采用阻锈剂、环氧树脂涂层钢筋或其他具有耐腐蚀性能的钢筋,采取阴极保护措施或采用可更换的构件等措施。

4. 结构在设计工作年限内的检测与维护要求

Inspection and maintenance requirements for structures in the design working life

(1) 建立定期检测、维修制度。
(2) 设计中可更换的混凝土构件应按规定更换。
(3) 构件表面的防护层应按规定维护或更换。
(4) 结构出现可见的耐久性缺陷时,应及时进行处理。

5. 结构处置

Structural disposal

出现下列情况之一时,应采取消除安全隐患的措施进行处理:
(1) 混凝土结构或结构构件的裂缝宽度或挠度超过限值。
(2) 混凝土结构或构件钢筋出现锈胀。
(3) 预应力混凝土构件锚固端的封端混凝土出现裂缝、剥落、渗漏、穿孔、预应力锚具暴露。
(4) 结构混凝土中氯离子含量超标或发现有碱骨料反应迹象。

(1) 钢筋混凝土构件的裂缝宽度及挠度验算是为了满足正常使用极限状态的要求,验算时要采用荷载标准值和可变荷载准永久值及材料强度标准值。同时,因为裂缝宽度及挠度都是随时间而增大的,所以,在验算时还要考虑荷载短期效应组合与荷载长期效应组合等问题。

(2) 减小纵向受拉钢筋直径,采用变形钢筋是减小裂缝宽度经济而有效的方法。加大截面高度是提高截面弯曲刚度最有效的方法。当梁、板截面高度满足一定的跨高比后,可以略去挠度验算。

(3) 钢筋混凝土受弯构件的挠度计算问题首先是如何确定其抗弯刚度的问题。在短期荷载作用下,根据平截面假定建立裂缝间距内的曲率与弯矩和短期刚度的关系式,再引入平衡关系和材料的物理关系及实测结果,就可得到短期刚度的计算表达式。同时还要考虑构件在长期荷载作用下,由于混凝土徐变等原因,抗弯刚度不断缓慢降低的情况,即考虑其长期刚度,用它可以计算钢筋混凝土梁在短期荷载效应作用下并考虑长期荷载影响(混凝土徐变等影响)的挠度。为了方便挠度验算,在构件的同号弯矩区段内的截面抗弯刚度近似取其最小值,即最小刚度原则。

(4) 混凝土结构耐久性的影响因素很多,混凝土碳化和钢筋锈蚀是两个最主要的影响因素,混凝土结构耐久性问题应得到更多重视,开展进一步研究。

第 9 章拓展知识和特别提示

视频：第 9 章小结讲解

习 题

9-1 选择题

1. 裂缝宽度验算是为了保证构件（　　）。
   A. 进入承载能力极限状态的概率足够小
   B. 进入正常使用极限状态的概率足够小
   C. 能在弹性阶段工作
   D. 能在带裂缝阶段工作

2. 抗裂度、裂缝宽度和变形验算中所采用的荷载及材料强度指标是（　　）。
   A. 荷载的设计值，材料强度的设计值
   B. 荷载的设计值，材料强度的标准值
   C. 荷载的标准值，材料强度的标准值
   D. 荷载的标准值，材料强度的设计值

3. 当验算受弯构件挠度时，出现超过挠度限值的情况，采取下列（　　）的措施是最有效的。
   A. 加大截面高度　　　　　　　　　　B. 加大截面宽度
   C. 提高混凝土强度等级　　　　　　　D. 提高钢筋强度等级

4. 有两根钢筋混凝土梁，甲梁的抗弯承载力比乙梁的抗弯承载力大，则甲梁的刚度（　　）。
   A. 比乙梁的刚度大
   B. 比乙梁的刚度小
   C. 可能比乙梁的刚度大，也可能比乙梁的刚度小

5. 长期荷载作用下，钢筋混凝土梁的挠度会随时间而增大，其主要原因是（　　）。
   A. 受拉钢筋产生塑性变形　　　　　　B. 受拉混凝土产生塑性变形
   C. 受压混凝土产生塑性变形　　　　　D. 受压混凝土产生徐变

9-2 思考题

1. 对于钢筋混凝土受弯构件，正常使用极限状态验算包括哪些内容？
2. 钢筋混凝土构件的裂缝和挠度过大都有哪些主要的危害？
3. 一根钢筋混凝土梁，从制作开始直至使用阶段，可能由于种种原因使构件产生裂缝，分析可能引起梁产生裂缝的各种原因；在钢筋混凝土构件中，大部分裂缝是由什么原因引起的？
4. 写出《混凝土结构设计标准》(GB/T 50010—2010)最大裂缝宽度计算公式及公式中

各个符号的含义。

5. 钢筋混凝土构件在正常使用阶段，若计算所得的最大裂缝宽度超过裂缝宽度限值时，应采取什么措施？

6. 裂缝对钢筋混凝土构件的耐久性有什么影响？

7. 设计人员在验算裂缝宽度时，发现超过《混凝土结构设计标准》(GB/T 50010—2010)要求，但超过不多，需要适度修改设计，该如何进行？

8. 钢筋混凝土受弯构件在荷载作用下开裂后，其刚度为什么不能直接用 EI 来计算？

9. 在长期荷载作用下，钢筋混凝土受弯构件的挠度为什么会增大？受弯构件挠度主要的影响因素有哪些？其中，最主要的影响因素是什么？

10. 在挠度和裂缝宽度验算公式中，是怎样体现"按荷载标准组合并考虑荷载准永久组合影响"进行计算的？

11. 混凝土结构耐久性的主要影响因素有哪些？

12. 如何进行混凝土结构耐久性概念设计？

9-3 计算题

1. 某矩形截面简支梁，结构安全等级二级，处于室内环境，截面尺寸为 $b \times h = 300\text{mm} \times 600\text{mm}$，计算跨度 $l_0 = 6\text{m}$，混凝土强度等级为 C25，纵向受拉钢筋 HRB400。使用期间承受均布荷载，其中永久荷载标准值 $g_k = 15\text{kN/m}$（包括自重），可变荷载标准值 $q_k = 7\text{kN/m}$，楼面可变荷载的准永久值系数 $\psi_q = 0.5$。求纵向受拉钢筋截面面积 A_s，验算最大裂缝宽度。

2. 某矩形截面简支梁，结构安全等级二级，处于室内环境，截面尺寸为 $b \times h = 200\text{mm} \times 500\text{mm}$，配置纵向受力钢筋 4$\Phi$16，混凝土强度等级为 C30，混凝土保护层厚度 $c = 25\text{mm}$，箍筋直径为 8mm，计算跨度 $l_0 = 5.6\text{m}$。使用期间承受均布荷载，其中永久荷载标准值 $g_k = 12\text{kN/m}$（包括自重），可变荷载标准值 $q_k = 6\text{kN/m}$，楼面可变荷载的准永久值系数 $\psi_q = 0.5$。验算梁的挠度。

3. 某矩形截面简支梁，截面尺寸为 $b \times h = 250\text{mm} \times 600\text{mm}$，计算跨度 $l_0 = 6\text{m}$。承受均布荷载，永久荷载标准值 $g_k = 8\text{kN/m}$，可变荷载标准值 $q_k = 10\text{kN/m}$，可变荷载的准永久值系数 $\psi_q = 0.5$。混凝土强度等级为 C25，在受拉区配置钢筋 2Φ20+2Φ16。混凝土保护层厚度为 $c = 25\text{mm}$，最大裂缝宽度限值为 0.3mm，梁的挠度限值为 $l_0/200$。验算最大裂缝宽度和梁的挠度。

参 考 文 献

[1] 中华人民共和国住房和城乡建设部.混凝土结构设计标准:GB/T 50010—2010[S].北京:中国建筑工业出版社,2024.
[2] 中华人民共和国住房和城乡建设部.工程结构通用规范:GB 55001—2021[S].北京:中国建筑工业出版社,2021.
[3] 中华人民共和国住房和城乡建设部.混凝土结构通用规范:GB 55008—2021[S].北京:中国建筑工业出版社,2022.
[4] 中华人民共和国住房和城乡建设部.建筑结构荷载规范:GB 50009—2012[S].北京:中国建筑工业出版社,2012.
[5] 中华人民共和国住房和城乡建设部.建筑抗震设计规范:GB 50011—2010(2016年版)[S].北京:中国建筑工业出版社,2016.
[6] 中国建筑标准设计研究院.混凝土结构施工图平面整体表示方法制图规则和构造详图(现浇混凝土框架、剪力墙、梁、板)22G101-1[S].北京:中国计划出版社,2022.
[7] 中国建筑科学研究院.高层建筑混凝土结构技术规程:JGJ 3—2010[S].北京:中国建筑工业出版社,2011.
[8] 中国建筑科学研究院.混凝土强度检验评定标准:GB/T 50107—2010[S].北京:中国建筑工业出版社,2010.
[9] 中华人民共和国住房和城乡建设部.工程结构设计基本术语标准:GB/T 50083—2014[S].北京:中国建筑工业出版社,2015.
[10] 中华人民共和国住房和城乡建设部.建筑结构可靠性设计统一标准:GB 50068—2018[S].北京:中国建筑工业出版社,2019.
[11] 清华大学.混凝土结构耐久性设计标准:GB/T 50476—2019[S].北京:中国建筑工业出版社,2019.
[12] 史美东.建筑结构原理及应用[M].北京:北京大学出版社,2012.
[13] 东南大学,天津大学,同济大学.混凝土结构设计原理[M].7版.北京:中国建筑工业出版社,2020.
[14] 梁兴文,史庆轩.混凝土结构设计原理[M].5版.北京:中国建筑工业出版社,2022.
[15] 戴维·达尔文,查尔斯 W.多兰,亚瑟 H.尼尔逊.混凝土结构基本原理(第15版)[M].影印版.北京:机械工业出版社,2018.
[16] 贺东青.混凝土结构基本原理[M].北京:中国建筑工业出版社,2017.
[17] 李章政,郝献华.混凝土结构基本原理[M].2版.武汉:武汉大学出版社,2017.
[18] 曹启坤.混凝土结构设计原理[M].北京:中国建材工业出版社,2016.
[19] 刘志钦,张玉新.混凝土结构设计原理[M].2版.重庆:重庆大学出版社,2015.
[20] 荣国能.混凝土结构设计原理[M].2版.成都:西南交通大学出版社,2015.
[21] 姚素玲,陈英杰.混凝土结构基本原理[M].北京:中国建材工业出版社,2015.
[22] 赵军,王新玲,楚留声,等.混凝土结构基本原理(英文版)[M].北京:中国建筑工业出版社,2015.
[23] 马芹永.混凝土结构基本原理[M].2版.北京:机械工业出版社,2013.
[24] 徐凤纯,王丽玫.钢筋混凝土与砌体结构[M].2版.北京:中国水利水电出版社,2013.
[25] 刘立新.混凝土结构基本原理[M].武汉:武汉理工大学出版社,2004.

附录 1

《混凝土结构设计标准》(GB/T 50010—2010)和《混凝土结构通用规范》(GB 55008—2021)规定的混凝土和钢筋的力学性能指标

附表 1-1 混凝土轴心抗压强度标准值　　　　　　　单位：N/mm²

强度	混凝土强度等级												
	C20	C25	C30	C35	C40	C45	C50	C55	C60	C65	C70	C75	C80
f_{ck}	13.4	16.7	20.1	23.4	26.8	29.6	32.4	35.5	38.5	41.5	44.5	47.4	50.2

附表 1-2 混凝土轴心抗拉强度标准值　　　　　　　单位：N/mm²

强度	混凝土强度等级												
	C20	C25	C30	C35	C40	C45	C50	C55	C60	C65	C70	C75	C80
f_{tk}	1.54	1.78	2.01	2.20	2.39	2.51	2.64	2.74	2.85	2.93	2.99	3.05	3.11

附表 1-3 混凝土轴心抗压强度设计值　　　　　　　单位：N/mm²

强度	混凝土强度等级												
	C20	C25	C30	C35	C40	C45	C50	C55	C60	C65	C70	C75	C80
f_c	9.6	11.9	14.3	16.7	19.1	21.1	23.1	25.3	27.5	29.7	31.8	33.8	35.9

附表 1-4 混凝土轴心抗拉强度设计值　　　　　　　单位：N/mm²

强度	混凝土强度等级												
	C20	C25	C30	C35	C40	C45	C50	C55	C60	C65	C70	C75	C80
f_t	1.10	1.27	1.43	1.57	1.71	1.80	1.89	1.96	2.04	2.09	2.14	2.18	2.22

附表 1-5　普通钢筋强度标准值

牌　号		符号	公称直径 d/mm	屈服强度标准值 f_{yk}/(N/mm²)	极限强度标准值 f_{stk}/(N/mm²)
热轧钢筋	HPB300	Φ	6～14	300	420
	HRB400 HRBF400 RRB400	Φ Φ^F Φ^R	6～50	400	540
	HRB500 HRBF500	Φ Φ^F	6～50	500	630
冷轧带肋钢筋	CRB550	Φ^R	5～12	500	550
	CRB600H	Φ^{RH}	5～12	540	600

附表 1-6　普通钢筋强度设计值　　　　单位：N/mm²

牌　号		抗拉强度设计值 f_y	抗压强度设计值 f'_y
热轧钢筋	HPB300	270	270
	HRB400、HRBF400、RRB400	360	360
	HRB500、HRBF500	435	410
冷轧带肋钢筋	CRB550	400	—
	CRB600H	430	—

附表 1-7　热轧钢筋、冷轧带肋钢筋及预应力筋的最大力总延伸率限值 δ_{gt}

钢筋牌号或种类	热轧钢筋				冷轧带肋钢筋		预应力筋	
	HPB300	HRB400、HRBF400、HRB500、HRBF500	HRB400E、HRB500E	RRB400	CRB550	CRB600H	中强度预应力钢丝、预应力冷轧带肋钢筋	消除应力钢丝、钢绞线、预应力螺纹钢筋
δ_{gt}/%	10.0	7.5	9.0	5.0	2.5	5.0	4.0	4.5

注：HRB400E、HRB500E 中的"E"表示对结构构件抗震设计有较高要求时可采用的钢筋品种。

附录 2

钢筋的公称直径、公称截面面积及理论质量

附表 2-1 钢筋的公称直径、公称截面面积及理论质量

公称直径/mm	不同根数钢筋的公称截面面积/mm²									单根钢筋理论质量/(kg/m)
	1	2	3	4	5	6	7	8	9	
6	28.3	57	85	113	142	170	198	226	255	0.222
8	50.3	101	151	201	252	302	352	402	453	0.395
10	78.5	157	236	314	393	471	550	628	707	0.617
12	113.1	226	339	452	565	678	791	904	1017	0.888
14	153.9	308	461	615	769	923	1077	1231	1385	1.21
16	201.1	402	603	804	1005	1206	1407	1608	1809	1.58
18	254.5	509	763	1017	1272	1527	1781	2036	2290	2.00(2.11)
20	314.2	628	942	1256	1570	1884	2199	2513	2827	2.47
22	380.1	760	1140	1520	1900	2281	2661	3041	3421	2.98
25	490.9	982	1473	1964	2454	2945	3436	3927	4418	3.85(4.10)
28	615.8	1232	1847	2463	3079	3695	4310	4926	5542	4.83
32	804.2	1609	2413	3217	4021	4826	5630	6434	7238	6.31(6.65)
36	1017.9	2036	3054	4072	5089	6107	7125	8143	9161	7.99

注：括号内为预应力螺纹钢筋的数值。

附表 2-2 不同钢筋间距时每米长度内钢筋截面面积　　单位：mm²

钢筋间距/mm	钢筋直径/mm									
	6	6/8	8	8/10	10	10/12	12	12/14	14	16
70	404	561	719	920	1121	1369	1616	1907	2199	2873
75	377	524	671	899	1047	1277	1508	1780	2052	2681
80	354	491	629	805	981	1198	1414	1669	1924	2514
85	333	462	592	758	924	1127	1331	1571	1811	2366
90	314	437	559	716	872	1064	1257	1438	1710	2234
95	298	414	529	678	826	1008	1190	1405	1620	2117
100	283	393	503	644	785	958	1131	1335	1539	2011
110	257	357	457	585	714	871	1028	1214	1399	1828

续表

钢筋间距/mm	钢筋直径/mm									
	6	6/8	8	8/10	10	10/12	12	12/14	14	16
120	236	327	419	537	654	798	942	1113	1283	1676
125	226	314	402	515	628	766	905	1068	1231	1608
130	218	302	387	495	604	737	870	1027	1184	1547
140	202	281	359	460	561	684	808	954	1099	1436
150	189	262	335	429	523	639	754	890	1026	1341
160	177	246	314	403	491	599	707	834	962	1257
170	166	231	296	379	462	564	665	785	905	1183
180	157	218	279	358	436	532	628	742	855	1117
190	149	207	265	339	413	504	595	703	810	1058
200	141	196	251	322	393	479	565	668	770	1005
220	129	176	229	293	357	436	514	607	700	914
240	118	164	210	268	327	399	471	556	641	838
250	113	157	201	258	314	383	452	534	616	804
260	109	151	193	248	302	368	435	514	592	773
280	101	140	180	230	281	342	404	477	550	718
300	94	131	168	215	262	320	377	445	513	670
320	88	123	157	201	245	299	353	417	481	628

注：表中"/"表示两种不同直径的钢筋间隔放置。

《混凝土结构设计标准》(GB/T 50010—2010)和《混凝土结构通用规范》(GB 55008—2021)的有关规定

附表 3-1　混凝土结构的环境类别

环境类别	条　件
一	室内干燥环境；无侵蚀性静水浸没环境
二 a	室内潮湿环境；非严寒和非寒冷地区的露天环境；非严寒和非寒冷地区与无侵蚀性的水或土壤直接接触的环境；严寒和寒冷地区的冰冻线以下与无侵蚀性的水或土壤直接接触的环境
二 b	干湿交替环境；水位频繁变动环境；严寒和寒冷地区的露天环境；严寒和寒冷地区的冰冻线以上与无侵蚀性的水或土壤直接接触的环境
三 a	严寒和寒冷地区冬季水位变动区环境；受除冰盐影响环境；海风环境
三 b	盐渍土环境；受除冰盐作用环境；海岸环境
四	海水环境
五	受人为或自然的侵蚀性物质影响的环境

注：混凝土的环境类别是指混凝土结构暴露表面所处的环境条件。

附表 3-2　混凝土保护层的最小厚度　　　　单位：mm

环境类别	板、墙、壳	梁、柱
一	15	20
二 a	20	25
二 b	25	35
三 a	30	40
三 b	40	50

注：①表中混凝土保护层厚度指最外层钢筋外边缘至混凝土表面的距离，适用于设计工作年限为 50 年的混凝土结构；②混凝土强度等级为 C25 时，表中保护层厚度数值应增加 5mm；③基础底面钢筋的保护层厚度，有混凝土垫层时应从垫层顶面算起，且不应小于 40mm。

附表 3-3 纵向受力普通钢筋的最小配筋率 ρ_{min}

受力构件类型			最小配筋率
受压构件	全部纵向钢筋	强度等级 500MPa	0.5%
		强度等级 400MPa	0.55%
		强度等级 300MPa	0.6%
	一侧纵向钢筋		0.2%
受弯构件、偏心受拉构件、轴心受拉构件一侧的受拉钢筋			0.2% 和 $0.45f_t/f_y$ 中的较大值

注：①当采用 C60 以上强度等级的混凝土时,受压构件全部纵向普通钢筋最小配筋率应按表中的规定值增加 0.10%采用；②除悬臂板、柱支承板之外的板类受弯构件,当纵向受拉钢筋采用强度等级 500MPa 的钢筋时,其最小配筋率应允许采用 0.15% 和 $0.45f_t/f_y$ 中的较大值；③对于卧置于地基上的钢筋混凝土板,板中受拉普通钢筋的最小配筋率不应小于 0.15%。

附表 3-4 梁纵向受拉钢筋最小配筋率

抗震等级	位 置	
	支座（取较大值）	跨中（取较大值）
一级	0.40% 和 $0.80f_t/f_y$	0.30% 和 $0.65f_t/f_y$
二级	0.30% 和 $0.65f_t/f_y$	0.25% 和 $0.55f_t/f_y$
三级、四级	0.25% 和 $0.55f_t/f_y$	0.20% 和 $0.45f_t/f_y$

附表 3-5 柱全部纵向受力钢筋最小配筋率　　　　　单位：%

柱类型	抗 震 等 级			
	一级	二级	三级	四级
中柱、边柱	0.90(1.00)	0.70(0.80)	0.60(0.70)	0.50(0.60)
角柱、框支柱	1.10	0.90	0.80	0.70

注：①表中括号内数值用于房屋建筑纯框架结构柱；②柱截面每一侧纵向普通钢筋配筋率不应小于 0.20%；③当柱的混凝土强度等级为 C60 以上时,应按表中规定值增加 0.10%采用；④当采用 400MPa 级纵向受力钢筋时,应按表中规定值增加 0.05%采用。

附表 3-6 梁端箍筋加密区的长度、箍筋最大间距和最小直径　　　　　单位：mm

抗震等级	加密区长度（取较大值）	箍筋最大间距（取较小值）	箍筋最小直径
一级	$2.0h_b$, 500	$h_b/4$, 6d, 100	10
二级	$1.5h_b$, 500	$h_b/4$, 8d, 100	8
三级	$1.5h_b$, 500	$h_b/4$, 8d, 150	8
四级	$1.5h_b$, 500	$h_b/4$, 8d, 150	6

注：表中 d 为纵向钢筋直径,mm；h_b 为梁截面高度,mm。

附表 3-7 柱箍筋加密区的箍筋最大间距和最小直径　　　　　单位：mm

抗 震 等 级	箍筋最大间距	箍筋最小直径
一级	6d 和 100 的较小值	10
二级	8d 和 100 的较小值	8
三级、四级	8d 和 150（柱根 100）的较小值	8

注：表中 d 为纵向普通钢筋的直径,mm；柱根指柱底部嵌固部位的箍筋加密区范围。